PROBLEM-SOLVING CASES IN MICROSOFT® ACCESS™ AND EXCEL®

PROBLEM-SOLVING CASES IN MICROSOFT® ACCESS™ AND EXCEL®

Annual Twelfth Edition

Ellen F. Monk

Joseph A. Brady

Gerard S. Cook

CENGAGE
Learning®

Australia • Brazil • Mexico • Singapore • United Kingdom • United States

**Problem-Solving Cases in Microsoft®
Access™ and Excel®, Annual 12th Edition**
Ellen F. Monk, Joseph A. Brady,
Gerard S. Cook

Product Director: Joe Sabatino

Product Manager: Clara Goosman

Senior Content Developer: Kate Mason

Development Editor: Dan Seiter

Product Assistant: Brad Sullender

Market Development Manager:
Heather Mooney

Marketing Coordinator: Eileen Corcoran

Art and Cover Direction, Production
Management, and Composition:
PreMediaGlobal

Manufacturing Planner: Ron Montgomery

Senior Rights Acquisition Specialist:
Christine Myaskovsky

Cover Images: Image # 166288815 (colorful
background) Copyright: © iStock/Thinkstock;
Image # 84468995 (two machinists)
Copyright: © Monkey Business/Thinkstock;
Image # 80404208 (woman working on flip
chart) Copyright: © Fuse/Thinkstock; Image #
164125422 (graphs and charts) Copyright:
© iStock/Thinkstock; Image # 64464652
(business people and computer) Copyright:
© Golden Pixels LLC/Shutterstock; Image #
163869640 (man with tablet in warehouse)
Copyright: © Catherine Yeulet/Photos.com

For product information and technology assistance, contact us at
Cengage Learning Customer & Sales Support, 1-800-354-9706.
For permission to use material from this text or product,
submit all requests online at **www.cengage.com/permissions.**
Further permissions questions can be e-mailed to
permissionrequest@cengage.com.

Some of the product names and company names used in this book have been
used for identification purposes only and may be trademarks or registered
trademarks of their respective manufacturers and sellers.

Library of Congress Control Number: 2013952103

ISBN-13: 978-1-285-86719-9

ISBN-10: 1-285-86719-X

Cengage Learning
200 First Stamford Place, 4th Floor
Stamford, CT 06902
USA

Screenshots for this book were created using Microsoft Access and Excel®,
and were used with permission from Microsoft.

Microsoft and the Office logo are either registered trademarks or trademarks
of Microsoft Corporation in the United States and/or other countries.
Cengage Learning is an independent entity from the Microsoft Corporation,
and is not affiliated with Microsoft in any manner.

The programs in this book are for instructional purposes only. They have
been tested with care, but are not guaranteed for any particular intent
beyond educational purposes. The author and the publisher do not offer any
warranties or representations, nor do they accept any liabilities with respect
to the programs.

Cengage Learning reserves the right to revise this publication and make
changes from time to time in its content without notice.

Cengage Learning is a leading provider of customized learning solutions with
office locations around the globe, including Singapore, the United Kingdom,
Australia, Mexico, Brazil, and Japan. Locate your local office at:
www.cengage.com/global.

Cengage Learning products are represented in Canada by Nelson Education, Ltd.

To learn more about Cengage Learning, visit **www.cengage.com.**

Purchase any of our products at your local college store or at our preferred
online store, **www.cengagebrain.com.**

Printed in the United States of America
1 2 3 4 5 6 7 17 16 15 14 13

To Kate Mason, our Senior Content Developer.
For many years, Kate has been our contact at Cengage.
It has been a pleasure to work with such a capable and pleasant person.

BRIEF CONTENTS

For two decades, we have taught MIS courses at the university level. From the start, we wanted to use good computer-based case studies for the database and decision-support portions of our courses.

At first, we could not find a casebook that met our needs! This surprised us because we thought our requirements were not unreasonable. First, we wanted cases that asked students to think about real-world business situations. Second, we wanted cases that provided students with hands-on experience, using the kind of software that they had learned to use in their computer literacy courses—and that they would later use in business. Third, we wanted cases that would strengthen students' ability to analyze a problem, examine alternative solutions, and implement a solution using software. Undeterred by the lack of casebooks, we wrote our own, and Cengage Learning published it.

This is the twelfth casebook we have written for Cengage Learning. The cases are all new, and the tutorials have been updated using Microsoft Office 2013.

As with our prior casebooks, we include tutorials that prepare students for the cases, which are challenging but doable. The cases are organized to help students think about the logic of each case's business problem and then about how to use the software to solve the business problem. The cases fit well in an undergraduate MIS course, an MBA information systems course, or a computer science course devoted to business-oriented programming.

BOOK ORGANIZATION

The book is organized into seven parts:

- Database cases using Access
- Decision support cases using the Excel Scenario Manager
- Decision support cases using the Excel Solver
- A decision support case using basic Excel functionality
- Integration cases using Access and Excel
- Advanced Excel skills
- Presentation skills

Part 1 begins with two tutorials that prepare students for the Access case studies. Parts 2 and 3 each begin with a tutorial that prepares students for the Excel case studies. All four tutorials provide students with hands-on practice in using the software's more advanced features—the kind of support that other books about Access and Excel do not provide. Part 4 asks students to use Excel's basic functionality for decision support. Part 5 challenges students to use both Access and Excel to find a solution to a business problem. Part 6 is a tutorial that teaches advanced skills students might need to complete some of the Excel cases. Part 7 is a tutorial that hones students' skills in creating and delivering an oral presentation to business managers. The next sections explore these parts of the book in more depth.

Part 1: Database Cases Using Access

This section begins with two tutorials and then presents five case studies.

Tutorial A: Database Design

This tutorial helps students understand how to set up tables to create a database, without requiring students to learn formal analysis and design methods, such as data normalization.

Tutorial B: Microsoft Access

The second tutorial teaches students the more advanced features of Access queries and reports—features that students will need to know to complete the cases.

Cases 1–5

Five database cases follow Tutorials A and B. The students must use the Access database in each case to create forms, queries, and reports that help management. The first case is an easier "warm-up" case. The next four cases require more effort to design the database and implement the results.

Part 2: Decision Support Cases Using Excel Scenario Manager

This section has one tutorial and two decision support cases that require the use of the Excel Scenario Manager.

Tutorial C: Building a Decision Support System in Excel

This section begins with a tutorial that uses Excel to explain decision support and fundamental concepts of spreadsheet design. The case emphasizes the use of Scenario Manager to organize the output of multiple "what-if" scenarios.

Cases 6–7

Students can perform these two cases with or without Scenario Manager, although it is nicely suited to both cases. In each case, students must use Excel to model two or more solutions to a problem. Students then use the model outputs to identify and document the preferred solution in a memorandum. The instructor might also require students to summarize their solutions in oral presentations.

Part 3: Decision Support Cases Using Microsoft Excel Solver

This section has one tutorial and two decision support cases that require the use of Excel Solver.

Tutorial D: Building a Decision Support System Using Microsoft Excel Solver

This section begins with a tutorial for using Excel Solver, a powerful decision support tool for solving optimization problems.

Cases 8–9

Once again, students use Excel and the Solver tool in each case to analyze alternatives and identify and document the preferred solution.

Part 4: Decision Support Case Using Basic Excel Functionality
Case 10

The book continues with a case that uses basic Excel functionality. (In other words, the case does not require Scenario Manager or the Solver.) Excel is used to test students' analytical skills in "what-if" analyses.

Part 5: Integration Cases Using Access and Excel
Cases 11 and 12

These cases integrate Access and Excel. The cases show students how to share data between Access and Excel to solve problems.

Part 6: Advanced Skills Using Excel

This part contains one tutorial that focuses on using advanced techniques in Excel.

Tutorial E: Guidance for Excel Cases

Some cases require the use of Excel techniques that are not discussed in other tutorials or cases in this casebook. For example, techniques for using data tables and pivot tables are explained in Tutorial E rather than in the cases themselves.

Part 7: Presentation Skills

Tutorial F: Giving an Oral Presentation

Each case includes an optional assignment that lets students practice making a presentation to management to summarize the results of their case analysis. This tutorial gives advice for creating oral presentations. It also includes technical information on charting, a technique that is useful in case analyses or as support for presentations. This tutorial will help students to organize their recommendations, to present their solutions both in words and graphics, and to answer questions from the audience. For larger classes, instructors may want to have students work in teams to create and deliver their presentations, which would model the team approach used by many corporations.

INDIVIDUAL CASE DESIGN

The format of the cases uses the following template:

- Each case begins with a *Preview* and an overview of the tasks.
- The next section, *Preparation*, tells students what they need to do or know to complete the case successfully. Again, the tutorials also prepare students for the cases.
- The third section, *Background*, provides the business context that frames the case. The background of each case models situations that require the kinds of thinking and analysis that students will need in the business world.
- The *Assignment* sections are generally organized to help students develop their analyses.
- The last section, *Deliverables*, lists the finished materials that students must hand in: printouts, a memorandum, a presentation, and files. The list is similar to the deliverables that a business manager might demand.

USING THE CASES

We have successfully used cases like these in our undergraduate MIS courses. We usually begin the semester with Access database instruction. We assign the Access database tutorials and then a case to each student. Then, to teach students how to use the Excel decision support system, we do the same thing: we assign a tutorial and then a case.

Some instructors have asked for access to extra cases, especially in the second semester of a school year. For example, they assigned the integration case in the fall, and they need another one for the spring. To meet this need, we have set up an online "Hall of Fame" that features some of our favorite cases from prior editions. These password-protected cases are available to instructors on the Cengage Learning Web site. Go to *www.cengage.com/* and search for this textbook by title, author, or ISBN. Note that the cases are in Microsoft Office 2013 format.

TECHNICAL INFORMATION

The cases in this textbook were written using Microsoft Office 2013, and the textbook was tested for quality assurance using the Windows 7 operating system, Microsoft Access 2013, and Microsoft Excel 2013.

Data Files and Solution Files

We have created "starter" data files for the Excel cases, so students need not spend time typing in the spreadsheet skeleton. Cases 11 and 12 also ask students to load Access and Excel starter files. All these files are on the Cengage Learning Web site, which is available both to students and instructors. Instructors should go to *www.cengage.com* and search for this textbook by title, author, or ISBN. Students will find the files at *www.cengagebrain.com*. You are granted a license to copy the data files to any computer or computer network used by people who have purchased this textbook.

Solutions to the material in the text are available to instructors at *login.cengage.com/sso*. Search for this textbook by title, author, or ISBN. The solutions are password protected.

ACKNOWLEDGMENTS

We would like to give many thanks to the team at Cengage Learning, including our Development Editor, Dan Seiter; Senior Content Developer, Kate Hennessy Mason; and our Content Project Manager, Arul Joseph Raj. As always, we acknowledge our students' diligent work.

PART 1

DATABASE CASES USING ACCESS

DATABASE DESIGN

This tutorial has three sections. The first section briefly reviews basic database terminology. The second section teaches database design. The third section features a database design problem for practice.

REVIEW OF TERMINOLOGY

You will begin by reviewing some basic terms that will be used throughout this textbook. In Access, a **database** is a group of related objects that are saved in one file. An Access **object** can be a table, form, query, or report. You can identify an Access database file by its suffix, .accdb.

A **table** consists of data that is arrayed in rows and columns. A **row** of data is called a **record**. A **column** of data is called a **field**. Thus, a record is a set of related fields. The fields in a table should be related to one another in some way. For example, a company might want to keep its employee data together by creating a database table called Employee. That table would contain data fields about employees, such as their names and addresses. It would not have data fields about the company's customers; that data would go in a Customer table.

A field's values have a **data type** that is declared when the table is defined. Thus, when data is entered into the database, the software knows how to interpret each entry. Data types in Access include the following:

- Text for words
- Integer for whole numbers
- Double for numbers that have a decimal value
- Currency for numbers that represent dollars and cents
- Yes/No for variables that have only two values (such as 1/0, on/off, yes/no, and true/false)
- Date/Time for variables that are dates or times

Each database table should have a **primary key** field—a field in which each record has a *unique* value. For example, in an Employee table, a field called Employee Identification Number (EIN) could serve as a primary key. (This assumes that each employee is given a number when hired, and that these numbers are not reused later.) Sometimes a table does not have a single field whose values are all different. In that case, two or more fields are combined into a **compound primary key**. The combination of the fields' values is unique.

Database tables should be logically related to one another. For example, suppose a company has an Employee table with fields for EIN, Name, Address, and Telephone Number. For payroll purposes, the company has an Hours Worked table with a field that summarizes Labor Hours for individual employees. The relationship between the Employee table and Hours Worked table needs to be established in the database so you can determine the number of hours worked by any employee. To create this relationship, you include the primary key field from the Employee table (EIN) as a field in the Hours Worked table. In the Hours Worked table, the EIN field is then called a **foreign key** because it's from a "foreign" table.

In Access, data can be entered directly into a table or it can be entered into a form, which then inserts the data into a table. A **form** is a database object that is created from an existing table to make the process of entering data more user-friendly.

A **query** is the database equivalent of a question that is posed about data in a table (or tables). For example, suppose a manager wants to know the names of employees who have worked for the company for more than five years. A query could be designed to search the Employee table for the information. The query would be run, and its output would answer the question.

Queries can be designed to search multiple tables at a time. For this to work, the tables must be connected by a **join** operation, which links tables on the values in a field that they have in common. The common field acts as a "hinge" for the joined tables; when the query is run, the query generator treats the joined tables as one large table.

In Access, queries that answer a question are called *select queries* because they select relevant data from the database records. Queries also can be designed to change data in records, add a record to the end of a table, or delete entire records from a table. These queries are called **update**, **append**, and **delete** queries, respectively.

Access has a **report** generator that can be used to format a table's data or a query's output.

DATABASE DESIGN

Designing a database involves determining which tables belong in the database and then creating the fields that belong in each table. This section begins with an introduction to key database design concepts, then discusses design rules you should use when building a database. First, the following key concepts are defined:

- Entities
- Relationships
- Attributes

Database Design Concepts

Computer scientists have highly formalized ways of documenting a database's logic. Learning their notations and mechanics can be time-consuming and difficult. In fact, doing so usually takes a good portion of a systems analysis and design course. This tutorial will teach you database design by emphasizing practical business knowledge; the approach should enable you to design serviceable databases quickly. Your instructor may add more formal techniques.

A database models the logic of an organization's operation, so your first task is to understand the operation. You can talk to managers and workers, make your own observations, and look at business documents such as sales records. Your goal is to identify the business's "entities" (sometimes called *objects*). An **entity** is a thing or event that the database will contain. Every entity has characteristics, called **attributes**, and one or more **relationships** to other entities. Take a closer look.

Entities

As previously mentioned, an entity is a tangible thing or an event. The reason for identifying entities is that *an entity eventually becomes a table in the database*. Entities that are things are easy to identify. For example, consider a video store. The database for the video store would probably need to contain the names of DVDs and the names of customers who rent them, so you would have one entity named Video and another named Customer.

In contrast, entities that are events can be more difficult to identify, probably because they are more conceptual. However, events are real, and they are important. In the video store example, one event would be Video Rental and another event would be Hours Worked by employees.

In general, your analysis of an organization's operations is made easier when you realize that organizations usually have physical entities such as these:

- Employees
- Customers
- Inventory (products or services)
- Suppliers

Thus, the database for most organizations would have a table for each of these entities. Your analysis also can be made easier by knowing that organizations engage in transactions internally (within the company) and externally (with the outside world). Such transactions are explained in an introductory accounting course, but most people understand them from events that occur in daily life. Consider the following examples:

- Organizations generate revenue from sales or interest earned. Revenue-generating transactions include event entities called Sales and Interest Earned.
- Organizations incur expenses from paying hourly employees and purchasing materials from suppliers. Hours Worked and Purchases are event entities in the databases of most organizations.

Thus, identifying entities is a matter of observing what happens in an organization. Your powers of observation are aided by knowing what entities exist in the databases of most organizations.

Relationships

As an analyst building a database, you should consider the relationship of each entity to the other entities you have identified. For example, a college database might contain entities for Student, Course, and Section to contain data about each. A relationship between Student and Section could be expressed as "Students enroll in sections."

An analyst also must consider the **cardinality** of any relationship. Cardinality can be one-to-one, one-to-many, or many-to-many:

- In a one-to-one relationship, one instance of the first entity is related to just one instance of the second entity.
- In a one-to-many relationship, one instance of the first entity is related to many instances of the second entity, but each instance of the second entity is related to only one instance of the first.
- In a many-to-many relationship, one instance of the first entity is related to many instances of the second entity, and one instance of the second entity is related to many instances of the first.

For a more concrete understanding of cardinality, consider again the college database with the Student, Course, and Section entities. The university catalog shows that a course such as Accounting 101 can have more than one section: 01, 02, 03, 04, and so on. Thus, you can observe the following relationships:

- The relationship between the entities Course and Section is one-to-many. Each course has many sections, but each section is associated with just one course.
- The relationship between Student and Section is many-to-many. Each student can be in more than one section, because each student can take more than one course. Also, each section has more than one student.

Thinking about relationships and their cardinalities may seem tedious to you. However, as you work through the cases in this text, you will see that this type of analysis can be valuable in designing databases. In the case of many-to-many relationships, you should determine the tables a given database needs; in the case of one-to-many relationships, you should decide which fields the tables need to share.

Attributes

An attribute is a characteristic of an entity. You identify attributes of an entity because *attributes become a table's fields*. If an entity can be thought of as a noun, an attribute can be considered an adjective that describes the noun. Continuing with the college database example, consider the Student entity. Students have names, so Last Name would be an attribute of the Student entity and therefore a field in the Student table. First Name would be another attribute, as well as Address, Phone Number, and other descriptive fields.

Sometimes it can be difficult to tell the difference between an attribute and an entity, but one good way is to ask whether more than one attribute is possible for each entity. If more than one instance is possible, but you do not know the number in advance, you are working with an entity. For example, assume that a student could have a maximum of two addresses—one for home and one for college. You could specify attributes Address 1 and Address 2. Next, consider that you might not know the number of student addresses in advance, meaning that all addresses have to be recorded. In that case, you would not know how many fields to set aside in the Student table for addresses. Therefore, you would need a separate Student Addresses table (entity) that would show any number of addresses for a given student.

Database Design Rules

As described previously, your first task in database design is to understand the logic of the business situation. Once you understand this logic, you are ready to build the database. To create a context for learning about database design, look at a hypothetical business operation and its database needs.

Example: The Talent Agency

Suppose you have been asked to build a database for a talent agency that books musical bands into nightclubs. The agent needs a database to keep track of the agency's transactions and to answer day-to-day questions. For example, a club manager often wants to know which bands are available on a certain date at a certain time, or wants to know the agent's fee for a certain band. The agent may want to see a list of all band members and the instrument each person plays, or a list of all bands that have three members.

Suppose that you have talked to the agent and have observed the agency's business operation. You conclude that your database needs to reflect the following facts:

1. A booking is an event in which a certain band plays in a particular club on a particular date, starting and ending at certain times, and performing for a specific fee. A band can play more than once a day. The Heartbreakers, for example, could play at the East End Cafe in the afternoon and then at the West End Cafe on the same night. For each booking, the club pays the talent agent. The agent keeps a five percent fee and then gives the remainder of the payment to the band.

2. Each band has at least two members and an unlimited maximum number of members. The agent notes a telephone number of just one band member, which is used as the band's contact number. No two bands have the same name or telephone number.

3. Band member names are not unique. For example, two bands could each have a member named Sally Smith.

4. The agent keeps track of just one instrument that each band member plays. For the purpose of this database, "vocals" are considered an instrument.

5. Each band has a desired fee. For example, the Lightmetal band might want $700 per booking, and would expect the agent to try to get at least that amount.

6. Each nightclub has a name, an address, and a contact person. The contact person has a telephone number that the agent uses to call the club. No two clubs have the same name, contact person, or telephone number. Each club has a target fee. The contact person will try to get the agent to accept that fee for a band's appearance.

7. Some clubs feed the band members for free; others do not.

Before continuing with this tutorial, you might try to design the agency's database on your own. Ask yourself: What are the entities? Recall that business databases usually have Customer, Employee, and Inventory entities, as well as an entity for the event that generates revenue transactions. Each entity becomes a table in the database. What are the relationships among the entities? For each entity, what are its attributes? For each table, what is the primary key?

Six Database Design Rules

Assume that you have gathered information about the business situation in the talent agency example. Now you want to identify the tables required for the database and the fields needed in each table. Observe the following six rules:

Rule 1: You do not need a table for the business. The database represents the entire business. Thus, in the example, Agent and Agency are not entities.

Rule 2: Identify the entities in the business description. Look for typical things and events that will become tables in the database. In the talent agency example, you should be able to observe the following entities:

- *Things*: The product (inventory for sale) is Band. The customer is Club.
- *Events*: The revenue-generating transaction is Bookings.

You might ask yourself: Is there an Employee entity? Isn't Instrument an entity? Those issues will be discussed as the rules are explained.

Rule 3: Look for relationships among the entities. Look for one-to-many relationships between entities. The relationship between those entities must be established in the tables, using a foreign key. For details, see the following discussion in Rule 4 about the relationship between Band and Band Member.

Look for many-to-many relationships between entities. Each of these relationships requires a third entity that associates the two entities in the relationship. Recall the many-to-many relationship from the college database scenario that involved Student and Section entities. To display the enrollment of specific students in specific sections, a third table would be required. The mechanics of creating such a table are described in Rule 4 during the discussion of the relationship between Band and Club.

Rule 4: Look for attributes of each entity and designate a primary key. As previously mentioned, you should think of the entities in your database as nouns. You should then create a list of adjectives that describe those nouns. These adjectives are the attributes that will become the table's fields. After you have identified fields for each table, you should check to see whether a field has unique values. If such a field exists, designate it as the primary key field; otherwise, designate a compound primary key.

In the talent agency example, the attributes, or fields, of the Band entity are Band Name, Band Phone Number, and Desired Fee, as shown in Figure A-1. Assume that no two bands have the same name, so the primary key field can be Band Name. The data type of each field is shown.

BAND	
Field Name	**Data Type**
Band Name (primary key)	Text
Band Phone Number	Text
Desired Fee	Currency

Source: © Cengage Learning 2015
FIGURE A-1 The Band table and its fields

Two Band records are shown in Figure A-2.

Band Name (primary key)	Band Phone Number	Desired Fee
Heartbreakers	981 831 1765	$800
Lightmetal	981 831 2000	$700

Source: © Cengage Learning 2015
FIGURE A-2 Records in the Band table

If two bands might have the same name, Band Name would not be a good primary key, so a different unique identifier would be needed. Such situations are common. Most businesses have many types of inventory, and duplicate names are possible. The typical solution is to assign a number to each product to use as the primary key field. A college could have more than one faculty member with the same name, so each faculty member would be assigned an employee identification number. Similarly, banks assign a personal identification number (PIN) for each depositor. Each automobile produced by a car manufacturer gets a unique Vehicle Identification Number (VIN). Most businesses assign a number to each sale, called an invoice number. (The next time you go to a grocery store, note the number on your receipt. It will be different from the number on the next customer's receipt.)

At this point, you might be wondering why Band Member would not be an attribute of Band. The answer is that, although you must record each band member, you do not know in advance how many members are in each band. Therefore, you do not know how many fields to allocate to the Band table for members. (Another way to think about band members is that they are the agency's employees, in effect. Databases for organizations usually have an Employee entity.) You should create a Band Member table with the attributes Member ID Number, Member Name, Band Name, Instrument, and Phone. A Member ID Number field is needed because member names may not be unique. The table and its fields are shown in Figure A-3.

BAND MEMBER	
Field Name	**Data Type**
Member ID Number (primary key)	Text
Member Name	Text
Band Name (foreign key)	Text
Instrument	Text
Phone	Text

Source: © Cengage Learning 2015
FIGURE A-3 The Band Member table and its fields

Note in Figure A-3 that the phone number is classified as a Text data type because the field values will not be used in an arithmetic computation. The benefit is that Text data type values take up fewer bytes than Numerical or Currency data type values; therefore, the file uses less storage space. You should also use the Text data type for number values such as zip codes.

Five records in the Band Member table are shown in Figure A-4.

Member ID Number	Member Name	Band Name	Instrument	Phone
0001	Pete Goff	Heartbreakers	Guitar	981 444 1111
0002	Joe Goff	Heartbreakers	Vocals	981 444 1234
0003	Sue Smith	Heartbreakers	Keyboard	981 555 1199
0004	Joe Jackson	Lightmetal	Sax	981 888 1654
0005	Sue Hoopes	Lightmetal	Piano	981 888 1765

Source: © Cengage Learning 2015

FIGURE A-4   Records in the Band Member table

You can include Instrument as a field in the Band Member table because the agent records only one instrument for each band member. Thus, you can use the instrument as a way to describe a band member, much like the phone number is part of the description. Phone could not be the primary key because two members might share a telephone and because members might change their numbers, making database administration more difficult.

You might ask why Band Name is included in the Band Member table. The common-sense reason is that you did not include the Member Name in the Band table. You must relate bands and members somewhere, and the Band Member table is the place to do it.

To think about this relationship in another way, consider the cardinality of the relationship between Band and Band Member. It is a one-to-many relationship: one band has many members, but each member in the database plays in just one band. You establish such a relationship in the database by using the primary key field of one table as a foreign key in the other table. In Band Member, the foreign key Band Name is used to establish the relationship between the member and his or her band.

The attributes of the Club entity are Club Name, Address, Contact Name, Club Phone Number, Preferred Fee, and Feed Band?. The Club table can define the Club entity, as shown in Figure A-5.

CLUB	
Field Name	**Data Type**
Club Name (primary key)	Text
Address	Text
Contact Name	Text
Club Phone Number	Text
Preferred Fee	Currency
Feed Band?	Yes/No

Source: © Cengage Learning 2015

FIGURE A-5   The Club table and its fields

Two records in the Club table are shown in Figure A-6.

Club Name (primary key)	Address	Contact Name	Club Phone Number	Preferred Fee	Feed Band?
East End	1 Duce St.	Al Pots	981 444 8877	$600	Yes
West End	99 Duce St.	Val Dots	981 555 0011	$650	No

Source: © Cengage Learning 2015

FIGURE A-6   Records in the Club table

You might wonder why Bands Booked into Club (or a similar name) is not an attribute of the Club table. There are two reasons. First, you do not know in advance how many bookings a club will have, so the value cannot be an attribute. Second, Bookings is the agency's revenue-generating transaction, an event entity, and you need a table for that business transaction. Consider the booking transaction next.

You know that the talent agent books a certain band into a certain club for a specific fee on a certain date, starting and ending at a specific time. From that information, you can see that the attributes of the Bookings entity are Band Name, Club Name, Date, Start Time, End Time, and Fee. The Bookings table and its fields are shown in Figure A-7.

BOOKINGS	
Field Name	**Data Type**
Band Name	Text
Club Name	Text
Date	Date/Time
Start Time	Date/Time
End Time	Date/Time
Fee	Currency

Source: © Cengage Learning 2015

FIGURE A-7   The Bookings table and its fields—and no designation of a primary key

Some records in the Bookings table are shown in Figure A-8.

Band Name	Club Name	Date	Start Time	End Time	Fee
Heartbreakers	East End	11/21/14	21:30	23:30	$800
Heartbreakers	East End	11/22/14	21:00	23:30	$750
Heartbreakers	West End	11/28/14	19:00	21:00	$500
Lightmetal	East End	11/21/14	18:00	20:00	$700
Lightmetal	West End	11/22/14	19:00	21:00	$750

Source: © Cengage Learning 2015

FIGURE A-8   Records in the Bookings table

Note that no single field is guaranteed to have unique values, because each band is likely to be booked many times and each club might be used many times. Furthermore, each date and time can appear more than once. Thus, no one field can be the primary key.

If a table does not have a single primary key field, you can make a compound primary key whose field values will be unique when taken together. Because a band can be in only one place at a time, one possible solution is to create a compound key from the Band Name, Date, and Start Time fields. An alternative solution is to create a compound primary key from the Club Name, Date, and Start Time fields.

If you don't want a compound key, you could create a field called Booking Number. Each booking would then have its own unique number, similar to an invoice number.

You can also think about this event entity in a different way. Over time, a band plays in many clubs, and each club hires many bands. Thus, Band and Club have a many-to-many relationship, which signals the need for a table between the two entities. A Bookings table would associate the Band and Club tables. You implement an associative table by including the primary keys from the two tables that are associated. In this case, the primary keys from the Band and Club tables are included as foreign keys in the Bookings table.

Rule 5: Avoid data redundancy. You should not include extra (redundant) fields in a table. Redundant fields take up extra disk space and lead to data entry errors because the same value must be entered in multiple tables, increasing the chance of a keystroke error. In large databases, keeping track of multiple instances of the same data is nearly impossible, so contradictory data entries become a problem.

Consider this example: Why wouldn't Club Phone Number be included in the Bookings table as a field? After all, the agent might have to call about a last-minute booking change and could quickly look up the number in the Bookings table. Assume that the Bookings table includes Booking Number as the primary key and Club Phone Number as a field. Figure A-9 shows the Bookings table with the additional field.

BOOKINGS	
Field Name	**Data Type**
Booking Number (primary key)	Text
Band Name	Text
Club Name	Text
Club Phone Number	Text
Date	Date/Time
Start Time	Date/Time
End Time	Date/Time
Fee	Currency

Source: © Cengage Learning 2015

FIGURE A-9   The Bookings table with an unnecessary field—Club Phone Number

The fields Date, Start Time, End Time, and Fee logically depend on the Booking Number primary key—they help define the booking. Band Name and Club Name are foreign keys and are needed to establish the relationship between the Band, Club, and Bookings tables. But what about Club Phone Number? It is not defined by the Booking Number. It is defined by Club Name—*in other words, it is a function of the club, not of the booking*. Thus, the Club Phone Number field does not belong in the Bookings table. It is already in the Club table.

Perhaps you can see the practical data-entry problem of including Club Phone Number in Bookings. Suppose a club changed its contact phone number. The agent could easily change the number one time, in the Club table. However, the agent would need to remember which other tables contained the field and change the values there too. In a small database, this task might not be difficult, but in larger databases, having redundant fields in many tables makes such maintenance difficult, which means that redundant data is often incorrect.

You might object by saying, "What about all of those foreign keys? Aren't they redundant?" In a sense, they are. But they are needed to establish the relationship between one entity and another, as discussed previously.

Rule 6: Do not include a field if it can be calculated from other fields. A **calculated field** is made using the query generator. Thus, the agent's fee is not included in the Bookings table because it can be calculated by query (here, five percent multiplied by the booking fee).

PRACTICE DATABASE DESIGN PROBLEM

Imagine that your town library wants to keep track of its business in a database, and that you have been called in to build the database. You talk to the town librarian, review the old paper-based records, and watch people use the library for a few days. You learn the following about the library:

1. Any resident of the town can get a library card simply by asking for one. The library considers each cardholder a member of the library.

2. The librarian wants to be able to contact members by telephone and by mail. She calls members when books are overdue or when requested materials become available. She likes to mail a thank-you note to each patron on his or her anniversary of becoming a member of the library. Without a database, contacting members efficiently can be difficult; for example, multiple members can have the same name. Also, a parent and a child might have the same first and last name, live at the same address, and share a phone.

3. The librarian tries to keep track of each member's reading interests. When new books come in, the librarian alerts members whose interests match those books. For example, long-time member Sue Doaks is interested in reading Western novels, growing orchids, and baking bread. There must be some way to match her interests with available books. One complication is that, although the librarian wants to track all of a member's reading interests, she wants to classify each book as being in just one category of interest. For example, the classic gardening book *Orchids of France* would be classified as a book about orchids or a book about France, but not both.

4. The library stocks thousands of books. Each book has a title and any number of authors. Also, more than one book in the library might have the same title. Similarly, multiple authors might have the same name.

5. A writer could be the author of more than one book.

6. A book will be checked out repeatedly as time goes on. For example, *Orchids of France* could be checked out by one member in March, by another member in July, and by another member in September.

7. The library must be able to identify whether a book is checked out.

8. A member can check out any number of books in one visit. Also, a member might visit the library more than once a day to check out books.

9. All books that are checked out are due back in two weeks, with no exceptions. The librarian would like to have an automated way of generating an overdue book list each day so she can telephone offending members.

10. The library has a number of employees. Each employee has a job title. The librarian is paid a salary, but other employees are paid by the hour. Employees clock in and out each day. Assume that all employees work only one shift per day and that all are paid weekly. Pay is deposited directly into an employee's checking account—no checks are hand-delivered. The database needs to include the librarian and all other employees.

Design the library's database, following the rules set forth in this tutorial. Your instructor will specify the format of your work. Here are a few hints in the form of questions:

- A book can have more than one author. An author can write more than one book. How would you describe the relationship between books and authors?
- The library lends books for free, of course. If you were to think of checking out a book as a sales transaction for zero revenue, how would you handle the library's revenue-generating event?
- A member can borrow any number of books at one checkout. A book can be checked out more than once. How would you describe the relationship between checkouts and books?

MICROSOFT ACCESS

Microsoft Access is a relational database package that runs on the Microsoft Windows operating system. There are many different versions of Access; this tutorial was prepared using Access 2013.

Before using this tutorial, you should know the fundamentals of Access and know how to use Windows. This tutorial explains advanced Access skills you will need to complete database case studies. The tutorial concludes with a discussion of common Access problems and how to solve them.

To prevent losing your work, always observe proper file-saving and closing procedures. To exit Access, click the File tab and select Close, then click the Close button in the upper-right corner. You can also simply select the Exit option to return to Windows. Always end your work with these steps. If you remove your USB key or other portable storage device when database forms and tables are shown on the screen, you will lose your work.

To begin this tutorial, you will create a new database called Employee.

AT THE KEYBOARD

Open Access. Click the Blank desktop database icon from the templates list. Name the database Employee. Click the file folder next to the filename to browse for the folder where you want to save the file. Otherwise, your file will be saved automatically in the Documents folder. Click the Create button.

A portion of your opening screen should resemble the screen shown in Figure B-1.

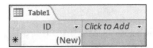

Source: Used with permission from Microsoft Corporation
FIGURE B-1   Entering data in Datasheet view

When you create a table, Access opens it in Datasheet view by default. Because you will use Design view to build your tables, close the new table by clicking the *X* in the upper-right corner of the table window that corresponds to Close 'Table 1.' You are now on the Home tab in the Database window of Access, as shown in Figure B-2. From this screen, you can create or change objects.

Source: Used with permission from Microsoft Corporation
FIGURE B-2   The Database window Home tab in Access

CREATING TABLES

Your database will contain data about employees, their wage rates, and the hours they worked.

Defining Tables

In the Database window, build three new tables using the following instructions.

AT THE KEYBOARD

Defining the Employee Table

This table contains permanent data about employees. To create the table, click the Create tab and then click Table Design in the Tables group. The table's fields are Last Name, First Name, Employee ID, Street Address, City, State, Zip, Date Hired, and US Citizen. The Employee ID field is the primary key field. Change the lengths of Short Text fields from the default 255 spaces to more appropriate lengths; for example, the Last Name field might be 30 spaces, and the Zip field might be 10 spaces. Your completed definition should resemble the one shown in Figure B-3.

Field Name	Data Type	Description (Optional)
Last Name	Short Text	
First Name	Short Text	
Employee ID	Short Text	
Street Address	Short Text	
City	Short Text	
State	Short Text	
Zip	Short Text	
Date Hired	Date/Time	
US Citizen	Yes/No	

Source: Used with permission from Microsoft Corporation

FIGURE B-3   Fields in the Employee table

When you finish, click the File tab, select Save As, select Save Object As, click the Save As button, and then enter a name for the table. In this example, the table is named Employee. (It is a coincidence that the Employee table has the same name as its database file.) After entering the name, click OK in the Save As window. Close the table by clicking the Close button (X) that corresponds to the Employee table.

Defining the Wage Data Table

This table contains permanent data about employees and their wage rates. The table's fields are Employee ID, Wage Rate, and Salaried. The Employee ID field is the primary key field. Use the data types shown in Figure B-4. Your definition should resemble the one shown in Figure B-4.

Field Name	Data Type	Description (Optional)
Employee ID	Short Text	
Wage Rate	Currency	
Salaried	Yes/No	

Source: Used with permission from Microsoft Corporation

FIGURE B-4   Fields in the Wage Data table

Click the File tab and then select Save As, select Save Object As, and click the Save As button to save the table definition. Name the table Wage Data.

Defining the Hours Worked Table

The purpose of this table is to record the number of hours that employees work each week during the year. The table's three fields are Employee ID (which has a Short Text data type), Week # (number–long integer), and Hours (number–double). The Employee ID and Week # are the compound keys.

In the following example, the employee with ID number 08965 worked 40 hours in Week 1 of the year and 52 hours in Week 2.

Employee ID	Week#	Hours
08965	1	40
08965	2	52

Note that no single field can be the primary key field because 08965 is an entry for each week. In other words, if this employee works each week of the year, 52 records will have the same Employee ID value at the end of the year. Thus, Employee ID values will not distinguish records. No other single field can distinguish these records either, because other employees will have worked during the same week number and some employees will have worked the same number of hours. For example, 40 hours—which corresponds to a full-time workweek—would be a common entry for many weeks.

All of this presents a problem because a table must have a primary key field in Access. The solution is to use a compound primary key; that is, use values from more than one field to create a combined field that will distinguish records. The best compound key to use for the current example consists of the Employee ID field and the Week # field, because as each person works each week, the week number changes. In other words, there is only *one* combination of Employee ID 08965 and Week # 1. Because those values *can occur in only one record*, the combination distinguishes that record from all others.

The first step of setting a compound key is to highlight the fields in the key. Those fields must appear one after the other in the table definition screen. (Plan ahead for that format.) As an alternative, you can highlight one field, hold down the Control key, and highlight the next field.

AT THE KEYBOARD

In the Hours Worked table, click the first field's left prefix area (known as the row selector), hold down the mouse button, and drag down to highlight the names of all fields in the compound primary key. Your screen should resemble the one shown in Figure B-5.

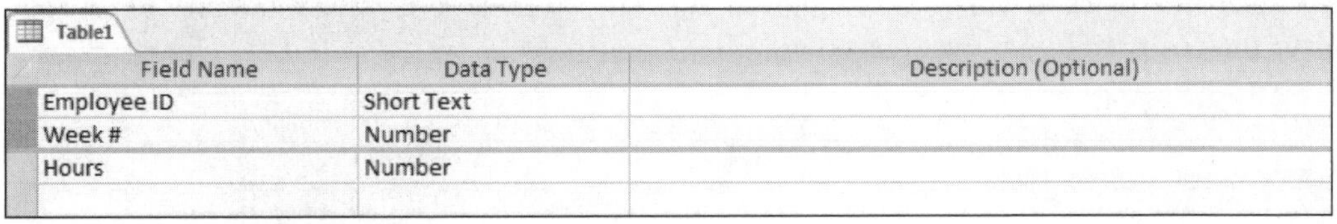

Field Name	Data Type	Description (Optional)
Employee ID	Short Text	
Week #	Number	
Hours	Number	

Source: Used with permission from Microsoft Corporation

FIGURE B-5 Selecting fields for the compound primary key for the Hours Worked table

Now click the Key icon. Your screen should resemble the one shown in Figure B-6.

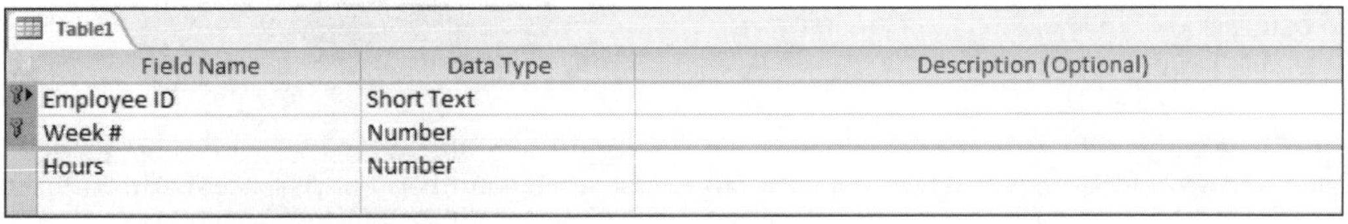

Field Name	Data Type	Description (Optional)
Employee ID	Short Text	
Week #	Number	
Hours	Number	

Source: Used with permission from Microsoft Corporation

FIGURE B-6 The compound primary key for the Hours Worked table

You have created the compound primary key and finished defining the table. Click the File tab and then select Save As, select Save Object As, and click the Save As button to save the table as Hours Worked.

Adding Records to a Table

At this point, you have set up the skeletons of three tables. The tables have no data records yet. If you printed the tables now, you would only see column headings (the field names). The most direct way to enter data into a table is to double-click the table's name in the navigation pane at the left side of the screen and then type the data directly into the cells.

NOTE

To display and open the database objects, Access 2013 uses a navigation pane, which is on the left side of the Access window.

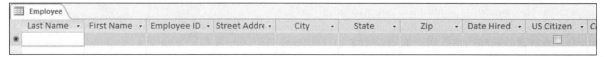

AT THE KEYBOARD

On the Home tab of the Database window, double-click the Employee table. Your data entry screen should resemble the one shown in Figure B-7.

Employee									
Last Name ▾	First Name ▾	Employee ID ▾	Street Addre ▾	City ▾	State ▾	Zip ▾	Date Hired ▾	US Citizen ▾	C
✱								☐	

Source: Used with permission from Microsoft Corporation

FIGURE B-7   The data entry screen for the Employee table

The Employee table has many fields, some of which may be off the screen to the right. Scroll to see obscured fields. (Scrolling happens automatically as you enter data.) Figure B-7 shows all of the fields on the screen.

Enter your data one field value at a time. Note that the first row is empty when you begin. Each time you finish entering a value, press Enter to move the cursor to the next cell. After you enter data in the last cell in a row, the cursor moves to the first cell of the next row *and* Access automatically saves the record. Thus, you do not need to click the File tab and then select Save after entering data into a table.

When entering data in your table, you should enter dates in the following format: 6/15/10. Access automatically expands the entry to the proper format in output.

Also note that Yes/No variables are clicked (checked) for Yes; otherwise, the box is left blank for No. You can change the box from Yes to No by clicking it.

Enter the data shown in Figure B-8 into the Employee table. If you make errors in data entry, click the cell, backspace over the error, and type the correction.

Employee									
Last Name ▾	First Name ▾	Employee ID ▾	Street Addre ▾	City ▾	State ▾	Zip ▾	Date Hired ▾	US Citizen ▾	Click to Add
Howard	Jane	11411	28 Sally Dr	Glasgow	DE	19702	6/1/2014	☑	
Smith	John	12345	30 Elm St	Newark	DE	19711	6/1/1996	☑	
Smith	Albert	14890	44 Duce St	Odessa	DE	19722	7/15/1987	☑	
Jones	Sue	22282	18 Spruce St	Newark	DE	19716	7/15/2004	☐	
Ruth	Billy	71460	1 Tater Dr	Baltimore	MD	20111	8/15/1999	☐	
Add	Your	Data	Here	Elkton	MD	21921		☑	
✱								☐	

Source: Used with permission from Microsoft Corporation

FIGURE B-8   Data for the Employee table

Note that the sixth record is *your* data record. Assume that you live in Elkton, Maryland, were hired on today's date (enter the date), and are a U.S. citizen. Make up a fictitious Employee ID number. For purposes of this tutorial, the sixth record has been created using the name of one of this text's authors and the employee ID 09911.

After adding records to the Employee table, open the Wage Data table and enter the data shown in Figure B-9.

Wage Data		
Employee ID ▾	Wage Rate ▾	Salaried ▾
11411	$10.00	☐
12345	$0.00	☑
14890	$12.00	☐
22282	$0.00	☑
71460	$0.00	☑
Your Employee ID	$8.00	☐
✱	$0.00	☐

Source: Used with permission from Microsoft Corporation

FIGURE B-9   Data for the Wage Data table

In this table, you are again asked to create a new entry. For this record, enter your own employee ID. Also assume that you earn $8 an hour and are not salaried. Note that when an employee's Salaried box is not checked (in other words, Salaried = No), the implication is that the employee is paid by the hour. Because salaried employees are not paid by the hour, their hourly rate is 0.00.

When you finish creating the Wage Data table, open the Hours Worked table and enter the data shown in Figure B-10.

Hours Worked		
Employee ID ▾	Week # ▾	Hours ▾
11411	1	40
11411	2	50
12345	1	40
12345	2	40
14890	1	38
14890	2	40
22282	1	40
22282	2	40
71460	1	40
71460	2	40
Your Employee ID	1	60
Your Employee ID	2	55
*	0	0

Source: Used with permission from Microsoft Corporation

FIGURE B-10 Data for the Hours Worked table

Notice that salaried employees are always given 40 hours. Nonsalaried employees (including you) might work any number of hours. For your record, enter your fictitious employee ID, 60 hours worked for Week 1, and 55 hours worked for Week 2.

CREATING QUERIES

Because you know how to create basic queries, this section explains the advanced queries you will create in the cases in this book.

Using Calculated Fields in Queries

A **calculated field** is an output field made up of *other* field values. A calculated field can be a field in a table; here it is created in the query generator. The calculated field here does not become part of the table—it is just part of the query output. The best way to understand this process is to work through an example.

AT THE KEYBOARD

Suppose you want to see the employee IDs and wage rates of hourly workers, and the new wage rates if all employees were given a 10 percent raise. To view that information, show the employee ID, the current wage rate, and the higher rate, which should be titled New Rate in the output. Figure B-11 shows how to set up the query.

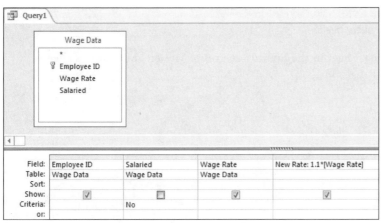

Field:	Employee ID	Salaried	Wage Rate	New Rate: 1.1*[Wage Rate]
Table:	Wage Data	Wage Data	Wage Data	
Sort:				
Show:	☑	☐	☑	☑
Criteria:		No		
or:				

Source: Used with permission from Microsoft Corporation

FIGURE B-11 Query setup for the calculated field

To set up this query, you need to select hourly workers by using the Salaried field with Criteria = No. Note in Figure B-11 that the Show box for the field is not checked, so the Salaried field values will not appear in the query output.

Note the expression for the calculated field, which you can see in the far-right field cell:

New Rate: 1.1 * [Wage Rate]

The term *New Rate:* merely specifies the desired output heading. (Don't forget the colon.) The rest of the expression, 1.1 * [Wage Rate], multiplies the old wage rate by 110 percent, which results in the 10 percent raise.

In the expression, the field name Wage Rate must be enclosed in square brackets. Remember this rule: *Any time an Access expression refers to a field name, the field name must be enclosed in square brackets.*

If you run this query, your output should resemble that in Figure B-12.

Query1		
Employee ID	Wage Rate	New Rate
11411	$10.00	11
14890	$12.00	13.2
09911	$8.00	8.8
*	$0.00	

Source: Used with permission from Microsoft Corporation

FIGURE B-12   Output for a query with calculated field

Notice that the calculated field output is not shown in Currency format, but as a Double—a number with digits after the decimal point. To convert the output to Currency format, select the output column by clicking the line above the calculated field expression. The column darkens to indicate its selection. Your data entry screen should resemble the one shown in Figure B-13.

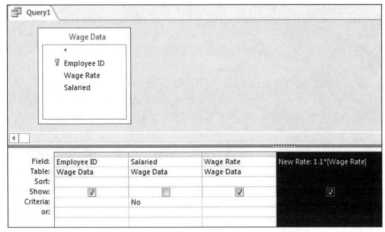

Source: Used with permission from Microsoft Corporation

FIGURE B-13   Activating a calculated field in query design

Then, on the Design tab, click Property Sheet in the Show/Hide group. The Field Properties sheet appears, as shown on the right in Figure B-14.

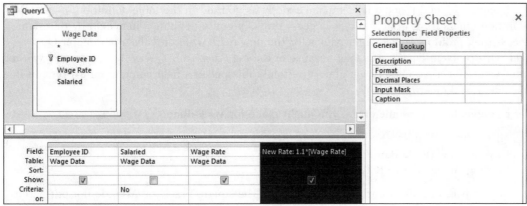

Source: Used with permission from Microsoft Corporation

FIGURE B-14   Field properties of a calculated field

Click Format and choose Currency, as shown in Figure B-15. Then click the *X* in the upper-right corner of the window to close it.

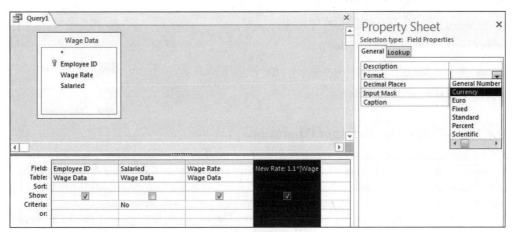

Source: Used with permission from Microsoft Corporation

FIGURE B-15   Currency format of a calculated field

When you run the query, the output should resemble that in Figure B-16.

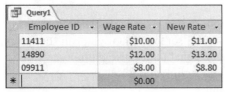

Source: Used with permission from Microsoft Corporation

FIGURE B-16   Query output with formatted calculated field

Next, you examine how to avoid errors when making calculated fields.

Avoiding Errors when Making Calculated Fields

Follow these guidelines to avoid making errors in calculated fields:

- Do not enter the expression in the *Criteria* cell as if the field definition were a filter. You are making a field, so enter the expression in the *Field* cell.

- Spell, capitalize, and space a field's name *exactly* as you did in the table definition. If the table definition differs from what you type, Access thinks you are defining a new field by that name. Access then prompts you to enter values for the new field, which it calls a Parameter Query field. This problem is easy to debug because of the tag *Parameter Query*. If Access asks you to enter values for a parameter, you almost certainly misspelled a field name in an expression in a calculated field or criterion.

For example, here are some errors you might make for Wage Rate:

> Misspelling: (Wag Rate)
> Case change: (wage Rate / WAGE RATE)
> Spacing change: (WageRate / Wage Rate)

- Do not use parentheses or curly braces instead of the square brackets. Also, do not put parentheses inside square brackets. You *can*, however, use parentheses outside the square brackets in the normal algebraic manner.

For example, suppose that you want to multiply Hours by Wage Rate to get a field called Wages Owed. This is the correct expression:

> Wages Owed: [Wage Rate] * [Hours]

The following expression also would be correct:

> Wages Owed: ([Wage Rate] * [Hours])

But it would *not* be correct to omit the inside brackets, which is a common error:

> Wages Owed: [Wage Rate * Hours]

"Relating" Two or More Tables by the Join Operation

Often, the data you need for a query is in more than one table. To complete the query, you must **join** the tables by linking the common fields. One rule of thumb is that joins are made on fields that have common *values*, and those fields often can be key fields. The names of the join fields are irrelevant; also, the names of the tables or fields to be joined may be the same, but it is not required for an effective join.

Make a join by bringing in (adding) the tables needed. Next, decide which fields you will join. Then click one field name and hold down the left mouse button while you drag the cursor over to the other field's name in its window. Release the button. Access inserts a line to signify the join. (If a relationship between two tables has been formed elsewhere, Access inserts the line automatically, and you do not have to perform the click-and-drag operation. Access often inserts join lines without the user forming relationships.)

You can join more than two tables. The common fields *need not* be the same in all tables; that is, you can daisy-chain them together.

A common join error is to add a table to the query and then fail to link it to another table. In that case, you will have a table floating in the top part of the QBE (query by example) screen. When you run the query, your output will show the same records over and over. The error is unmistakable because there is *so much* redundant output. The two rules are to add only the tables you need and to link all tables.

Next, you will work through an example of a query that needs a join.

AT THE KEYBOARD

Suppose you want to see the last names, employee IDs, wage rates, salary status, and citizenship only for U.S. citizens and hourly workers. Because the data is spread across two tables, Employee and Wage Data, you should add both tables and pull down the five fields you need. Then you should add the Criteria expressions. Set up your work to resemble that in Figure B-17. Make sure the tables are joined on the common field, Employee ID.

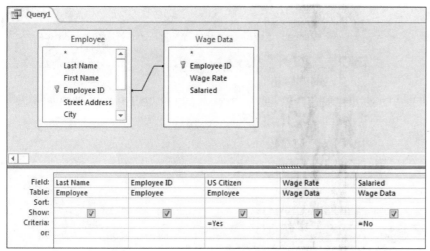

Source: Used with permission from Microsoft Corporation

FIGURE B-17   A query based on two joined tables

You should quickly review the criteria you will need to set up this join: If you want data for employees who are U.S. citizens *and* who are hourly workers, the Criteria expressions go in the *same* Criteria row. If you want data for employees who are U.S. citizens *or* who are hourly workers, one of the expressions goes in the second Criteria row (the one with the or: notation).

Now run the query. The output should resemble that in Figure B-18, with the exception of the name "Brady."

Last Name	Employee ID	US Citizen	Wage Rate	Salaried
Howard	11411	☑	$10.00	☐
Smith	14890	☑	$12.00	☐
Brady	09911	☑	$8.00	☐

Source: Used with permission from Microsoft Corporation

FIGURE B-18   Output of a query based on two joined tables

You do not need to print or save the query output, so return to Design view and close the query. Another practice query follows.

AT THE KEYBOARD

Suppose you want to see the wages owed to hourly employees for Week 2. You should show the last name, the employee ID, the salaried status, the week #, and the wages owed. Wages will have to be a calculated field ([Wage Rate] * [Hours]). The criteria are No for Salaried and 2 for the Week #. (This means that another "And" query is required.) Your query should be set up like the one in Figure B-19.

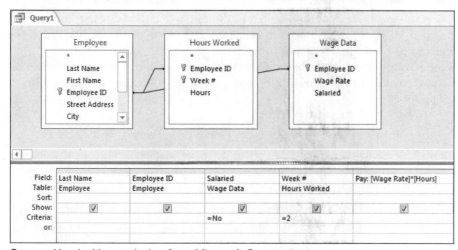

Source: Used with permission from Microsoft Corporation

FIGURE B-19   Query setup for wages owed to hourly employees for Week 2

NOTE

In the query in Figure B-19, the calculated field column was widened so you could see the whole expression. To widen a column, click the column boundary line and drag to the right.

Run the query. The output should be similar to that in Figure B-20, if you formatted your calculated field to Currency.

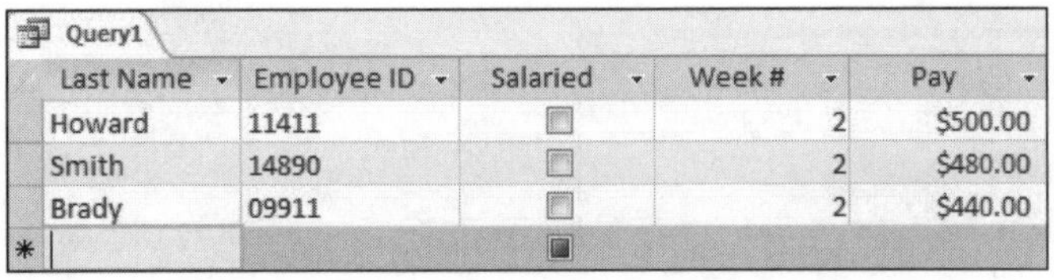

Query1				
Last Name ▾	Employee ID ▾	Salaried ▾	Week # ▾	Pay ▾
Howard	11411	☐	2	$500.00
Smith	14890	☐	2	$480.00
Brady	09911	☐	2	$440.00
*		◼		

Source: Used with permission from Microsoft Corporation

FIGURE B-20   Query output for wages owed to hourly employees for Week 2

Notice that it was not necessary to pull down the Wage Rate and Hours fields to make the query work. You do not need to save or print the query output, so return to Design view and close the query.

Summarizing Data from Multiple Records (Totals Queries)

You may want data that summarizes values from a field for several records (or possibly all records) in a table. For example, you might want to know the average hours that all employees worked in a week or the total (sum) of all of the hours worked. Furthermore, you might want data grouped or stratified in some way. For example, you might want to know the average hours worked, grouped by all U.S. citizens versus all non-U.S. citizens. Access calls such a query a **Totals query**. These queries include the following operations:

Sum	The total of a given field's values
Count	A count of the number of instances in a field—that is, the number of records. In the current example, you would count the number of employee IDs to get the number of employees.
Average	The average of a given field's values
Min	The minimum of a given field's values
Var	The variance of a given field's values
StDev	The standard deviation of a given field's values
Where	The field has criteria for the query output

AT THE KEYBOARD

Suppose you want to know how many employees are represented in the example database. First, bring the Employee table into the QBE screen. Because you will need to count the number of employee IDs, which is a Totals query operation, you must bring down the Employee ID field.

To tell Access that you want a Totals query, click the Design tab and then click the Totals button in the Show/Hide group. A new row called the Total row opens in the lower part of the QBE screen. At this point, the screen resembles that in Figure B-21.

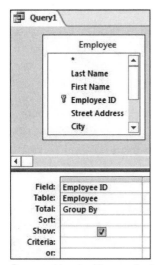

Source: Used with permission from Microsoft Corporation

FIGURE B-21   Totals query setup

Note that the Total cell contains the words *Group By*. Until you specify a statistical operation, Access assumes that a field will be used for grouping (stratifying) data.

To count the number of employee IDs, click next to Group By to display an arrow. Click the arrow to reveal a drop-down menu, as shown in Figure B-22.

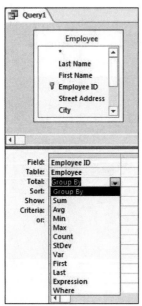

Source: Used with permission from Microsoft Corporation

FIGURE B-22   Choices for statistical operation in a Totals query

Select the Count operator. (You might need to scroll down the menu to see the operator you want.) Your screen should resemble the one shown in Figure B-23.

Source: Used with permission from Microsoft Corporation

FIGURE B-23   Count in a Totals query

Run the query. Your output should resemble that in Figure B-24.

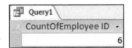

Source: Used with permission from Microsoft Corporation

FIGURE B-24   Output of Count in a Totals query

Notice that Access created a pseudo-heading, "CountOfEmployee ID," by splicing together the statistical operation (Count), the word Of, and the name of the field (Employee ID). If you wanted a phrase such as "Count of Employees" as a heading, you would go to Design view and change the query to resemble the one shown in Figure B-25.

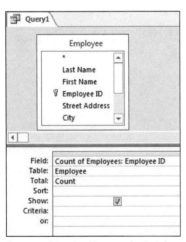

Source: Used with permission from Microsoft Corporation

FIGURE B-25   Heading change in a Totals query

When you run the query, the output should resemble that in Figure B-26.

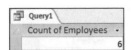

Source: Used with permission from Microsoft Corporation

FIGURE B-26   Output of heading change in a Totals query

You do not need to print or save the query output, so return to Design view and close the query.

AT THE KEYBOARD

As another example of a Totals query, suppose you want to know the average wage rate of employees, grouped by whether the employees are salaried. Figure B-27 shows how to set up your query.

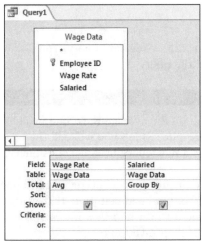

Source: Used with permission from Microsoft Corporation

FIGURE B-27   Query setup for average wage rate of employees

When you run the query, your output should resemble that in Figure B-28.

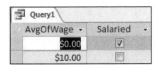

Source: Used with permission from Microsoft Corporation

FIGURE B-28   Output of query for average wage rate of employees

Recall the convention that salaried workers are assigned zero dollars an hour. Suppose you want to eliminate the output line for zero dollars an hour because only hourly-rate workers matter for the query. The query setup is shown in Figure B-29.

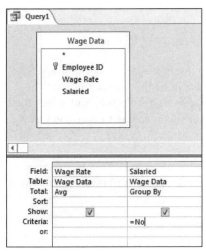

Source: Used with permission from Microsoft Corporation

FIGURE B-29   Query setup for nonsalaried workers only

When you run the query, you will get output for nonsalaried employees only, as shown in Figure B-30.

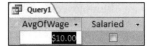

FIGURE B-30   Query output for nonsalaried workers only

Thus, it is possible to use Criteria in a Totals query, just as you would with a "regular" query. You do not need to print or save the query output, so return to Design view and close the query.

AT THE KEYBOARD

Assume that you want to see two pieces of information for hourly workers: (1) the average wage rate, which you will call Average Rate in the output; and (2) 110 percent of the average rate, which you will call the Increased Rate. To get this information, you can make a calculated field in a new query from a Totals query. In other words, you use one query as a basis for another query.

Create the first query; you already know how to perform certain tasks for this query. The revised heading for the average rate will be Average Rate, so type *Average Rate: Wage Rate* in the Field cell. Note that you want the average of this field. Also, the grouping will be by the Salaried field. (To get hourly workers only, enter *Criteria: No*.) Confirm that your query resembles that in Figure B-31, then save the query and close it.

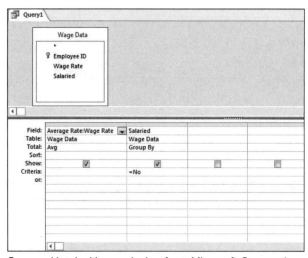

FIGURE B-31   A totals query with average

Now begin a new query. However, instead of bringing in a table to the query design, select a query. To start a new query, click the Create tab and then click the Query Design button in the Queries group. The Show Table dialog box appears. Click the Queries tab instead of using the default Tables tab, and select the query you just saved as a basis for the new query. The most difficult part of this query is to construct the expression for the calculated field. Conceptually, it is as follows:

Increased Rate: 1.1 * [The current average]

You use the new field name in the new query as the current average, and you treat the new name like a new field:

Increased Rate: 1.1 * [Average Rate]

The query within a query is shown in Figure B-32.

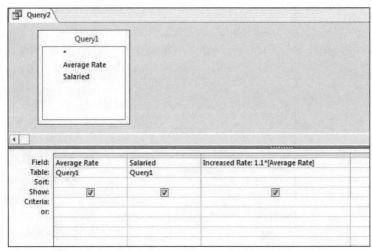

Source: Used with permission from Microsoft Corporation

FIGURE B-32   A query within a query

Figure B-33 shows the output of the new query. Note that the calculated field is formatted.

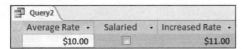

Source: Used with permission from Microsoft Corporation

FIGURE B-33   Output of an Expression in a Totals query

You do not need to print or save the query output, so return to Design view and close the query.

Using the Date() Function in Queries

Access has two important date function features:

- The built-in Date() function gives you today's date. You can use the function in query criteria or in a calculated field. The function "returns" the day on which the query is run; in other words, it inserts the value where the Date() function appears in an expression.
- Date arithmetic lets you subtract one date from another to obtain the difference—in number of days—between two calendar dates. For example, suppose you create the following expression:
10/9/2012 – 10/4/2012
Access would evaluate the expression as the integer 5 (9 minus 4 is 5).

As another example of how date arithmetic works, suppose you want to give each employee a one-dollar bonus for each day the employee has worked. You would need to calculate the number of days between the employee's date of hire and the day the query is run, and then multiply that number by $1.

You would find the number of elapsed days by using the following equation:

Date() – [Date Hired]

Also suppose that for each employee, you want to see the last name, employee ID, and bonus amount. You would set up the query as shown in Figure B-34.

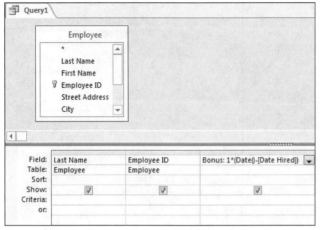

Source: Used with permission from Microsoft Corporation

FIGURE B-34   Date arithmetic in a query

Assume that you set the format of the Bonus field to Currency. The output will be similar to that in Figure B-35, although your Bonus data will be different because you used a different date.

Last Name	Employee ID	Bonus
Brady	09911	$0.00
Howard	11411	$25.00
Smith	12345	$6,234.00
Smith	14890	$9,478.00
Jones	22282	$3,268.00
Ruth	71460	$5,064.00

Source: Used with permission from Microsoft Corporation

FIGURE B-35   Output of query with date arithmetic

Using Time Arithmetic in Queries

Access also allows you to subtract the values of time fields to get an elapsed time. Assume that your database has a Job Assignments table showing the times that nonsalaried employees were at work during a day. The definition is shown in Figure B-36.

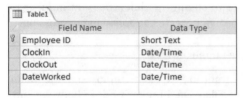

Field Name	Data Type
Employee ID	Short Text
ClockIn	Date/Time
ClockOut	Date/Time
DateWorked	Date/Time

Source: Used with permission from Microsoft Corporation

FIGURE B-36   Date/Time data definition in the Job Assignments table

Assume that the DateWorked field is formatted for Long Date and that the ClockIn and ClockOut fields are formatted for Medium Time. Also assume that for a particular day, nonsalaried workers were scheduled as shown in Figure B-37.

Employee ID	ClockIn	ClockOut	DateWorked	Click to Add
09911	8:30 AM	4:30 PM	Monday, September 29, 2014	
11411	9:00 AM	3:00 PM	Monday, September 29, 2014	
14890	7:00 AM	5:00 PM	Monday, September 29, 2014	

Source: Used with permission from Microsoft Corporation

FIGURE B-37   Display of date and time in a table

You want a query showing the elapsed time that your employees were on the premises for the day. When you add the tables, your screen may show the links differently. Click and drag the Job Assignments, Employee, and Wage Data table icons to look like those in Figure B-38.

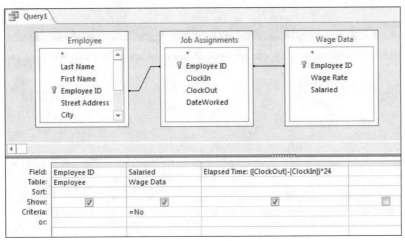

Source: Used with permission from Microsoft Corporation

FIGURE B-38   Query setup for time arithmetic

Figure B-39 shows the output, which looks correct. For example, employee 09911 was at work from 8:30 a.m. to 4:30 p.m., which is eight hours. But how does the odd expression that follows yield the correct answers?

Employee ID ▾	Salaried ▾	Elapsed Time ▾
11411	☐	6
14890	☐	10
09911	☐	8
*	■	

Source: Used with permission from Microsoft Corporation

FIGURE B-39   Query output for time arithmetic

([ClockOut] – [ClockIn]) * 24

Why wouldn't the following expression work?

[ClockOut] – [ClockIn]

Here is the answer: In Access, subtracting one time from the other yields the *decimal* portion of a 24-hour day. Returning to the example, you can see that employee 09911 worked eight hours, which is one-third of a day, so the time arithmetic function yields .3333. That is why you must multiply by 24—to convert from decimals to an hourly basis. Hence, for employee 09911, the expression performs the following calculation: $1/3 \times 24 = 8$.

Note that parentheses are needed to force Access to do the subtraction *first*, before the multiplication. Without parentheses, multiplication takes precedence over subtraction. For example, consider the following expression:

[ClockOut] – [ClockIn] * 24

In this example, ClockIn would be multiplied by 24, the resulting value would be subtracted from Clock-Out, and the output would be a nonsensical decimal number.

Deleting and Updating Queries

The queries presented in this tutorial so far have been Select queries. They select certain data from specific tables based on a given criterion. You also can create queries to update the original data in a database. Businesses use such queries often, and in real time. For example, when you order an item from a Web site, the company's database is updated to reflect your purchase through the deletion of that item from the company's inventory.

Consider an example. Suppose you want to give all nonsalaried workers a $0.50 per hour pay raise. Because you have only three nonsalaried workers, it would be easy to change the Wage Rate data in the table. However, if you had 3,000 nonsalaried employees, it would be much faster and more accurate to change the Wage Rate data by using an Update query that adds $0.50 to each nonsalaried employee's wage rate.

AT THE KEYBOARD

Now you will change each of the nonsalaried employees' pay via an Update query. Figure B-40 shows how to set up the query.

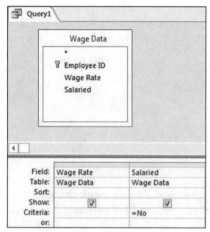

Source: Used with permission from Microsoft Corporation
FIGURE B-40   Query setup for an Update query

So far, this query is just a Select query. Click the Update button in the Query Type group, as shown in Figure B-41.

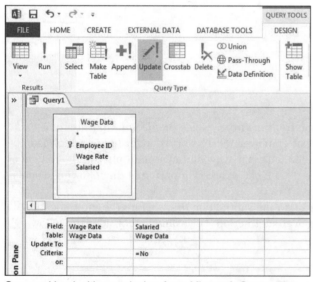

Source: Used with permission from Microsoft Corporation
FIGURE B-41   Selecting a query type

Notice that you now have another line on the QBE grid called Update To:, which is where you specify the change or update the data. Notice that you will update only the nonsalaried workers by using a filter under the Salaried field. Update the Wage Rate data to Wage Rate plus $0.50, as shown in Figure B-42. Note that the update involves the use of brackets [ ], as in a calculated field.

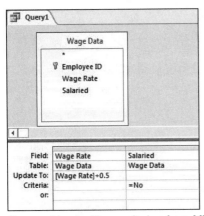

Source: Used with permission from Microsoft Corporation

FIGURE B-42 Updating the wage rate for nonsalaried workers

Now run the query by clicking the Run button in the Results group. If you cannot run the query because it is blocked by Disabled Mode, click the Database Tools tab, then click Message Bar in the Show/Hide group. Click the Options button, choose "enable this content," and then click OK. When you successfully run the query, the warning message in Figure B-43 appears.

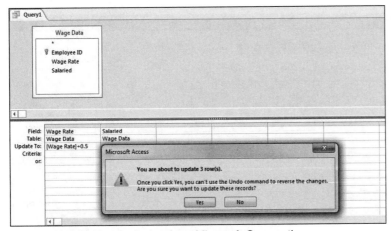

Source: Used with permission from Microsoft Corporation

FIGURE B-43 Update query warning

When you click Yes, the records are updated. Check the updated records by viewing the Wage Data table. Each nonsalaried wage rate should be increased by $0.50. You could add or subtract data from another table as well. If you do, remember to put the field name in square brackets.

Another type of query is the Delete query, which works like Update queries. For example, assume that your company has been purchased by the state of Delaware, which has a policy of employing only state residents. Thus, you must delete (or fire) all employees who are not exclusively Delaware residents. To do that, you would create a Select query. Using the Employee table, you would click the Delete button in the Query Type group, then bring down the State field and filter only those records that were not in Delaware (DE). Do not perform the operation, but note that if you did, the setup would look like the one in Figure B-44.

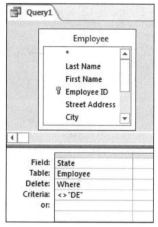

Source: Used with permission from Microsoft Corporation

FIGURE B-44   Deleting all employees who are not Delaware residents

Using Parameter Queries

A **Parameter query** is actually a type of Select query. For example, suppose your company has 5,000 employees and you want to query the database to find the same kind of information repeatedly, but about different employees each time. For example, you might want to know how many hours a particular employee has worked. You could run a query that you created and stored previously, but run it only for a particular employee.

AT THE KEYBOARD

Create a Select query with the format shown in Figure B-45.

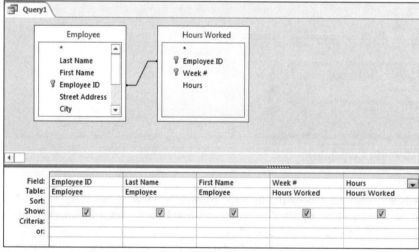

Source: Used with permission from Microsoft Corporation

FIGURE B-45   Design of a Parameter query beginning as a Select query

In the Criteria line of the QBE grid for the Employee ID field, type what is shown in Figure B-46.

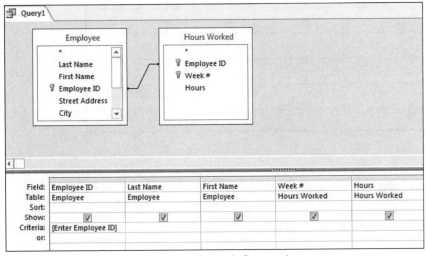

Source: Used with permission from Microsoft Corporation

FIGURE B-46   Design of a Parameter query, continued

Note that the Criteria line uses square brackets, as you would expect to see in a calculated field.

Now run the query. You will be prompted for the employee's ID number, as shown in Figure B-47.

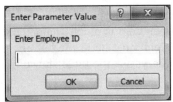

Source: Used with permission from Microsoft Corporation

FIGURE B-47   Enter Parameter Value dialog box

Enter your own employee ID. Your query output should resemble that in Figure B-48.

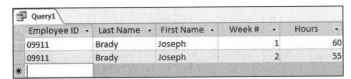

Source: Used with permission from Microsoft Corporation

FIGURE B-48   Output of a Parameter query

MAKING SEVEN PRACTICE QUERIES

This portion of the tutorial gives you additional practice in creating queries. Before making these queries, you must create the specified tables and enter the records shown in the "Creating Tables" section of this tutorial. The output shown for the practice queries is based on those inputs.

AT THE KEYBOARD

For each query that follows, you are given a problem statement and a "scratch area." You also are shown what the query output should look like. Set up each query in Access and then run the query. When you are satisfied with the results, save the query and continue with the next one. Note that you will work with the Employee, Hours Worked, and Wage Data tables.

1.   Create a query that shows the employee ID, last name, state, and date hired for employees who live in Delaware *and* were hired after 12/31/99. Perform an ascending sort by employee ID. First click the Sort cell of the field, and then choose Ascending or Descending. Before creating your query, use the table shown in Figure B-49 to work out your QBE grid on paper.

Field					
Table					
Sort					
Show					
Criteria					
Or:					

Source: © Cengage Learning 2015

FIGURE B-49   QBE grid template

Your output should resemble that in Figure B-50.

Employee ID	Last Name	State	Date Hired
11411	Howard	DE	6/1/2014
22282	Jones	DE	7/15/2004

Source: Used with permission from Microsoft Corporation

FIGURE B-50   Number 1 query output

2. Create a query that shows the last name, first name, date hired, and state for employees who live in Delaware *or* were hired after 12/31/99. The primary sort (ascending) is on last name, and the secondary sort (ascending) is on first name. The Primary Sort field must be to the left of the Secondary Sort field in the query setup. Before creating your query, use the table shown in Figure B-51 to work out your QBE grid on paper.

Field					
Table					
Sort					
Show					
Criteria					
Or:					

Source: © Cengage Learning 2015

FIGURE B-51   QBE grid template

If your name was Joseph Brady, your output would look like that in Figure B-52.

Last Name	First Name	Date Hired	State
Brady	Joseph	9/15/2014	MD
Howard	Jane	6/1/2014	DE
Jones	Sue	7/15/2004	DE
Smith	Albert	7/15/1987	DE
Smith	John	6/1/1996	DE

Source: Used with permission from Microsoft Corporation

FIGURE B-52   Number 2 query output

3. Create a query that sums the number of hours worked by U.S. citizens and the number of hours worked by non-U.S. citizens. In other words, create two sums, grouped on citizenship. The heading for total hours worked should be Total Hours Worked. Before creating your query, use the table shown in Figure B-53 to work out your QBE grid on paper.

Field					
Table					
Total					
Sort					
Show					
Criteria					
Or:					

Source: © Cengage Learning 2015

FIGURE B-53 QBE grid template

Your output should resemble that in Figure B-54.

Practice Query 3

Total Hours Worked	US Citizen
363	☑
160	☐

Source: Used with permission from Microsoft Corporation

FIGURE B-54 Number 3 query output

4. Create a query that shows the wages owed to hourly workers for Week 1. The heading for the wages owed should be Total Owed. The output headings should be Last Name, Employee ID, Week #, and Total Owed. Before creating your query, use the table shown in Figure B-55 to work out your QBE grid on paper.

Field					
Table					
Sort					
Show					
Criteria					
Or:					

Source: © Cengage Learning 2015

FIGURE B-55 QBE grid template

If your name was Joseph Brady, your output would look like that in Figure B-56.

Practice Query 4

Last Name	Employee ID	Week #	Total Owed
Howard	11411	1	$420.00
Smith	14890	1	$475.00
Brady	09911	1	$510.00

Source: Used with permission from Microsoft Corporation

FIGURE B-56 Number 4 query output

5. Create a query that shows the last name, employee ID, hours worked, and overtime amount owed for hourly employees who earned overtime during Week 2. Overtime is paid at 1.5 times the normal hourly rate for all hours worked over 40. Note that the amount shown in the query should be just the overtime portion of the wages paid. Also, this is not a Totals query—amounts should be shown for individual workers. Before creating your query, use the table shown in Figure B-57 to work out your QBE grid on paper.

Field					
Table					
Sort					
Show					
Criteria					
Or:					

Source: © Cengage Learning 2015

FIGURE B-57   QBE grid template

If your name was Joseph Brady, your output would look like that in Figure B-58.

Practice Query 5			
Last Name ▾	Employee ID ▾	Hours ▾	OT Pay ▾
Howard	11411	50	$157.50
Brady	09911	55	$191.25
*			

Source: Used with permission from Microsoft Corporation

FIGURE B-58   Number 5 query output

6. Create a Parameter query that shows the hours employees have worked. Have the Parameter query prompt for the week number. The output headings should be Last Name, First Name, Week #, and Hours. This query is for nonsalaried workers only. Before creating your query, use the table shown in Figure B-59 to work out your QBE grid on paper.

Field					
Table					
Sort					
Show					
Criteria					
Or:					

Source: © Cengage Learning 2015

FIGURE B-59   QBE grid template

Run the query and enter 2 when prompted for the week number. Your output should look like that in Figure B-60.

Practice Query 6			
Last Name ▾	First Name ▾	Week # ▾	Hours ▾
Howard	Jane	2	50
Smith	Albert	2	40
Brady	Joseph	2	55
*			

Source: Used with permission from Microsoft Corporation

FIGURE B-60   Number 6 query output

7. Create an Update query that gives certain workers a merit raise. First, you must create an additional table, as shown in Figure B-61.

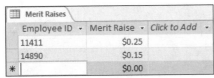

Merit Raises		
Employee ID ▾	Merit Raise ▾	Click to Add ▾
11411	$0.25	
14890	$0.15	
*	$0.00	

Source: Used with permission from Microsoft Corporation

FIGURE B-61   Merit Raises table

Create a query that adds the Merit Raise to the current Wage Rate for employees who will receive a raise. When you run the query, you should be prompted with *You are about to update two rows*. Check the original Wage Data table to confirm the update. Before creating your query, use the table shown in Figure B-62 to work out your QBE grid on paper.

Field					
Table					
Update to					
Criteria					
Or:					

Source: © Cengage Learning 2015

FIGURE B-62   QBE grid template

CREATING REPORTS

Database packages let you make attractive management reports from a table's records or from a query's output. If you are making a report from a table, the Access report generator looks up the data in the table and puts it into report format. If you are making a report from a query's output, Access runs the query in the background (you do not control it or see it happen) and then puts the output in report format.

There are different ways to make a report. One method is to create one from scratch in Design view, but this tedious process is not explained in this tutorial. A simpler way is to select the query or table on which the report is based and then click Create Report. This streamlined method of creating reports is explained in this tutorial.

Creating a Grouped Report

This tutorial assumes that you already know how to create a basic ungrouped report, so this section teaches you how to make a grouped report. If you do not know how to create an ungrouped report, you can learn by following the first example in the upcoming section.

AT THE KEYBOARD

Suppose you want to create a report from the Hours Worked table. Select the table by clicking it once. Click the Create tab, then click Report in the Reports group. A report appears, as shown in Figure B-63.

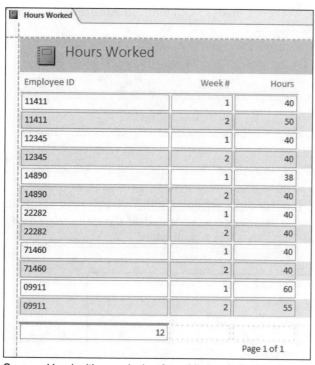

Source: Used with permission from Microsoft Corporation

FIGURE B-63   Initial report based on a table

On the Design tab, select the Group & Sort button in the Grouping & Totals group. Your report will have an additional selection at the bottom, as shown in Figure B-64.

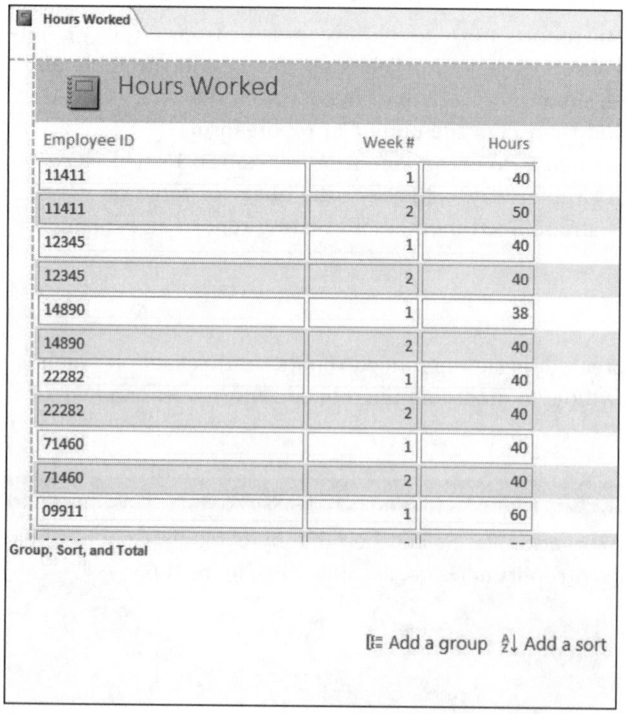

Source: Used with permission from Microsoft Corporation

FIGURE B-64   Report with grouping and sorting options

Click the Add a group button at the bottom of the report, and then select Employee ID. Your report will be grouped as shown in Figure B-65.

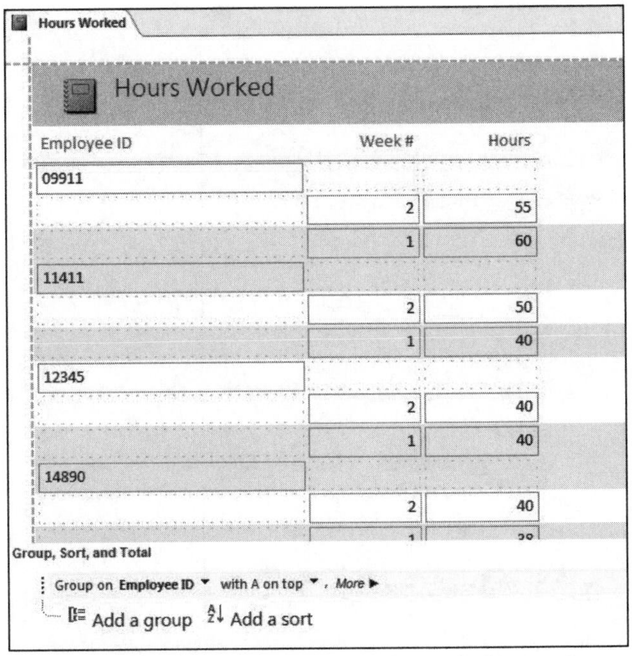

Source: Used with permission from Microsoft Corporation

FIGURE B-65   Grouped report

To complete this report, you need to total the hours for each employee by selecting the Hours column heading. Your report will show that the entire column is selected. On the Design tab, click the Totals button in the Grouping & Totals group, and then choose Sum from the menu, as shown in Figure B-66.

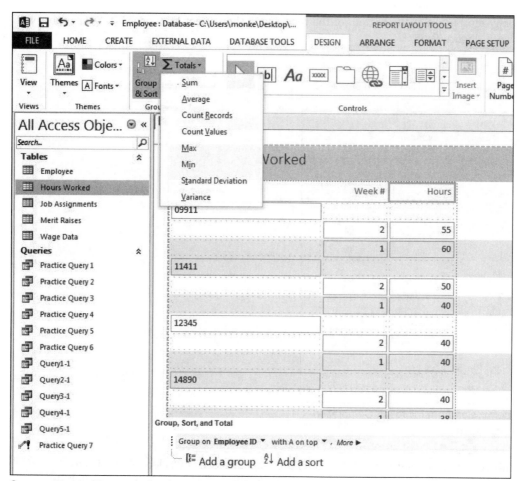

Source: Used with permission from Microsoft Corporation

FIGURE B-66   Totaling the hours

Your report will look like the one in Figure B-67.

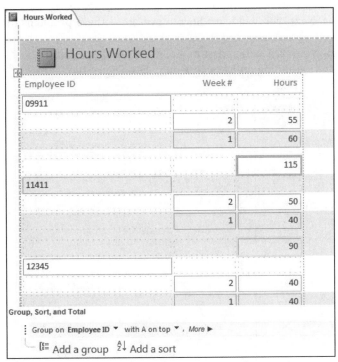

Source: Used with permission from Microsoft Corporation

FIGURE B-67   Completed report

Your report is currently in Layout view. To see how the final report looks when printed, click the Design tab and select Report View from the Views group. Your report looks like the one in Figure B-68, although only a portion is shown in the figure.

Source: Used with permission from Microsoft Corporation

FIGURE B-68   Report in Report view

NOTE

To change the picture or logo in the upper-left corner of the report when in Layout view, click the notebook symbol and press the Delete key. You can insert a logo in place of the notebook by clicking the Design tab and then clicking the Insert Image button in the Controls group.

Moving Fields in Layout View

If you group records based on more than one field in a report, the report will have an odd "staircase" look or display repeated data, or it will have both problems. Next, you will learn how to overcome these problems in Layout view.

Suppose you make a query that shows an employee's last name, first name, week number, and hours worked, and then you make a report from that query, grouping on last name only. See Figure B-69.

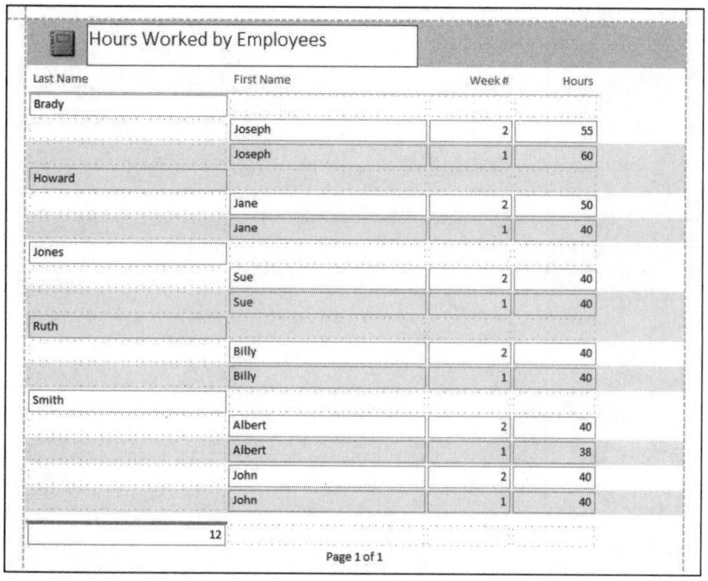

Source: Used with permission from Microsoft Corporation

FIGURE B-69   Query-based report grouped on last name

As you preview the report, notice the repeating data from the First Name field. In the report shown in Figure B-69, notice that the first name repeats for each week worked—hence, the staircase effect. The Week # and Hours fields are shown as subordinate to Last Name, as desired.

Suppose you want the last name and first name to appear on the same line. If so, take the report into Layout view for editing. Click the first record for the First Name (in this case, Joseph), and drag the name up to the same line as the Last Name (in this case, Brady). Your report will now show the First Name on the same line as Last Name, thereby eliminating the staircase look, as shown in Figure B-70.

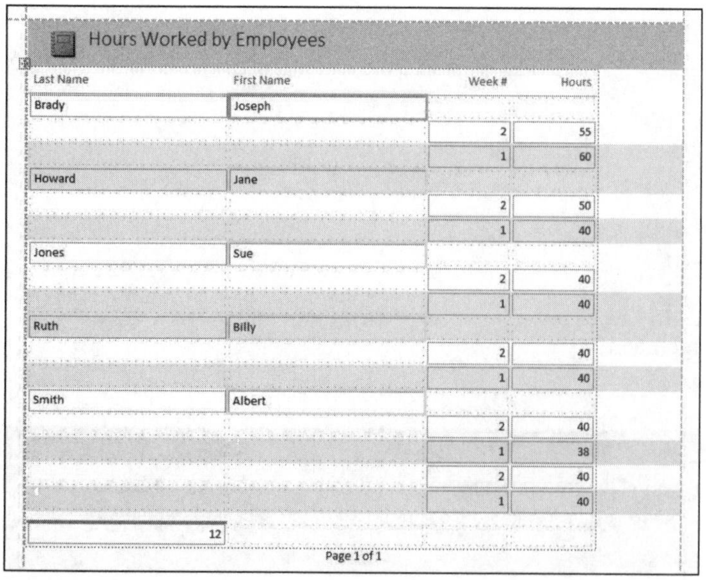

Source: Used with permission from Microsoft Corporation

FIGURE B-70   Report in Layout view with Last Name and First Name on the same line

You can now add the sum of Hours for each group. Also, if you want to add more fields to your report, such as Street Address and Zip, you can repeat the preceding procedure.

IMPORTING DATA

Text or spreadsheet data is easy to import into Access. In business, it is often necessary to import data because companies use disparate systems. For example, assume that your healthcare coverage data is on the human resources manager's computer in a Microsoft Excel spreadsheet. Open the Excel application and then create a spreadsheet using the data shown in Figure B-71.

	A	B	C
1	Employee ID	Provider	Level
2	11411	BlueCross	family
3	12345	BlueCross	family
4	14890	Coventry	spouse
5	22282	None	none
6	71460	Coventry	single
7	Your ID	BlueCross	single

Source: Used with permission from Microsoft Corporation

FIGURE B-71   Excel data

Save the file and then close it. Now you can easily import the spreadsheet data into a new table in Access. With your Employee database open, click the External Data tab, then click Excel in the Import & Link group. Browse to find the Excel file you just created, and make sure the first radio button is selected to import the source data into a new table in the current database (see Figure B-72). Click OK.

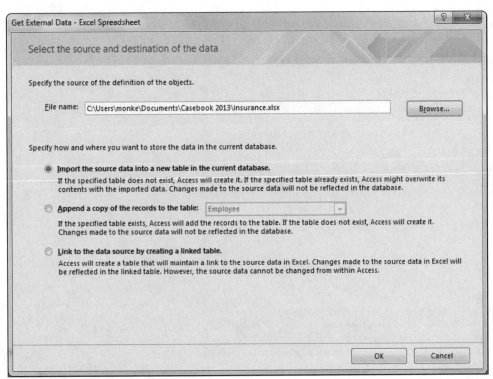

Source: Used with permission from Microsoft Corporation

FIGURE B-72   Importing Excel data into a new table

Choose the correct worksheet. Assuming that you have just one worksheet in your Excel file, your next screen should look like the one in Figure B-73.

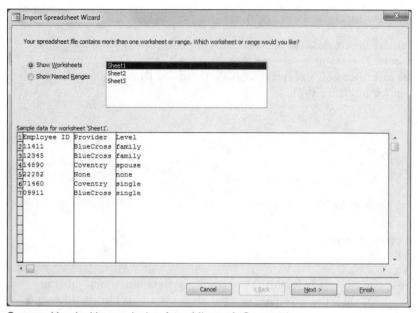

Source: Used with permission from Microsoft Corporation

FIGURE B-73   First screen in the Import Spreadsheet Wizard

Choose Next, and then make sure to select the First Row Contains Column Headings box, as shown in Figure B-74.

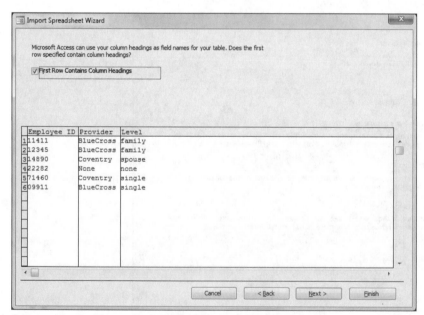

Source: Used with permission from Microsoft Corporation

FIGURE B-74   Choosing column headings in the Import Spreadsheet Wizard

Choose Next. Accept the default setting for each field you are importing on the screen. Each field is assigned a text data type, which is correct for this table. Your screen should look like the one in Figure B-75.

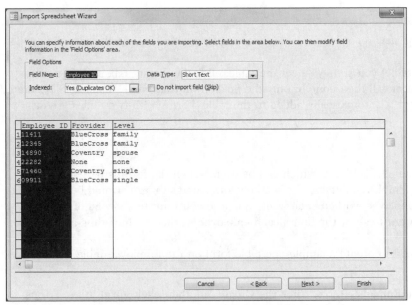

Source: Used with permission from Microsoft Corporation

FIGURE B-75   Choosing the data type for each field in the Import Spreadsheet Wizard

Choose Next. In the next screen of the wizard, you will be prompted to create an index—that is, to define a primary key. Because you will store your data in a new table, choose your own primary key (Employee ID), as shown in Figure B-76.

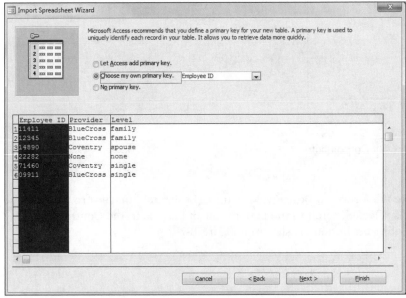

Source: Used with permission from Microsoft Corporation

FIGURE B-76   Choosing a primary key field in the Import Spreadsheet Wizard

Continue through the wizard, giving your table an appropriate name. After importing the table, take a look at its design by right-clicking the table and choosing Design View. Note that each field is very wide. Adjust the field properties as needed.

MAKING FORMS

Forms simplify the process of adding new records to a table. Creating forms is easy, and they can be applied to one or more tables.

When you base a form on one table, you simply select the table, click the Create tab, and then select Form from the Forms group. The form will then contain only the fields from that table. When data is entered into the form, a complete new record is automatically added to the table. Forms with two tables are discussed next.

Making Forms with Subforms

You also can create a form that contains a subform, which can be useful when the form is based on two or more tables. Return to the example Employee database to see how forms and subforms would be useful for viewing all of the hours that each employee worked each week. Suppose you want to show all of the fields from the Employee table; you also want to show the hours each employee worked by including all fields from the Hours Worked table as well.

To create the form and subform, first create a simple one-table form on the Employee table. Follow these steps:

1. Click once to select the Employee table. Click the Create tab, then click Form in the Forms group. After the main form is complete, it should resemble the one in Figure B-77.

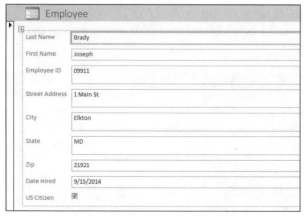

Source: Used with permission from Microsoft Corporation

FIGURE B-77   The Employee form

2. To add the subform, take the form into Design view. On the Design tab, make sure that the Use Control Wizards option is selected, scroll to the bottom row of buttons in the Controls group, and click the Subform/Subreport button, as shown in Figure B-78.

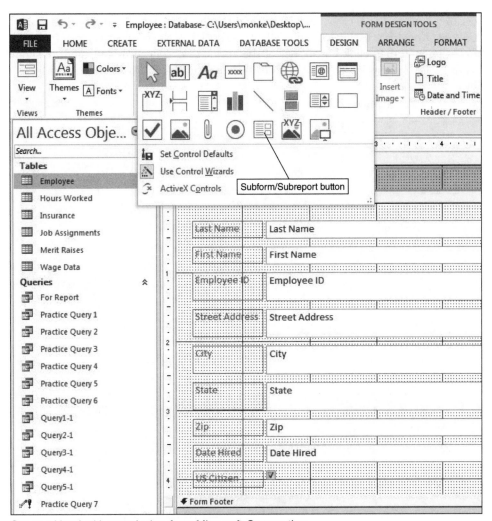

Source: Used with permission from Microsoft Corporation

FIGURE B-78   The Subform/Subreport button

3. Use your cursor to stretch out the box under your main form. You might need to expand the area beneath the main form. The dialog box shown in Figure B-79 appears.

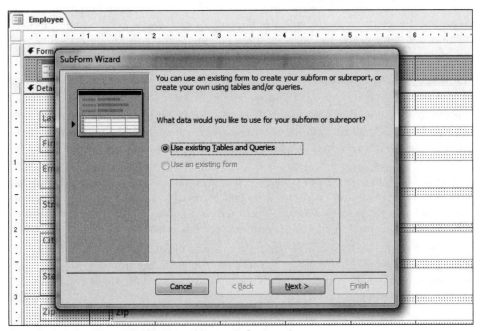

Source: Used with permission from Microsoft Corporation

FIGURE B-79   Adding a subform

4. Select Use existing Tables and Queries, click Next, and then select Table: Hours Worked from the Tables/Queries drop-down list. Select all available fields. Click Next, select Choose from a list, click Next again, and then click Finish. Select the Form view. Your form and subform should resemble Figure B-80. You may need to stretch out the subform box in Design view if all fields are not visible.

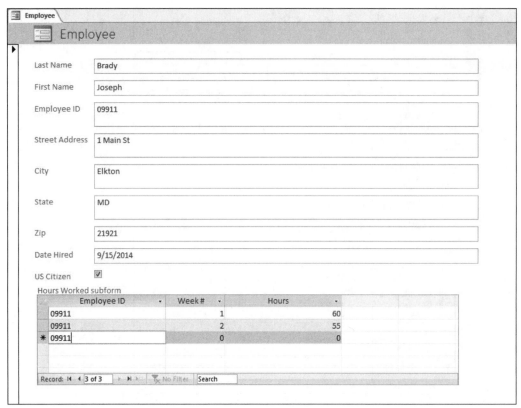

Source: Used with permission from Microsoft Corporation

FIGURE B-80   Form with subform

TROUBLESHOOTING COMMON PROBLEMS

Access is a powerful program, but it is complex and sometimes difficult for new users. People sometimes unintentionally create databases that have problems. Some of these common problems are described below, along with their causes and corrections.

1. *"I saved my database file, but I can't find it on my computer or my external secondary storage medium! Where is it?"*

   You saved your file to a fixed disk or a location other than the Documents folder. Use the Windows Search option to find all files ending in .accdb (search for *.accdb). If you saved the file, it is on the hard drive (C:\) or a network drive. Your site assistant can tell you the drive designators.

2. *"What is a 'duplicate key field value'? I'm trying to enter records into my Sales table. The first record was for a sale of product X to customer 101, and I was able to enter that one. But when I try to enter a second sale for customer #101, Access tells me I already have a record with that key field value. Am I allowed to enter only one sale per customer?"*

   Your primary key field needs work. You may need a compound primary key—a combination of the customer number and some other field(s). In this case, the customer number, product number, and date of sale might provide a unique combination of values, or you might consider using an invoice number field as a key.

3. *"My query reads 'Enter Parameter Value' when I run it. What is that?"*

This problem almost always indicates that you have misspelled a field name in an expression in a Criteria field or calculated field. Access is very fussy about spelling; for example, it is case-sensitive. Access is also "space-sensitive," meaning that when you insert a space in a field name when defining a table, you must also include a space in the field name when you reference it in a query expression. Fix the typo in the query expression.

4. *"I'm getting an enormous number of rows in my query output—many times more than I need. Most of the rows are duplicates!"*

This problem is usually caused by a failure to link all of the tables you brought into the top half of the query generator. The solution is to use the manual click-and-drag method to link the common fields between tables. The spelling of the field names is irrelevant because the link fields need not have the same spelling.

5. *"For the most part, my query output is what I expected, but I am getting one or two duplicate rows or not enough rows."*

You may have linked too many fields between tables. Usually, only a single link is needed between two tables. It is unnecessary to link each common field in all combinations of tables; it is usually sufficient to link the primary keys. A simplistic explanation for why overlinking causes problems is that it causes Access to "overthink" and repeat itself in its answer.

On the other hand, you might be using too many tables in the query design. For example, you brought in a table, linked it on a common field with some other table, but then did not use the table. In other words, you brought down none of its fields, and/or you used none of its fields in query expressions. In this case, if you got rid of the table, the query would still work. Click the unneeded table's header at the top of the QBE area, and press the Delete key to see if you can make the few duplicate rows disappear.

6. *"I expected six rows in my query output, but I got only five. What happened to the other one?"*

Usually, this problem indicates a data entry error in your tables. When you link the proper tables and fields to make the query, remember that the linking operation joins records from the tables *on common values* (*equal* values in the two tables). For example, if a primary key in one table has the value "123," the primary key or the linking field in the other table should be the same to allow linking. Note that the text string "123" is not the same as the text string " 123"—the space in the second string is considered a character too. Access does not see unequal values as an error. Instead, Access moves on to consider the rest of the records in the table for linking. The solution is to examine the values entered into the linked fields in each table and fix any data entry errors.

7. *"I linked fields correctly in a query, but I'm getting the empty set in the output. All I get are the field name headings!"*

You probably have zero common (equal) values in the linked fields. For example, suppose you are linking on Part Number, which you declared as text. In one field, you have part numbers "001", "002", and "003"; in the other table, you have part numbers "0001," "0002," and "0003." Your tables have no common values, which means that no records are selected for output. You must change the values in one of the tables.

8. *"I'm trying to count the number of today's sales orders. A Totals query is called for. Sales are denoted by an invoice number, and I made that a text field in the table design. However, when I ask the Totals query to 'Sum' the number of invoice numbers, Access tells me I cannot add them up! What is the problem?"*

Text variables are words! You cannot add words, but you can count them. Use the Count Totals operator (not the Sum operator) to count the number of sales, each being denoted by an invoice number.

9. *"I'm doing time arithmetic in a calculated field expression. I subtracted the Time In from the Time Out and got a decimal number! I expected eight hours, and I got the number .33333. Why?"*

[Time Out] – [Time In] yields the decimal percentage of a 24-hour day. In your case, eight hours is one-third of a day. You must complete the expression by multiplying by 24: ([Time Out] – [Time In]) * 24. Don't forget the parentheses.

10. *"I formatted a calculated field for Currency in the query generator, and the values did show as currency in the query output; however, the report based on the query output does not show the dollar sign in its output. What happened?"*

Go to the report Design view. A box in one of the panels represents the calculated field's value. Click the box and drag to widen it. That should give Access enough room to show the dollar sign as well as the number in the output.

11. *"I told the Report Wizard to fit all of my output to one page. It does print to one page, but some of the data is missing. What happened?"*

Access fits all the output on one page by leaving data out. If you can tolerate having the output on more than one page, deselect the Fit to a Page option in the wizard. One way to tighten output is to enter Design view and remove space from each box that represents output values and labels. Access usually provides more space than needed.

12. *"I grouped on three fields in the Report Wizard, and the wizard prints the output in a staircase fashion. I want the grouping fields to be on one line. How can I do that?"*

Make adjustments in Design view and Layout view. See the "Creating Reports" section of this tutorial for instructions on making these adjustments.

13. *"When I create an Update query, Access tells me that zero rows are updating or more rows are updating than I want. What is wrong?"*

If your Update query is not set up correctly (for example, if the tables are not joined properly), Access will either try not to update anything, or it will update all of the records. Check the query, make corrections, and run it again.

14. *"I made a Totals query with a Sum in the Group By row and saved the query. Now when I go back to it, the Sum field reads 'Expression,' and 'Sum' is entered in the field name box. Is that wrong?"*

Access sometimes changes the Sum field when the query is saved. The data remains the same, and you can be assured your query is correct.

15. *"I cannot run my Update query, but I know it is set up correctly. What is wrong?"*

Check the security content of the database by clicking the Security Content button. You may need to enable certain actions.

PRELIMINARY CASE: THE PERSONAL TRAINING DATABASE

Setting Up a Relational Database to Create Tables, Forms, Queries, and Reports

PREVIEW

In this case, you will create a relational database for a gym that arranges personal training sessions. First, you will create four tables and populate them with data. Next, you will create a form and subform for recording students and their training sessions. You will create five queries: a select query, a parameter query, an update query, a totals query, and a query used as the basis for a report. Finally, you will create the report from the fifth query.

PREPARATION

- Before attempting this case, you should have some experience using Microsoft Access.
- Complete any part of Tutorial B that your instructor assigns, or refer to the tutorial as necessary.

BACKGROUND

You are interested in fitness and you work out at the gym of your university, so your sister bought you five sessions with a personal trainer at the gym as your birthday present. You used those sessions and became hooked on working out with a trainer. Unfortunately, being a college student, you do not have the money to continue using a trainer. You decide to contact the owner of the gym and offer to help computerize its personal training system in exchange for more time with a personal trainer. The gym owner accepts your offer because he realizes that his business is growing and that the trainers should not be responsible for scheduling clients on their own smartphones. You suggest creating an Access database to track and schedule all training sessions; you have experience using the program in your information systems coursework.

Your first tasks are to design the database, create the tables, and populate them with current data. You have decided to begin in a simple fashion, so your database design includes only four tables, as shown in Figures 1-1, 1-2, 1-3, and 1-4:

- Students, which keeps track of student ID numbers, students' names, their full addresses and telephone numbers, and the date that each student joined the gym
- Trainers, which records each trainer's ID number, name, address, and phone number
- Trainer Assignments, which identifies the trainer assigned to each student
- Training Sessions, which records the date and time of each student training session

After the database tables are complete and populated with data, you want to computerize several common tasks. First, you need a streamlined way to have student customers sign up for training sessions and then to record the date and starting time. All training sessions last one hour, so there is no need to record the ending time of the session. You recognize that a form and subform would be ideal for recording information about training sessions.

To reward the gym's longtime members, the owner would like to mail discount coupons to anyone who has been training there for more than two years. This listing will be easy to complete using a basic select query.

Students sometimes leave their belongings in the gym after a session. Therefore, the owner would like to have a way to know which students were in the gym on any particular day so he can contact them about missing items. You suggest a parameter query that prompts for an input date to satisfy his request. Also, students occasionally don't like the personal trainer assigned to them, so the owner has requested a way to change the trainer for individual students. You realize that an update query will accomplish this task.

The gym owner tells you that he likes to encourage occasional customers to book more personal training sessions. The owner has found that if a student's attendance begins to decline, the student is more likely to leave the gym entirely. Therefore, the owner would like a listing of students who have attended fewer than three sessions over a specified time period. You are confident that a totals query will satisfy this request.

Finally, the owner would like a comprehensive list of training sessions that is subdivided into sessions with each personal trainer. The list should include the name of each student and the date of each session. You decide that the best solution is to use a query that feeds into a report. The report will show each trainer's customers and information about their sessions.

ASSIGNMENT 1: CREATING TABLES

Use Microsoft Access to create tables that contain the fields shown in Figures 1-1 through 1-4; you learned about these tables in the Background section. Populate the database tables as shown. Add your name to the Students table with a fictitious ID number; complete the entry by adding your address and phone number, and then enter today's date as your join date.

This database contains the following four tables:

Student ID	Student Name	Street Address	City	State	Zip	Phone	Join Date	Click to Add
B-10	Joe Breskley	5 Fifth Ave	Newark	DE	19711	(302)738-1220	9/1/2014	
F-5	Mary Beth Fairnham	1 Main St	Newark	DE	19713	(302)454-0009	8/15/2014	
H-10	George Hammer	6 Stewart St	Elsmere	DE	19832	(302)998-7682	6/5/2011	
J-11	Lou Jones	10 Lindon Rd	Wilmington	DE	19808	(302)454-3399	5/10/2012	
J-6	Petra Jordan	89 Oak Ave	Newark	DE	19711	(302)731-9821	1/15/2011	
L-16	David Luber	90 Harbor Ct	Wilmington	DE	19808	(302)998-3562	1/30/2013	
S-59	Ann Sweeney	6 Elm St	Wilmington	DE	19808	(302)998-1114	2/12/2013	
S-62	Pat Short	515 Oak Ave	Newark	DE	19713	(302)733-8732	9/28/2012	
W-9	Jackie Weblos	78 Wayne Ave	Newark	DE	19711	(302)733-9812	9/28/2012	

Source: Used with permission from Microsoft Corporation

FIGURE 1-1   The Students table

Trainer ID	Trainer First Name	Trainer Last Name	Street Address	City	State	Zip	Phone	Click to Add
12	Craig	Peterson	6 Pike Lane	Newark	DE	19711	(302)738-0011	
15	Susan	Braun	1001-A Littlewood Ct	Wilmington	DE	19808	(302)998-4556	
18	Betsy	Sandia	15 Maple Pl	Hockessin	DE	19801	(302)856-9899	
20	Betsy	Zeeba	25 West Ave	Newark	DE	19713	(302)731-4380	

Source: Used with permission from Microsoft Corporation

FIGURE 1-2   The Trainers table

Trainer Assignments

Trainer ID	Student ID	Click to Add
12	B-10	
12	F-5	
12	H-10	
15	J-11	
15	J-6	
18	L-16	
18	S-59	
20	S-62	
20	W-9	

Source: Used with permission from Microsoft Corporation

FIGURE 1-3 The Trainer Assignments table

Training Sessions

Training Session Date	Training Session Time	Student ID	Click to Add
10/15/2014	1:00 PM	B-10	
10/15/2014	5:00 PM	F-5	
10/16/2014	6:30 PM	H-10	
10/17/2014	10:00 AM	J-11	
10/17/2014	11:30 AM	J-6	
10/18/2014	7:15 PM	L-16	
10/18/2014	8:00 AM	S-59	
10/18/2014	6:00 AM	S-62	
10/19/2014	5:00 AM	W-9	
10/20/2014	1:00 PM	B-10	
10/20/2014	5:00 PM	F-5	
10/20/2014	6:30 PM	H-10	
10/20/2014	10:00 AM	J-11	
10/20/2014	5:00 AM	W-9	
10/23/2014	1:00 PM	B-10	
10/23/2014	5:00 PM	F-5	
10/23/2014	6:30 PM	H-10	
10/23/2014	10:00 AM	J-11	
10/23/2014	4:45 PM	S-59	
10/23/2014	1:00 PM	S-62	

Source: Used with permission from Microsoft Corporation

FIGURE 1-4 The Training Sessions table

ASSIGNMENT 2: CREATING A FORM, QUERIES, AND A REPORT

Assignment 2A: Creating a Form

Create a form for easy recording of students and their training sessions. The main form should be based on the Students table, and the subform should be inserted with the fields from the Training Sessions table. Save the form as Students. View one record; if required by your instructor, print the record. Your output should resemble that in Figure 1-5.

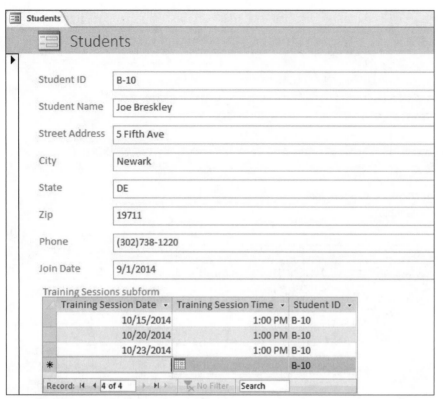

Source: Used with permission from Microsoft Corporation

FIGURE 1-5   The Students form with subform

Assignment 2B: Creating a Select Query

Create a query that lists all students who have trained at the gym for more than two years. Include columns that display the Student Name, Street Address, City, State, Zip, and Join Date. Save the query as Students Who Have Trained for More than 2 Years. Your output should resemble that shown in Figure 1-6. Print the output if desired.

Student Name	Street Address	City	State	Zip	Join Date
George Hammer	6 Stewart St	Elsmere	DE	19832	6/5/2011
Lou Jones	10 Lindon Rd	Wilmington	DE	19808	5/10/2012
Petra Jordan	89 Oak Ave	Newark	DE	19711	1/15/2011

Source: Used with permission from Microsoft Corporation

FIGURE 1-6   Students Who Have Trained for More than 2 Years query

Assignment 2C: Creating a Parameter Query

Create a parameter query that prompts for a specific date and then lists all students with sessions scheduled for that date. The query should include columns for Student Name, Phone, and Training Session Date. Save the query as Students by Training Session Date. Your output should resemble Figure 1-7 after you enter a date at the prompt.

Student Name	Phone	Training Session Date
Joe Breskley	(302)738-1220	10/15/2014
Mary Beth Fairnham	(302)454-0009	10/15/2014

Source: Used with permission from Microsoft Corporation

FIGURE 1-7   Students by Training Session Date query

Assignment 2D: Creating an Update Query

Create a query that updates Pat Short's student record so that her personal trainer is changed to trainer 20. Click the Run button to test the query. When prompted to change the record, answer "Yes." Save the query as Trainer Reassignment.

Assignment 2E: Creating a Totals Query

Create a totals query that lists all students who have attended fewer than three personal training sessions. To create this query, you must count the number of sessions that each student has attended. In the output, display columns for Student Name, Street Address, City, State, and Zip code. Save the query as Students with Few Sessions. Your output should resemble that shown in Figure 1-8. Print the output if desired.

Student Name	Street Address	City	State	Zip
Ann Sweeney	6 Elm St	Wilmington	DE	19808
David Luber	90 Harbor Ct	Wilmington	DE	19808
Jackie Weblos	78 Wayne Ave	Newark	DE	19711
Pat Short	515 Oak Ave	Newark	DE	19713
Petra Jordan	89 Oak Ave	Newark	DE	19711

Source: Used with permission from Microsoft Corporation

FIGURE 1-8   Students With Few Sessions query

Assignment 2F: Generating a Report

Generate a report based on a query that lists all training sessions. The query should display columns for Trainer First Name, Trainer Last Name, Student Name, and Training Session Date. Save the query as For Report. From that query, create a report that groups session information by each trainer's name. Make any needed adjustments to the output to avoid repeating names and to ensure that all fields and data are visible. Enter "Training Sessions by Trainer" as the title at the top of the report, and save the report under the same name. Your report output should resemble that in Figure 1-9.

Training Sessions by Trainer

Trainer First Name	Trainer Last Name	Student Name	Training Session Date
Betsy	Zeeba		
		Jackie Weblos	10/20/2014
		Jackie Weblos	10/19/2014
		Pat Short	10/23/2014
		Pat Short	10/18/2014
		Ann Sweeney	10/23/2014
		Ann Sweeney	10/18/2014
		David Luber	10/18/2014
Craig	Peterson		
		Mary Beth Fairnham	10/23/2014
		George Hammer	10/16/2014
		George Hammer	10/20/2014
		Joe Breskley	10/15/2014
		Mary Beth Fairnham	10/15/2014
		Mary Beth Fairnham	10/20/2014

Source: Used with permission from Microsoft Corporation

FIGURE 1-9   Training Sessions by Trainer report

If you are working with a portable storage disk or USB thumb drive, make sure that you remove it *after* closing the database file.

DELIVERABLES

Assemble the following deliverables for your instructor, either electronically or in printed form:

1. Four tables
2. Form and subform: Students
3. Query 1: Students Who Have Trained for More than 2 Years
4. Query 2: Students by Training Session Date
5. Query 3: Trainer Reassignment
6. Query 4: Students with Few Sessions
7. Query 5: For Report
8. Report: Training Sessions by Trainer
9. Any other required printouts or electronic media

Staple all the pages together. Write your name and class number at the top of the page. If required, make sure that your electronic media are labeled.

RENT-A-DRESS DATABASE

Designing a Relational Database to Create Tables, Forms, Queries, and Reports

PREVIEW

In this case, you will design a relational database for a business that rents designer dresses. After your design is completed and correct, you will create database tables and populate them with data. Then you will produce one form with a subform, three queries, and two reports. The queries will address the following questions: Which dresses are rented for less than a specified price limit? Which dresses have been kept over the seven-day limit? How many dresses has each customer rented? Your reports will summarize dress rentals for each customer and will display the savings that customers enjoy by renting designer dresses instead of buying them.

PREPARATION

- Before attempting this case, you should have some experience in database design and in using Microsoft Access.
- Complete any part of Tutorial A that your instructor assigns.
- Complete any part of Tutorial B that your instructor assigns, or refer to the tutorial as necessary.
- Refer to Tutorial F as necessary.

BACKGROUND

Your older sister and her friend recently graduated with MBA degrees from a prestigious university. Kate and Malia's capstone project was to create a business, and now they want to put that project into action. Their business plan calls for renting expensive designer dresses to customers using a model that is similar to renting DVDs through the mail. Kate and Malia have completed a test run of the company during the past six months and are ready to launch a full-scale venture. Over your university break, you visited your sister and her friend and told them you had learned to design and implement databases in your information systems class. Kate and Malia decide to hire you to create a prototype system that could eventually be scaled up for a larger business.

You talk to Kate and Malia to see how the business works. The women have purchased many designer dresses in various colors and patterns. Through local advertising in the Philadelphia area, they have attracted customers to rent these dresses. The rental cost of a dress is approximately 10 percent of its value. Shipping or courier delivery is included in the rental cost, as is the cost of dry cleaning. A prepaid mailing box is included with each rental to avoid an additional charge to the customer and streamline the process of receiving returned dresses. Customers rent each dress for seven days.

Because you have learned about database design and have worked with Microsoft Access, you will create a prototype database to help you understand the business. Kate and Malia will use this prototype during their presentation to a banker when they request funds for the business start-up. You decide to focus on a few key parts of the business for the presentation.

First, you need to keep track of all the dresses that are available to rent. Your sister's business partner has a listing of each dress, which includes a unique identifying number, a designer, a name (dresses are named by designers), color, rental cost, retail cost, and size. Customer essentials need to be recorded as well. Your idea is to organize dress information so that eventually it can be transferred to a secure Web site where customers can browse the available dresses.

When customers rent dresses, they provide their e-mail address, regular mailing address, and credit card number. (All transactions are conducted with credit cards.) Your database must record each dress rental along with the date rented and date returned.

As customers rent dresses and return them, Kate and Malia would like to have a convenient way of recording these transactions. A form with a subform is needed. Again, this form eventually will be made available on the Web so that customers can reserve all their rentals online.

You can imagine that customers will often call and ask which dresses are available in certain price ranges. Your sister would like to be able to enter an upper limit for a dress rental and see which dresses are available. From your discussions with Kate and Malia about the business, you know that the company needs a way to find out which customers are keeping their dresses over the seven-day limit. Kate and Malia would also like to know how many dresses each customer has rented. This information will be used for marketing purposes.

Finally, you need to create two reports. The first report lists the dresses in stock, the rental cost and retail value of each dress, and the savings that the customer will realize by renting a dress instead of buying it. This information will be used for advertising purposes. The second report will summarize dress rentals by each customer, including rental costs.

ASSIGNMENT 1: CREATING THE DATABASE DESIGN

In this assignment, you design your database tables using a word-processing program. Pay close attention to the logic and structure of the tables. Do not start developing your Access database in Assignment 2 before getting feedback from your instructor on Assignment 1. Keep in mind that you need to examine the requirements in Assignment 2 to design your fields and tables properly. It is good programming practice to look at the required outputs before beginning your design. When designing the database, observe the following guidelines:

- First, determine the tables you will need by listing the name of each table and the fields it should contain. Avoid data redundancy. Do not create a field if it can be created by a calculated field in a query.
- You will need transaction tables. Think about the business events that occur with each customer's actions. Avoid duplicating data.
- Document your tables using the table feature of your word processor. The format of your tables should resemble that shown in Figure 2-1.
- You must mark the appropriate key field(s) by entering an asterisk (*) next to the field name. Keep in mind that some tables might need a compound primary key to uniquely identify a record within a table.
- Print the database design.

Table Name	
Field Name	Data Type (text, numeric, currency, etc.)
...	...
...	...

Source: © Cengage Learning 2015

FIGURE 2-1   Table design

NOTE

Have your design approved before beginning Assignment 2; otherwise, you may need to redo Assignment 2.

ASSIGNMENT 2: CREATING THE DATABASE, QUERIES, AND REPORTS

In this assignment, you first create database tables in Access and populate them with data. Next, you create a form, three queries, and two reports.

Assignment 2A: Creating Tables in Access

In this part of the assignment, you create your tables in Access. Use the following guidelines:

- Enter at least 20 records for dresses from different designers. Use the Internet to find dress designers and ideas for dress names and colors.
- Enter records for at least eight customers, including their names, addresses, telephone numbers, e-mail addresses, and fictional credit card numbers. Enter your own name and information as an additional customer.
- Each dress should be rented at least once. Each customer should rent a dress twice, and at least two customers should have more than two rental transactions.
- Make sure that a few customers have kept their dresses longer than the seven-day rental period.
- Appropriately limit the size of the text fields; for example, a telephone number does not need the default length of 255 characters.
- Print all tables if your instructor requires it.

Assignment 2B: Creating Forms, Queries, and Reports

You must generate one form with a subform, three queries, and two reports, as outlined in the Background section of this case.

Form

Create a form and subform based on your Customers table and Rentals table (or whatever you named these tables). Save the form as Customers. Your form should resemble the one shown in Figure 2-2.

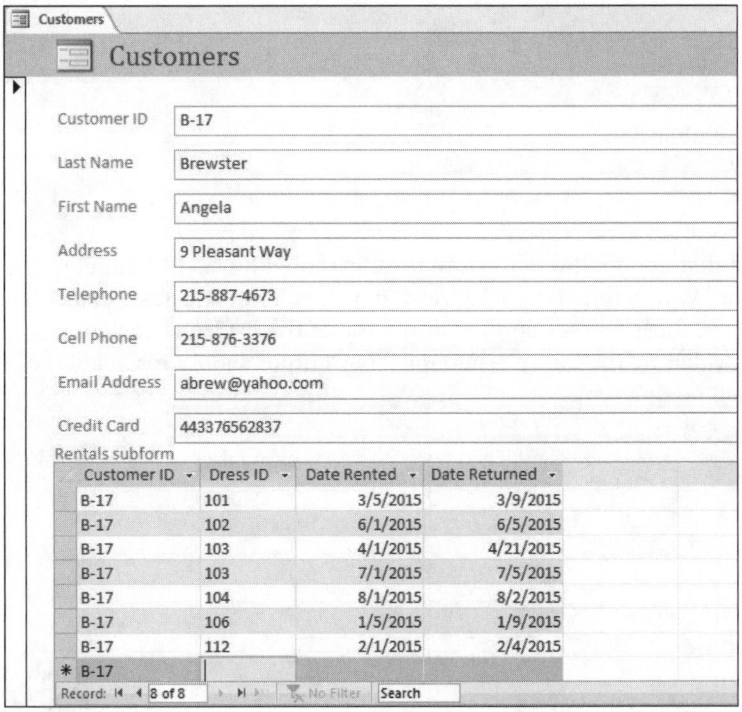

Source: Used with permission from Microsoft Corporation

FIGURE 2-2   Customers form and subform

Query 1

Create a parameter query called Upper Limit of Price Range that prompts for an upper limit to the rental price of a dress and then displays columns for Dress ID, Designer, Dress Name, Color, and Rental Price. In the example shown in Figure 2-3, the upper price limit is $100. Your output should resemble that in Figure 2-3, although your data will be different.

Upper Limit of Price Range

Dress ID	Designer	Dress Name	Color	Rental Price
101	Vera Wang	Paisley Lace	Pink	$75.00
104	Temperley London	Dot 'n Dot	Camel	$75.00
109	Herve Leger	Pebbles	Mauve	$65.00
113	Donna Karan	Plain Sheath	Black	$50.00
115	Calvin Klein	Wow	Orange	$65.00
119	Calvin Klein	Vixen	Grey	$75.00
*				$0.00

Source: Used with permission from Microsoft Corporation

FIGURE 2-3 Upper Limit of Price Range query

Query 2

Create a query called Dresses Kept Over 7 Days that calculates the total number of days each dress has been held by a customer and then lists dresses that have been kept longer than the seven-day limit. The query should include columns for Last Name, First Name, Dress ID, Designer, and Total Number of Days Held. Note that the Total Number of Days Held heading is a column heading change from the default setting provided by the query generator. Your output should look like that in Figure 2-4, although your data will be different.

Dresses Kept Over 7 Days

Last Name	First Name	Dress ID	Designer	Total Number of Days Held
Brewster	Angela	103	Shoshanna	20
Freeny	Geraldine	102	Tracy Reese	8
May	Allie	101	Vera Wang	18
May	Allie	107	Missoni	13
May	Allie	115	Calvin Klein	8
Pettyjon	Jane	111	Elie Tahari	10
Quinn	Samantha	117	Kate Spade	9
Signorelli	Patricia	109	Herve Leger	28
Zecon	Joyce	108	Kate Spade	31
*				

Source: Used with permission from Microsoft Corporation

FIGURE 2-4 Dresses Kept Over 7 Days query

Query 3

Create a query called Total Dresses Rented that counts the number of dresses rented by each customer. The query should include columns for Last Name, First Name, Email Address, and Number of Dresses Rented. Sort the output so that the customer who rents the most dresses appears at the top of the list. Note the column heading change from the default setting provided by the query generator. Your output should resemble the format shown in Figure 2-5, but the data will be different.

Total Dresses Rented

Last Name	First Name	Email Address	Number of Dresses Rented
Brewster	Angela	abrew@yahoo.com	7
Zecon	Joyce	zeconj@comcast.net	5
May	Allie	belle@comcast.net	5
Pettyjon	Jane	jp82@comcast.net	4
Freeny	Geraldine	gf59@gmail.com	4
Signorelli	Patricia	patti1@gmail.com	2
Quinn	Samantha	quinn45@gmail.com	2
Lopez	Maria	mrl@yahoo.com	2

Source: Used with permission from Microsoft Corporation

FIGURE 2-5 Total Dresses Rented query

Report 1

Create a report named Savings by Rental that summarizes the amounts of money customers saved by renting designer dresses instead of buying them. The report's output should show headings for Dress ID, Designer, Dress Name, Color, Rental Price, Retail Price, and Savings. The Savings column shows the difference between the rental price and retail price. You need to create a query first to calculate the values in the Savings field. Depending on your data, the output should resemble that in Figure 2-6. Note that only a portion of the report appears in the figure.

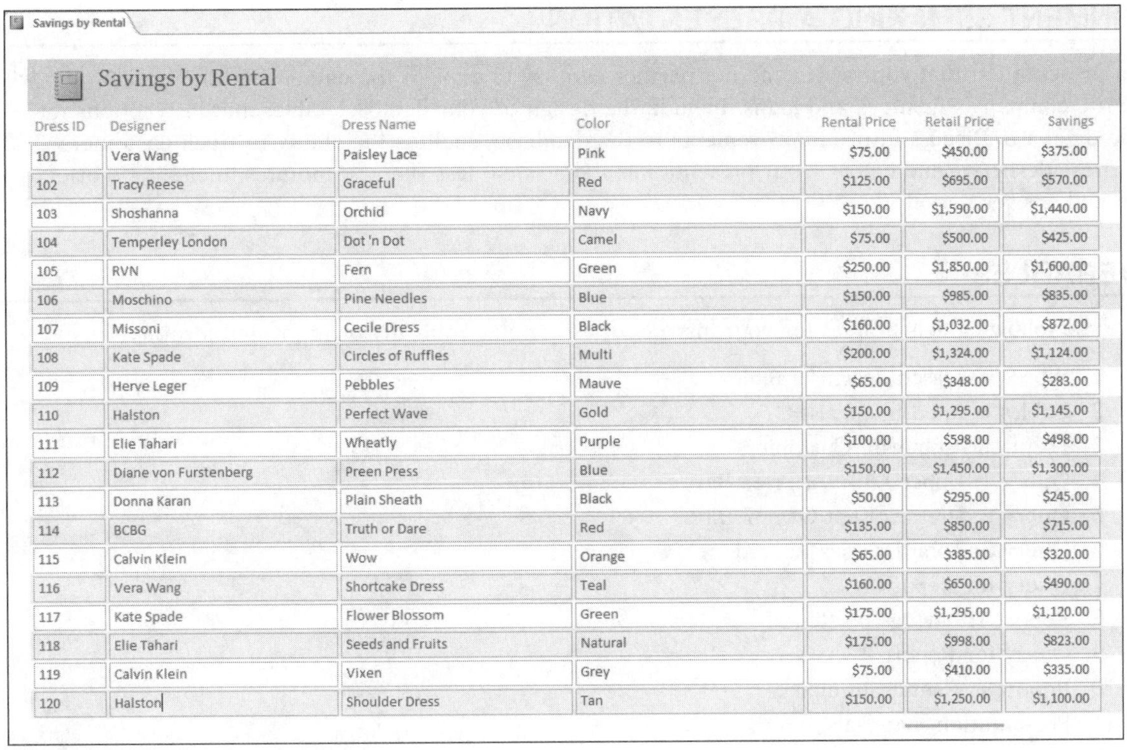

Savings by Rental

Dress ID	Designer	Dress Name	Color	Rental Price	Retail Price	Savings
101	Vera Wang	Paisley Lace	Pink	$75.00	$450.00	$375.00
102	Tracy Reese	Graceful	Red	$125.00	$695.00	$570.00
103	Shoshanna	Orchid	Navy	$150.00	$1,590.00	$1,440.00
104	Temperley London	Dot 'n Dot	Camel	$75.00	$500.00	$425.00
105	RVN	Fern	Green	$250.00	$1,850.00	$1,600.00
106	Moschino	Pine Needles	Blue	$150.00	$985.00	$835.00
107	Missoni	Cecile Dress	Black	$160.00	$1,032.00	$872.00
108	Kate Spade	Circles of Ruffles	Multi	$200.00	$1,324.00	$1,124.00
109	Herve Leger	Pebbles	Mauve	$65.00	$348.00	$283.00
110	Halston	Perfect Wave	Gold	$150.00	$1,295.00	$1,145.00
111	Elie Tahari	Wheatly	Purple	$100.00	$598.00	$498.00
112	Diane von Furstenberg	Preen Press	Blue	$150.00	$1,450.00	$1,300.00
113	Donna Karan	Plain Sheath	Black	$50.00	$295.00	$245.00
114	BCBG	Truth or Dare	Red	$135.00	$850.00	$715.00
115	Calvin Klein	Wow	Orange	$65.00	$385.00	$320.00
116	Vera Wang	Shortcake Dress	Teal	$160.00	$650.00	$490.00
117	Kate Spade	Flower Blossom	Green	$175.00	$1,295.00	$1,120.00
118	Elie Tahari	Seeds and Fruits	Natural	$175.00	$998.00	$823.00
119	Calvin Klein	Vixen	Grey	$75.00	$410.00	$335.00
120	Halston	Shoulder Dress	Tan	$150.00	$1,250.00	$1,100.00

Source: Used with permission from Microsoft Corporation

FIGURE 2-6 Savings by Rental report

Report 2

Create a report called Rental Summary that lists dress rentals by each customer. The report includes columns for Last Name, First Name, Designer, Dress Name, and Rental Price. You need to create a query for this report first. Group the report by the Last Name column and include subtotals for the rental price charged to each customer. Adjust your output so that the Last Name and First Name columns are on the same line and all fields are formatted and visible, as shown in Figure 2-7. Note that only a portion of the report appears in the figure and that your data will vary.

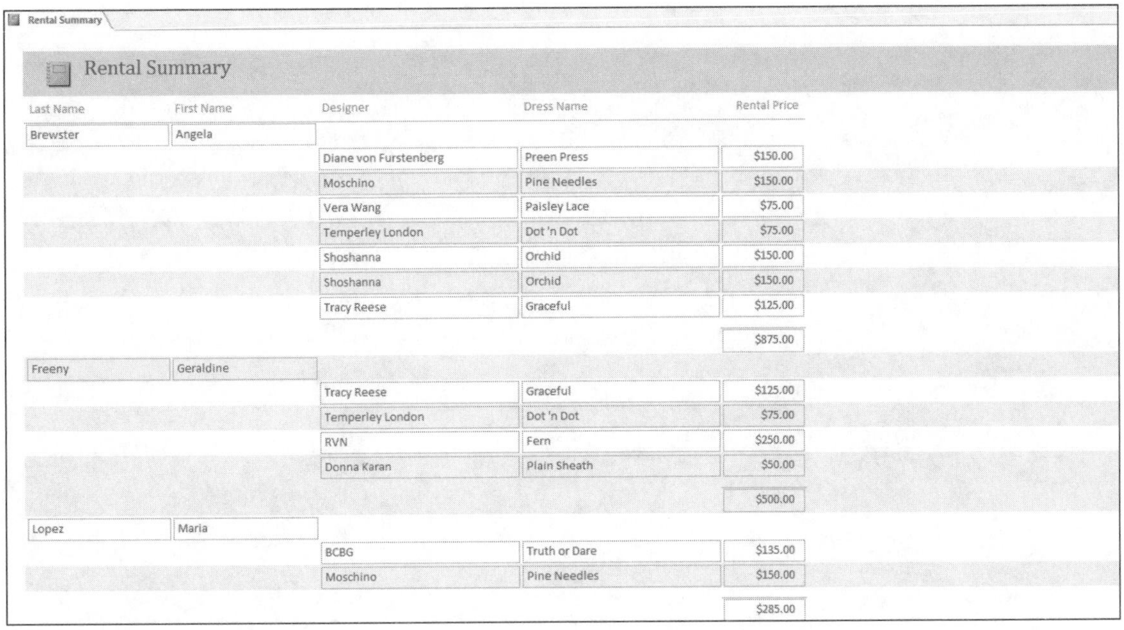

Rental Summary

Last Name	First Name	Designer	Dress Name	Rental Price
Brewster	Angela			
		Diane von Furstenberg	Preen Press	$150.00
		Moschino	Pine Needles	$150.00
		Vera Wang	Paisley Lace	$75.00
		Temperley London	Dot 'n Dot	$75.00
		Shoshanna	Orchid	$150.00
		Shoshanna	Orchid	$150.00
		Tracy Reese	Graceful	$125.00
				$875.00
Freeny	Geraldine			
		Tracy Reese	Graceful	$125.00
		Temperley London	Dot 'n Dot	$75.00
		RVN	Fern	$250.00
		Donna Karan	Plain Sheath	$50.00
				$500.00
Lopez	Maria			
		BCBG	Truth or Dare	$135.00
		Moschino	Pine Needles	$150.00
				$285.00

Source: Used with permission from Microsoft Corporation

FIGURE 2-7 Rental Summary report

ASSIGNMENT 3: MAKING A PRESENTATION

Create a presentation that your sister and her partner can use to explain the database to the banker who will approve the company's finances and loans. Include the design of your database tables and instructions for using the database. Discuss future improvements to the database, such as the ability to track customers' favorite dress designers and colors. Your presentation should take less than 10 minutes, including a brief question-and-answer period.

DELIVERABLES

Assemble the following deliverables for your instructor, either electronically or in printed form:

1. Word-processed design of tables
2. Tables created in Access
3. Form and subform: Customers
4. Query 1: Upper Limit of Price Range
5. Query 2: Dresses Kept Over 7 Days
6. Query 3: Total Dresses Rented
7. Query for Report 1
8. Report 1: Savings by Rental
9. Query for Report 2
10. Report 2: Rental Summary
11. Presentation materials
12. Any other required printouts or electronic media

Staple all the pages together. Write your name and class number at the top of each page. Make sure that your electronic media are labeled, if required.

THE IMPORT FOOD MARKET DATABASE

Designing a Relational Database to Create Tables, Forms, Queries, and Reports

PREVIEW

In this case, you will design a relational database for a market that sells and delivers imported food to customers. After your design is completed and correct, you will create database tables and populate them with data. You will produce one form with a subform that allows you to record incoming orders. You will also produce seven queries and a report. Five queries will address the following questions: Which products contain sauce? Which orders are arriving within a specified time frame? Which products cost less than a specified price? Which products are the most popular? What is the average delivery time? Another query will update the prices of the market's products. Your report, based on a query, will display all the orders for a specified date and the price of the orders.

PREPARATION

- Before attempting this case, you should have some experience in database design and in using Microsoft Access.
- Complete any part of Tutorial A that your instructor assigns.
- Complete any part of Tutorial B that your instructor assigns, or refer to the tutorial as necessary.
- Refer to Tutorial F as necessary.

BACKGROUND

The Import Food Market (IFM) in Austin, Texas, sells food from Asian countries such as China, Thailand, and Korea. The owners of the IFM are Kim and Joe Lee. Because you are a good friend of their son, they have asked you to design and create a database that allows them to keep track of their customers more efficiently.

The Lees explain that all customers are very valuable to them and that they pride themselves on customer service. For example, customers who call in orders have the groceries delivered directly to their homes for no additional charge. Timely delivery is so important to the Lees that they record the time each order was taken and when it was delivered. Kim and Joe want a good way to keep track of customers, where they live, and how to get in touch with them, either by phone or by e-mail. A typical customer order includes items such as sriracha, bamboo shoots, and rice. All money for grocery orders is collected upon delivery on a cash-only basis. Joe Lee tells you to ignore payment information for the database project.

You are currently taking a class on databases in which you have learned about their design and implementation with Microsoft Access. You plan to create a simple database for the Lees' business using Access. Then, after a trial run, you plan to replace the database with a larger one that can be made available on the Internet and accessed by multiple users simultaneously.

For now, you need to concentrate on designing the database. You have already been told about the customer information that is needed. Kim explains that all products and their prices will also have to be tracked. Order information must be recorded as well. You will need to keep track of all order dates and the time each order was received and delivered.

After you create and populate the database tables, you must think about how the market can best use the information in them. First, you need to create a form that lets the Lees easily record all incoming orders. When the company grows, the form can be made available on the market's Web site so customers can order quickly and easily.

The database should be able to answer several questions, so the next step is to create sample queries to show the Lees how the system works. For example, because customers often call and ask for recommendations about different sauces that the market sells, you can create a query that lists all the sauce products in the store's inventory.

Another issue involves your friend, the Lees' son, who makes all late-night deliveries. The Lees often wonder whether the market has enough late-night orders to justify paying their son for a full night's work. You will create a query that lists all deliveries after 11 p.m. on a specified day to support your friend's argument that he deserves a full night's pay.

Another useful query would list the products that are most popular on a specified day. Also, the Lees say that customers often call and ask for the least expensive products, so you want to create a query to list all products that cost less than a specified price. In addition, the market often offers discounted prices at certain times of the year, so you will set up a query that allows the Lees to decrease prices for products in their inventory.

Finally, you need to create an attractive sales report that lists all customers, what they ordered, and how much money they owe for a specified day's orders. This information will be important for tracking revenue and tax records.

ASSIGNMENT 1: CREATING THE DATABASE DESIGN

In this assignment, you design your database tables using a word-processing program. Pay close attention to the logic and structure of the tables. Do not start developing your Access database in Assignment 2 before getting feedback from your instructor on Assignment 1. Keep in mind that you need to examine the requirements in Assignment 2 to design your fields and tables properly. It is good programming practice to look at the required outputs before beginning your design. When designing the database, observe the following guidelines:

- First, determine the tables you will need by listing the name of each table and the fields it should contain. Avoid data redundancy. Do not create a field if it can be created by a calculated field in a query.
- You will need transaction tables. Think about the business events that occur with each customer's order. Record orders for only one specific date. Avoid duplicating data.
- Keep in mind that customers typically order more than one product within a single order.
- Document your tables using the table feature of your word processor. The format of your tables should resemble that shown in Figure 3-1.
- You must mark the appropriate key field(s) by entering an asterisk (*) next to the field name.
- Print the database design if your instructor requires it.

Table Name	
Field Name	Data Type (text, numeric, currency, etc.)
...	...
...	...

Source: © Cengage Learning 2015

FIGURE 3-1   Table design

N O T E

Have your design approved before beginning Assignment 2; otherwise, you may need to redo Assignment 2.

ASSIGNMENT 2: CREATING THE DATABASE, QUERIES, AND REPORT

In this assignment, you first create database tables in Access and populate them with data. Next, you create a form, queries, and a report.

Assignment 2A: Creating Tables in Access

In this part of the assignment, you create your tables in Access. Use the following guidelines:

- Create at least nine customer records and add yourself as the tenth customer.
- Create 10 different types of products to sell.
- Create at least 10 orders on a single day. Make sure that most customers order more than one product and that they order multiple quantities of some products.
- Appropriately limit the size of the text fields; for example, a telephone number does not need the default length of 255 characters.
- Print all tables if your instructor requires it.

Assignment 2B: Creating Forms, Queries, and a Report

You must generate one form with a subform, seven queries, and one report, as outlined in the Background section of this case.

Form

Create a form and subform based on the table of orders you developed. Save the form as Orders. Your form should resemble the one in Figure 3-2.

Source: Used with permission from Microsoft Corporation

FIGURE 3-2   Orders form and Order Line Item subform

Query 1

Create a query called Sauce Products that displays the name and price of all products that contain the word "sauce" in their names. Consider using a wildcard to filter data for this query. Your output should resemble that in Figure 3-3, although your data will be different.

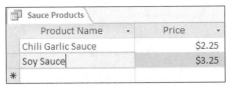

Source: Used with permission from Microsoft Corporation

FIGURE 3-3   Sauce Products query

Query 2

Create a query called Orders At or After 11 p.m. that lists late-night orders. Your query should filter orders for a specified date and include columns for the customers' Last Name, Address, and Telephone as well as for Product Name and Quantity. Make sure that you include the filter in the query. Your output should look like that in Figure 3-4, although your data will be different.

Last Name	Address	Telephone	Product Name	Quantity
Smith	34 Redback Road	(512)998-0675	Chili Garlic Sauce	3
Smith	34 Redback Road	(512)998-0675	Red Wine Vinegar	2
O'Hara	20 Cheswold Blvd	(307)887-0176	Cornstarch	2
Isaacs	10010 N. Barrett Dr	(201)584-1028	Soy Sauce	2
Isaacs	10010 N. Barrett Dr	(201)584-1028	Szechuan Pepper	2
Isaacs	10010 N. Barrett Dr	(201)584-1028	Black Beans	2
Downing	200 Mac Duff Rd	(801)912-6564	Rice	2
Downing	200 Mac Duff Rd	(801)912-6564	Tofu	3

Source: Used with permission from Microsoft Corporation

FIGURE 3-4   Orders At or After 11 p.m. query

Query 3

Create a query called Products by Price that prompts for a maximum price a customer wants to spend on a particular product. Include columns for Product Name and Price. If you run this query and enter $2 as the maximum price, the format of your output should resemble that shown in Figure 3-5, but the data will be different.

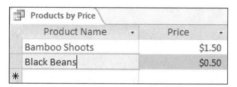

Product Name	Price
Bamboo Shoots	$1.50
Black Beans	$0.50

Source: Used with permission from Microsoft Corporation

FIGURE 3-5   Products by Price query

Query 4

Create a query named Updated Prices that revises the prices of all products in the market's inventory. For example, assume that you want to decrease each price by 2 percent. Test the query by running it and checking the updated prices in your product table.

Query 5

Create a query named Popular Products that lists the number of products sold on a specific date. Include columns for Product Name and Number Ordered. Prompt for a specific date, but do not show the date in the output. Sort your output so that the most popular product is listed first and the least popular product is shown last. Note the column heading change from the default setting provided by the query generator. The format of your output should resemble that shown in Figure 3-6, but the data will be different.

Product Name	Number Ordered
Tofu	5
Soy Sauce	5
Chili Garlic Sauce	5
Red Wine Vinegar	4
Szechuan Pepper	3
Sriracha	3
Rice	3
Cornstarch	2
Black Beans	2
Bamboo Shoots	1

Source: Used with permission from Microsoft Corporation

FIGURE 3-6   Popular Products query

Query 6

Create a query that calculates the average delivery time in minutes for all grocery orders on a specified date. Display columns for Date Ordered and Delivery Time in Minutes. Note the column heading change from the default setting provided by the query generator. Save the query as Delivery Time. Your output should resemble that shown in Figure 3-7, but the data will be different.

Date Ordere ⌄	Delivery Time in Minutes ⌄
12/14/2014	20.00

Source: Used with permission from Microsoft Corporation

FIGURE 3-7   Delivery Time query

Report

Create a report named Sales Report based on a query that lists all grocery sales on a specific date. Display columns for the customers' Last Name, First Name, and Address as well as for Product Name, Quantity, and Total Price, which is a calculated field. Use the Date field to compile your query, but do not display a Date column in the report output. Instead, enter "Sales Report for *current date*" as the title at the top of the report, where *current date* reflects the date for the data shown in the report. Make sure that all data and column headings are visible, and format your output appropriately, such as including currency signs. Depending on the data, your output should resemble that in Figure 3-8; only a portion of the report is shown.

Sales Report for 12/14/2014

Last Name	First Name	Address	Product Name	Quantity	Total Price
Bevans	Beverly	24 Raway Rd			
			Chili Garlic Sauce	2	$4.50
			Soy Sauce	3	$9.75
					$14.25
Chavez	Mary	9 Elm Arch Road			
			Sriracha	1	$5.00
			Bamboo Shoots	1	$1.50
			Rice	1	$8.25
					$14.75
DeVivo	Paul	625 Dawson Rd			
			Red Wine Vinegar	2	$5.00
			Sriracha	2	$10.00
					$15.00
Downing	Charlie	200 Mac Duff Rd			
			Tofu	3	$9.00
			Rice	2	$16.50
					$25.50

Source: Used with permission from Microsoft Corporation

FIGURE 3-8   Sales report

ASSIGNMENT 3: MAKING A PRESENTATION

Create a presentation that explains the database to the Lees. Demonstrate how they can use the database by running the queries and generating a report. Discuss future improvements and additions to the database, and how the Lees might use it to expand their market. Your presentation should take less than 10 minutes.

DELIVERABLES

Assemble the following deliverables for your instructor, either electronically or in printed form:

1. Word-processed design of tables
2. Tables created in Access
3. Form and subform: Orders
4. Query 1: Sauce Products
5. Query 2: Orders At or After 11 p.m.
6. Query 3: Products by Price
7. Query 4: Updated Prices
8. Query 5: Popular Products
9. Query 6: Delivery Time
10. Query 7: For Report
11. Report: Sales Report
12. Presentation materials
13. Any other required printouts or electronic media

Staple all the pages together. Include your name and class number at the top of each page. Make sure that your electronic media are labeled, if required.

THE MUSIC FESTIVAL DATABASE

Designing a Relational Database to Create Tables, Forms, Queries, and Reports

PREVIEW

In this case, you will design a relational database for an organizer of a music festival. After your design is completed and correct, you will create database tables and populate them with data. Then you will produce one form with a subform, six queries, and two reports. The queries will answer questions such as which bands fall into a certain category, when a particular band is scheduled to play, what bands are playing on a specific date, and how long each band is scheduled to play. You will produce two reports based on queries that display the festival schedule by band category and the volunteer hours worked.

PREPARATION

- Before attempting this case, you should have some experience in database design and in using Microsoft Access.
- Complete any part of Tutorial A that your instructor assigns.
- Complete any part of Tutorial B that your instructor assigns, or refer to the tutorial as necessary.
- Refer to Tutorial F as necessary.

BACKGROUND

You love music and have attended many summer music festivals. A large three-day music festival is scheduled soon in a town near your university, and you definitely want to attend. However, you don't think you have the money to pay for a three-day ticket, or even for a single day. Instead, you contact the manager of the festival, Pat "PJ" Jordan, about the possibility of volunteering to work for him in exchange for free admission. When PJ learns that you can design and implement databases, he tells you that he needs help organizing the festival and agrees to give you free admission if you can assist him with his new database.

To begin, you interview PJ to find out what information he needs from the database. You explain that because he will be the only user of the database, Microsoft Access is a perfect software solution for his needs. PJ has the entire Microsoft Office Suite on his computer, so he will be ready to use the system once it is implemented and he has been trained.

PJ explains that he needs help in three areas. First, he needs to keep track of each band that is scheduled to play at the festival. Each band plays a certain category of music, such as folk or "indie" (independent). Second, PJ assigns a unique number to each band for record-keeping purposes. Third, each band has a slot in the schedule that cannot be duplicated. For example, a certain band might be scheduled to play at the West stage on Saturday night between 6 p.m. and 9 p.m. (Note that the festival has two stages that are identified as East and West.) All of this information needs to be recorded in the database.

The database must also keep track of the festival's luxurious camping facilities. These so-called "glamping" sites (short for *glamorous camping*) offer fully outfitted tents that include air conditioning and beds with linens. A limited number of glamping sites are available; each site accommodates two, four, six, or eight people. PJ needs to know who has reserved the sites so he can sign up more people for the remaining available sites. He also has several volunteers who help to set up the glamping sites. The database must record the names, telephone numbers, and e-mail addresses of those volunteers, along with the number of hours they work to set up the sites.

After you have set up the data, PJ requires several outputs from the database system. First, he wants to be able to view the list of glamping sites and see which ones are reserved. You tell him that a form and subform would be ideal because he can easily scroll through the various sites.

Festival patrons often want to know when a particular band or a type of band is playing. In response, PJ wants to be able to produce lists that address the patrons' requests. For example, patrons might want to know which indie bands are booked and when they are scheduled to perform. Patrons might also want to know when a specific band is scheduled to play or which bands are playing on a specific date. You explain to PJ that you can develop queries to answer such questions easily.

PJ knows that each band booked at the festival has requested a particular time slot and has agreed to play for a particular length of time. He needs to be able to calculate these times to ensure that a set doesn't overlap subsequent sets on any stage. He would also like a list that summarizes the length of each band's set. Again, you describe how a query can calculate these times easily.

Finally, PJ would like to be able to generate reports. The first report, which will be made available on the festival's Web site, will contain the festival schedule broken down by category of music so patrons can plan which shows to attend. The second report is for internal purposes; it will list all volunteers and calculate their hours worked so that PJ can award free festival tickets to the top volunteers.

ASSIGNMENT 1: CREATING THE DATABASE DESIGN

In this assignment, you design your database tables using a word-processing program. Pay close attention to the logic and structure of the tables. Do not start developing your Access database in Assignment 2 before getting feedback from your instructor on Assignment 1. Keep in mind that you need to examine the requirements in Assignment 2 to design your fields and tables properly. It is good programming practice to look at the required outputs before beginning your design. When designing the database, observe the following guidelines:

- First, determine the tables you will need by listing the name of each table and the fields it should contain. Avoid data redundancy. Do not create a field if it can be created by a calculated field in a query.
- You will need transaction tables. Think about the business events that occur with band scheduling, glamping reservations, and volunteers' work shifts. Avoid duplicating data.
- Document your tables using the table feature of your word processor. The format of your tables should resemble that shown in Figure 4-1.
- You must mark the appropriate key field(s) by entering an asterisk (*) next to the field name. Keep in mind that some tables might need a compound primary key to uniquely identify a record within a table.
- Print the database design.

Table Name	
Field Name	Data Type (text, numeric, currency, etc.)
...	...
...	...

Source: © Cengage Learning 2015
FIGURE 4-1   Table design

N O T E

Have your design approved before beginning Assignment 2; otherwise, you may need to redo Assignment 2.

ASSIGNMENT 2: CREATING THE DATABASE, QUERIES, AND REPORTS

In this assignment, you first create database tables in Access and populate them with data. Next, you create a form, six queries, and two reports.

Assignment 2A: Creating Tables in Access

In this part of the assignment, you create your tables in Access. Use the following guidelines:

- Enter data for at least 10 of your favorite bands and create at least three different categories of music that apply to the bands. If you need help identifying categories, see Figure 4-7. Make sure that each band is scheduled to play on a specific date, and that each performance has starting and ending times and is assigned to a particular stage.
- Create at least 10 glamping sites and reserve at least half of them to ticket holders.
- Enter data for at least 10 volunteers, and use your name as one of the volunteers. Make up hours worked for some of the volunteers.
- Appropriately limit the size of the text fields; for example, a phone number does not need the default length of 255 characters.
- Print all tables if your instructor requires it.

Assignment 2B: Creating Forms, Queries, and Reports

You will generate one form with a subform, six queries, and two reports, as outlined in the Background section of this case.

Form

Create a form and subform based on your Glamping Spots table and Reservations table (or whatever you named the tables). Save the form as Glamping Spots. Your form should resemble that in Figure 4-2. Note that only a portion of the Reservations subform appears; the rest of the fields become visible when you scroll.

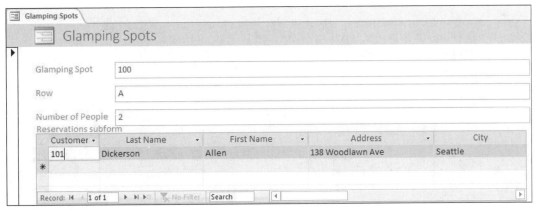

Source: Used with permission from Microsoft Corporation

FIGURE 4-2   Glamping Spots form and subform

Query 1

Create a select query called Indie Bands that lists the names of all participating bands in the Indie category. Your output should resemble that in Figure 4-3, although your data will be different.

Source: Used with permission from Microsoft Corporation

FIGURE 4-3   Indie Bands query

Query 2

Create a parameter query called When is Band Playing. The query prompts for a band name and then displays headings for Date, Start Time, Stage, and Band Name. For example, if you enter "Passion Pit" when prompted, the output should resemble that in Figure 4-4, although your data will be different.

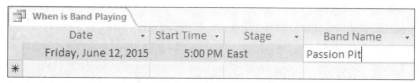

Source: Used with permission from Microsoft Corporation
FIGURE 4-4   When is Band Playing query

Query 3

Create a parameter query called Bands on Specific Date that displays headings for Band Name, Stage, Start Time, and Date. This query should prompt the user to specify the date on which bands are scheduled to play. Your output should resemble that shown in Figure 4-5, but the data will be different.

Bands on Specific Date			
Band Name	Stage	Start Time	Date
Vampire Weekend	West	11:00 AM	Saturday, June 13, 2015
Kopecky Family Band	West	5:00 PM	Saturday, June 13, 2015
Alabama Shakes	East	8:00 PM	Saturday, June 13, 2015
Yeah Yeah Yeahs	West	9:00 PM	Saturday, June 13, 2015

Source: Used with permission from Microsoft Corporation
FIGURE 4-5   Bands on Specific Date query

Query 4

Create a query called Length of Play that lists bands at the festival and how long they are scheduled to play. The query displays headings for Band Name and Time Expected to Play in hours, which is a calculation. Your output should resemble that shown in Figure 4-6, but the data will be different.

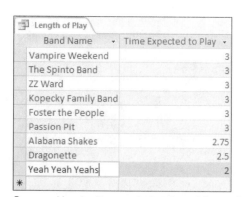

Length of Play	
Band Name	Time Expected to Play
Vampire Weekend	3
The Spinto Band	3
ZZ Ward	3
Kopecky Family Band	3
Foster the People	3
Passion Pit	3
Alabama Shakes	2.75
Dragonette	2.5
Yeah Yeah Yeahs	2

Source: Used with permission from Microsoft Corporation
FIGURE 4-6   Length of Play query

Report 1

Create a report called Schedule by Category that lists scheduled bands at the festival by their category of music. First, you need to create a query to amass the required data. The report should include headings for Category Type, Band Name, Date, Start Time, and Stage. Group the report by the Category Type column. Sort the data alphabetically by Category Type in the query. Depending on your data, the output should resemble that in Figure 4-7.

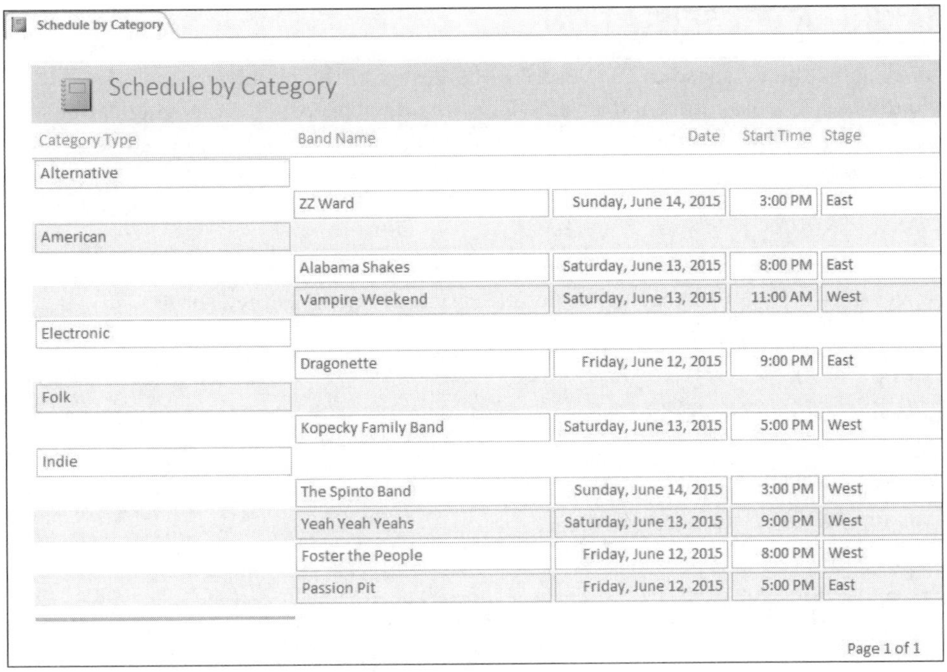

Source: Used with permission from Microsoft Corporation

FIGURE 4-7   Schedule by Category report

Report 2

Create a report called Volunteer Hours Worked that includes headings for Last Name and First Name of the volunteers as well as headings for Date Worked and Time Worked. First, you need to create a query to calculate the amount of time each volunteer worked. Bring the query into a report and group the data by the Last Name column. Adjust the output so that the First Name and Last Name are on the same line. Include subtotals that report the total time worked by each volunteer. Make sure that all column headings and data are visible. Depending on your data, the report should resemble the one in Figure 4-8.

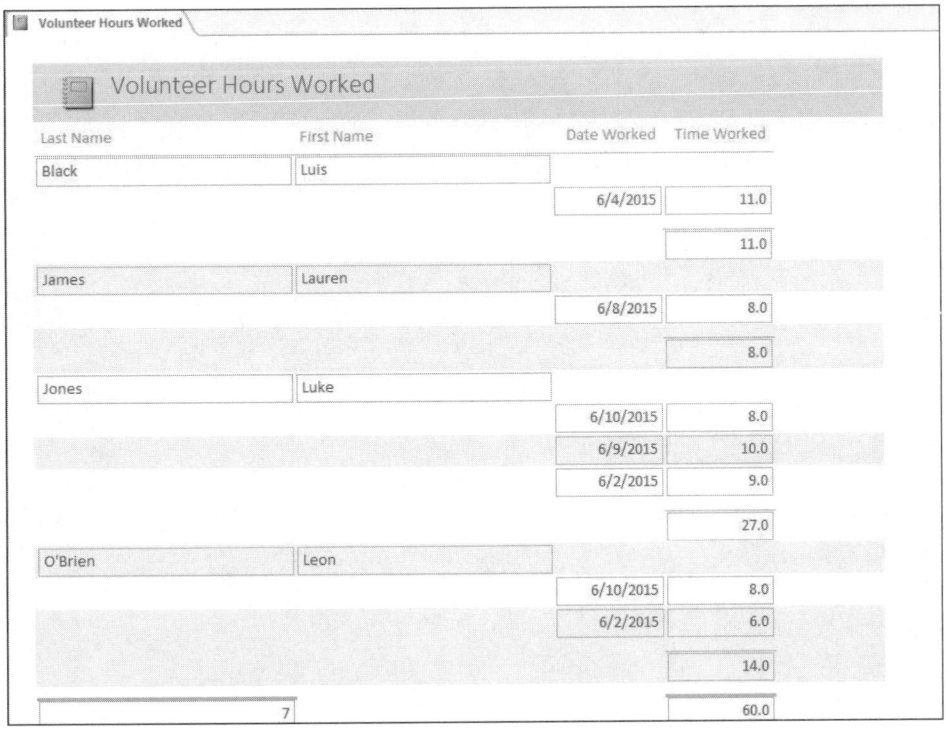

Source: Used with permission from Microsoft Corporation

FIGURE 4-8   Volunteer Hours Worked report

ASSIGNMENT 3: MAKING A PRESENTATION

Create a presentation that explains the database to PJ. Describe the design of the tables and include instructions for using the database. Discuss future improvements to the database, such as moving various parts of the system online. Your presentation should take less than 10 minutes, including a question-and-answer period.

DELIVERABLES

Assemble the following deliverables for your instructor, either electronically or in printed form:

1. Word-processed design of tables
2. Tables created in Access
3. Form and subform: Glamping Spots
4. Query 1: Indie Bands
5. Query 2: When is Band Playing
6. Query 3: Bands on Specific Date
7. Query 4: Length of Play
8. Queries 5 and 6: For reports
9. Report 1: Schedule by Category
10. Report 2: Volunteer Hours Worked
11. Presentation materials
12. Any other required printouts or electronic media

Staple all the pages together. Include your name and class number at the top of the page. Make sure that your electronic media are labeled, if required.

CASE **5**

CITY BICYCLE RENTAL DATABASE

Designing a Relational Database to Create Tables,
Queries, and Reports

PREVIEW

In this case, you will design a relational database for a citywide bicycle rental service. After your tables are designed and created, you will populate the database and create a form with a subform, eight queries, and a report. The form and subform will allow for easy recording of bike rentals. The queries will address the following questions: How many times has each bicycle been rented? What are the most popular stations for dropping off and picking up bicycles? What is the average amount of time that a customer uses a bicycle? How often does each customer rent a bicycle? What are the total charges for each customer? The report will summarize all customer rentals.

PREPARATION

- Before attempting this case, you should have some experience in database design and in using Microsoft Access.
- Complete any part of Tutorial A that your instructor assigns.
- Complete any part of Tutorial B that your instructor assigns, or refer to the tutorial as necessary.
- Refer to Tutorial F as necessary.

BACKGROUND

While visiting Europe during a study-abroad program last year, you noticed that many cities had racks of bicycles available for short-term rentals to enable customers to get around easily. When you returned to the United States for the summer, you discovered that your hometown is beginning a bicycle rental service similar to those in Europe. You do not have a summer job, so you immediately contact the manager of the new program, Maria Trejo, to see if you can help with the start-up. Maria interviews you on the telephone and discovers your talent for database design and implementation. She hires you on the spot to create a prototype database system that will be the basis for the new service.

On your first day of work, Maria explains to you how the business will operate: "Customers can go to any bicycle kiosk in the city and enter their credit card number, name, and address to register for rentals. After they register, they immediately should be assigned a customer identification number at the kiosk, and they can then request a bicycle at any rental location throughout the city. The kiosk will display a bicycle number and an unlocking code. The customer has 10 minutes to locate the bike on the rack and unlock it for use. It's easy to adjust the seat and mirrors and then take off to explore our beautiful city." She further explains, "Customers usually rent for a period of a few days. During that time, they can pick up bikes from any kiosk in the city and drop them off in different locations."

You ask about the pricing scheme of a bicycle rental. Maria explains: "There are several tiers of charges. If someone rents a bicycle for less than 24 hours, it's $3. If they rent time on bikes for 10 days or less, it's $15. Anything over 10 days is $150. The prices are low to encourage people to rent bikes and not drive, because the city has a lot of traffic and pollution. During the rental period, customers can drop off and pick up bikes as many times as necessary. For example, if a vacationer is here for three days, she can pick up and drop off bikes at multiple kiosks during that period."

Maria continues to explain how the kiosks or stations work. "Ten stations are strategically located around the city. They each have a unique number and street address. When a customer rents a bike, we need to record who takes it, from which station, and when. We also need to record the time the bike is returned and the station where it is returned. Customers can return the bikes to any station in the city. So, you could pick up a bike near your hotel, ride it to your favorite restaurant, drop the bike off at the local kiosk near the restaurant, and eat your meal. Then you could pick up a bike and reverse the journey for the way back to the hotel. We think this rental service will be a huge hit with tourists, especially with Europeans."

At this point, you think you know how you will design the tables in the database. Now you need to know what additional information Maria needs from the database. When you meet with her the following week, Maria says: "First of all, I'd like an easy way to record customer rental information on some sort of form. Next, the sponsors of this service are very interested in numbers. They'd like to know how many times each bicycle has been rented so they can plan for replacements as bikes become worn out. Our city planners are also interested in numbers. They want to know which stations have the most drop-offs and pick-ups around the city. Based on usage, they might need to increase capacity at some stations, decrease capacity at others, and even close stations that don't get enough use. Personally, I'd like to know how often each customer rents a bicycle. Also, I'd like to know the average time of each bicycle ride. Our accountants need to know the total price charged to each customer. Finally, I'd like to have a report that lists each segment of all customer bicycle rentals. This could be a great report to validate our hard work."

You explain to Maria that you can meet all her requirements by creating queries and a report, although the query that calculates the total price will be more challenging. It will require research and a preliminary query that will flow into the final query that calculates the total charges. However, you are confident that you are up to the task.

ASSIGNMENT 1: CREATING THE DATABASE DESIGN

In this assignment, you design your database tables using a word-processing program. Pay close attention to the logic and structure of the tables. Do not start developing your Access database in Assignment 2 before getting feedback from your instructor on Assignment 1. Keep in mind that you need to examine the requirements in Assignment 2 to design your fields and tables properly. It is good programming practice to look at the required outputs before beginning your design. When designing the database, observe the following guidelines:

- First, determine the tables you will need by listing the name of each table and the fields it should contain. Avoid data redundancy. Do not create a field if it can be created by a calculated field in a query.
- You will need a transaction table to record all rental information.
- Document your tables using the table feature of your word processor. The format of your tables should resemble that shown in Figure 5-1.
- You must mark the appropriate key field(s) by entering an asterisk (*) next to the field name. Keep in mind that some tables might need a compound primary key to uniquely identify a record within a table.
- Print the database design.

Table Name	
Field Name	Data Type (text, numeric, currency, etc.)
...	...
...	...

Source: © Cengage Learning 2015
FIGURE 5-1   Table design

N O T E

Have your design approved before beginning Assignment 2; otherwise, you may need to redo Assignment 2.

ASSIGNMENT 2: CREATING THE DATABASE, FORM, QUERIES, AND REPORT

In this assignment, you first create database tables in Access and populate them with data. Next, you create a form and subform, eight queries, and a report.

Assignment 2A: Creating Tables in Access

In this part of the assignment, you create your tables in Access. Use the following guidelines:

- Create at least 10 customer records with fictitious data such as names, addresses, phone numbers, e-mail addresses, and credit card numbers.
- Create 10 different stations for the bicycles, as outlined in the Background section of this case.
- Create approximately 100 rental transactions with different pick-up and drop-off stations. Consider using Microsoft Excel to generate random data and thus reduce the amount of data you need to enter. You can use the RANDBETWEEN function in Excel to create random data.
- Appropriately limit the size of the text fields; for example, a customer ID number does not need the default length of 255 characters.
- Print all tables if your instructor requires it.

Assignment 2B: Creating a Form, Queries, and Report

You will create a form and subform, eight queries, and a report, as outlined in the Background section of this case.

Form

Create a form and subform based on your Customers table and Rentals table (or whatever you named these tables). Save the form as Customers, as shown in Figure 5-2.

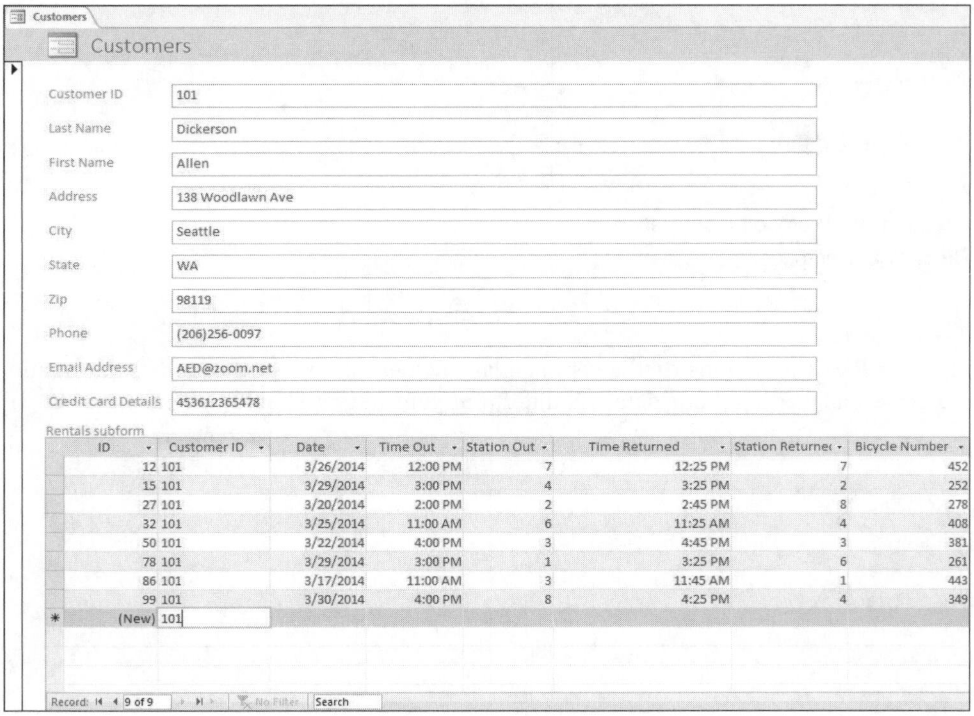

Source: Used with permission from Microsoft Corporation

FIGURE 5-2 Customers form and subform

Query 1

Create a query called Number of Times Rented that counts how many times each bicycle has been rented. Display columns for Bicycle Number and Number of Times Rented; sort the query so that the bicycle with the most rentals appears at the top of the list. Note the column heading change from the default setting provided by the query generator. Your data will vary, but the output should resemble that in Figure 5-3.

Number of Times Rented

Bicycle Number	Number of Times Rented
312	3
252	3
297	3
462	2
313	2
451	2
452	2
453	2
321	2
461	2
307	2
342	2
467	2

Source: Used with permission from Microsoft Corporation

FIGURE 5-3 Number of Times Rented query

Query 2

Create a query called Popular Drop Off Stations that displays columns for Station Returned and Location and then calculates the number of times that bikes were dropped off at each location. Make sure to show the most popular station at the top of the list. Note the column heading change from the default setting provided by the query generator. Your data will differ, but your output should resemble that in Figure 5-4.

Popular Drop Off Stations

Station Returned	Location	Number of Times Dropped Off
6	1000 10th Street	15
5	417 Oak Avenue	13
2	1600 West 5th Avenue	13
9	15 Albermarle Road	12
10	29 Maple Blvd	9
8	10 South Main Street	9
7	18 Square Circle	9
3	89 Reese Place	9
4	290 Elm Avenue	8
1	283 North Main Street	3

Source: Used with permission from Microsoft Corporation

FIGURE 5-4 Popular Drop Off Stations query

Query 3

Create a query called Popular Pick Up Stations that is very similar to Query 2, except that it lists information for stations where bicycles were picked up. Your data will differ, but your output should resemble that in Figure 5-5.

Popular Pick Up Stations

Station Out	Location	Number of Times Rented From
9	15 Albermarle Road	14
8	10 South Main Street	14
7	18 Square Circle	13
6	1000 10th Street	11
3	89 Reese Place	10
1	283 North Main Street	10
4	290 Elm Avenue	9
2	1600 West 5th Avenue	8
5	417 Oak Avenue	6
10	29 Maple Blvd	5

Source: Used with permission from Microsoft Corporation

FIGURE 5-5 Popular Pick Up Stations query

Query 4

Create a query called Average Elapsed Time that calculates the average amount of time in minutes for all bicycle rentals. Your data will differ, but your output should resemble that in Figure 5-6. Note the column heading change from the default setting provided by the query generator.

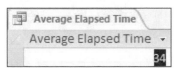

Source: Used with permission from Microsoft Corporation

FIGURE 5-6   Average Elapsed Time query

Query 5

Create a query called Rental Frequency by Customer. Display headings for Last Name, First Name, Email Address, and Number of Times Rented. The last field is a counted field that has been sorted to show customers with the most rentals at the top of the list. Note the column heading change from the default setting provided by the query generator. Your data will differ, but your output should resemble that in Figure 5-7.

Last Name	First Name	Email Address	Number of Times Rented
Sunzar	Sam	55SS@zoom.net	15
Wills	Billy	BillyW@aol.com	13
Lavelle	Shirley	Shirl121@hotmail.com	11
Schwartz	Byron	BS599@aol.com	10
Nelson	Janice	NelsJ@hotmail.com	10
Hearn	Arthur	Ahearn@aol.com	10
Trapp	John	Trapp334@hotmail.com	9
Faber	Dale	Oldie@brandywine.net	8
Dickerson	Allen	AED@zoom.net	8
Turner	Cynthia	CT@brandywine.net	6

Source: Used with permission from Microsoft Corporation

FIGURE 5-7   Rental Frequency by Customer query

Query 6

Create a query called Total Price that shows the bike rental charges billed to each customer. Display headings for the customer details, as shown in Figure 5-8, and for the Total Number of Days and Price. To make this query work properly, you must create an intermediate query as a basis for the final one. To create the intermediate query, you need to determine which days the customers began and ended their rentals. Consider using a totals query with MAX and MIN functions. Next, make a new query to calculate the total number of days rented. That calculated field can then be used to determine how much each customer owes: $3 for under 24 hours, $15 for more than 24 hours but 10 days or less, or $150 for more than 10 days. Your data will differ, but your output should resemble that in Figure 5-8.

Last Name	First Name	Address	City	State	Zip	Total Number of Days	Price
Dickerson	Allen	138 Woodlawn Ave	Seattle	WA	98119	13	$150.00
Faber	Dale	121 Chaucer Lane	Bronx	NY	10463	8	$15.00
Hearn	Arthur	26 Julie Court	Media	PA	19063	19	$150.00
Lavelle	Shirley	4001 Birch Street	Peekskill	NY	10566	20	$150.00
Nelson	Janice	23 Geneva Blvd	Piscataway	NJ	08854	13	$150.00
Schwartz	Byron	2 Waverly Rd	Deep River	CT	09776	16	$150.00
Sunzar	Sam	103 Chadd Rd	Owings Mills	MD	21117	20	$150.00
Turner	Cynthia	1502 Valley Stream Lane	Salt Lake City	UT	84109	9	$15.00
Trapp	John	220 E Main Street	Scotch Plains	NJ	07076	19	$150.00
Wills	Billy	25 Brown Lane	Centerville	OH	45459	18	$150.00

Source: Used with permission from Microsoft Corporation

FIGURE 5-8   Total Price query

Report

Create a report called Customer Rentals. First create a query that includes columns for Last Name, First Name, and Date of all customer bicycle rentals. Bring all three fields into a report and group it by the Last Name column. In the report generator, make sure that the First Name and Last Name columns are on the same line. Make sure that all field headings are visible and all data is formatted correctly. Your data will differ, but your output should resemble that in Figure 5-9. Note that only a portion of the output is shown here.

Source: Used with permission from Microsoft Corporation

FIGURE 5-9   Customer Rentals report

ASSIGNMENT 3: MAKING A PRESENTATION

Create a presentation that explains all the useful functions of your database. Maria can show this presentation to the program's sponsors, city planners, and accountants. Consider discussing how your database could be used to develop a more comprehensive system that controls all city bike rentals. For example, you may want to discuss ideas for developing a larger system that is shared by multiple users, using GPS to track the location of each bicycle, or developing apps that customers can download to their smartphones. Your presentation should take less than 15 minutes, including a brief question-and-answer period.

DELIVERABLES

Assemble the following deliverables for your instructor, either electronically or in printed form:

1. Word-processed design of tables
2. Tables created in Access
3. Query 1: Number of Times Rented
4. Query 2: Popular Drop Off Stations
5. Query 3: Popular Pick Up Stations
6. Query 4: Average Elapsed Time
7. Query 5: Rental Frequency by Customer
8. Query 6: Total Price (includes an additional query)
9. Report: Customer Rentals (includes a query)

Staple all the pages together. Include your name and class number at the top of the page. Make sure that your electronic media are labeled, if required.

PART 2

DECISION SUPPORT CASES USING EXCEL SCENARIO MANAGER

BUILDING A DECISION SUPPORT SYSTEM IN EXCEL

Decision Support Systems (DSS) are computer programs used to help managers solve complex business problems. DSS programs are commonly found in large, integrated packages called enterprise resource planning software that provide information services to an organization. Software packages such as SAP™, Microsoft Dynamics™, and PeopleSoft™ offer sophisticated DSS capabilities. However, many business problems can be modeled for solutions using less complex tools, such as Visual Basic, Microsoft Access, and Microsoft Excel.

A DSS program is actually a model representing a quantitative business problem. The problem can range from finding a desired product mix to sales forecasts to risk analysis, but almost all of the problems examine *financial outcomes*. The model itself contains the data and the algorithms (mathematical processes) needed to solve the problem.

In a DSS program, the user manually inputs data or the program accesses data from a file in the system. The program runs the data through its algorithms and displays output formatted as information; the manager uses this data to decide what action to take to solve the problem. Some sophisticated DSS programs display multiple solutions and recommend one based on predefined parameters.

Managers often find the Excel spreadsheet program particularly useful for their DSS needs. Excel contains hundreds of built-in arithmetic, statistical, logical, and financial functions. It can import data in numerous formats from large database programs, and it can be set up to display well-organized, visually appealing tables and graphs from the output.

This tutorial is organized into four sections:

1. **Spreadsheet and DSS Basics**—This section lets you "get your feet wet" by creating a DSS program in Excel. The program is a cash flow model for a small business looking to expand. You will get an introduction to spreadsheet design, building a DSS, and using financial functions.
2. **Scenario Manager**—Here you will learn how to use the Excel Scenario Manager. A DSS typically gives you one set of answers based on one set of inputs—the real value of the tool lies in its ability to play "what if" and take a comparative look at all the solutions based on all combinations of the inputs. Rather than inputting and running the DSS several times manually, you can use Scenario Manager to run and display the outputs from all possible combinations of the inputs. The output is summarized on a separate worksheet in the Excel workbook.
3. **Practice Using Scenario Manager**—Next, you will be given a new problem to model as a DSS, using Scenario Manager to display your solutions.
4. **Review of Excel Basics**—This section reviews additional information that will help you complete the spreadsheet cases that follow this tutorial. You will learn some basic operations, logical functions, and cash flow calculations.

SPREADSHEET AND DSS BASICS

You are the owner of a thrift shop that resells clothing and housewares in a university town. Many of your customers are college students. Your business is unusual in that sales actually increase during an economic recession. Your cost of obtaining used items basically follows the consumer price index. It is the end of 2014, and business has been very good due to the continuing recession. You are thinking of expanding your business to an adjacent storefront that is for sale, but you will have to apply for a business loan to finance the purchase. The bank requires a projection of your profit and cash flows for the next two years before it will loan you the money to expand, so you have to determine your net income (profit) and cash flows for 2015 and 2016. You decide that your forecast should be based on four factors: your 2014 sales dollars, your cost of goods sold per sales dollar, your estimates of the underlying economy, and the business loan payment amount and interest rate.

Because you will present this model to your prospective lenders, you decide to use an Income and Cash Flow Statements framework. You will input values for two possible states of the economy for 2015 and 2016: R for a continuing recession and B for a "boom" (recovery). Your sales in the recession were growing at 20% per year. If the recession continues and you expand the business, you expect sales to continue growing at 30% per year. However, if the economy recovers, some of your customers will switch to buying "new," so you expect sales growth for your thrift shop to be 15% above the previous year (only 5% growth plus 10% for the business expansion). If you do not expand, your recession or boom growth percentages will only be 20% and 5%, respectively. To determine the cost of goods sold for purchasing your merchandise, which is currently 70% of your sales, you will input values for two possible consumer price outlooks: H for high inflation (1.06 multiplied by the average cost of goods sold) and L for low inflation (1.02 multiplied by the cost of goods sold).

You currently own half the storefront and will need to borrow $100,000 to buy and renovate the other half. The bank has indicated that, depending on your forecast, it may be willing to loan you the money for your expansion at 5% interest during the current recession with a 10-year repayment compounded annually ("R"). However, if the prime rate drops at the start of 2015 because of an economic turnaround ("B"), the bank can drop your interest rate to 4% with the same repayment terms.

As an entrepreneur, an item of immediate interest is your cash flow position with the additional burden of a loan payment. After all, one of your main objectives is to make a profit (Net Income After Taxes). You can use the DSS model to determine if it is more profitable *not* to expand the business.

Organization of the DSS Model

A well-organized spreadsheet will make the design of your DSS model easier. Your spreadsheet should have the following sections:

- Constants
- Inputs
- Summary of Key Results
- Calculations (with separate calculations for Expansion vs. No Expansion)
- Income and Cash Flow Statements (with separate statements for Expansion vs. No Expansion)

You can also download the spreadsheet skeleton if you prefer; it will save you time. To access this skeleton file, select Tutorial C from your data files, and then select **Thrift Shop Expansion Skeleton.xlsx**.

Figures C-1 and C-2 illustrate the spreadsheet setup for the DSS model you want to build.

	A	B	C	D
1	**Tutorial Exercise--Collegetown Thrift Shop**			
2				
3	**Constants**	**2014**	**2015**	**2016**
4	Tax Rate	NA	33%	35%
5	Loan Amount for Store Expansion	NA	$100,000	NA
6				
7	**Inputs**	**2014**	**2015**	**2016**
8	Economic Outlook (R=Recession, B=Boom)	NA		NA
9	Inflation Outlook (H=High, L=Low)	NA		NA
10				
11	**Summary of Key Results**	**2014**	**2015**	**2016**
12	Net Income after Taxes (Expansion)	NA		
13	End-of-year Cash on Hand (Expansion)	NA		
14	Net Income after Taxes (No Expansion)	NA		
15	End-of-year Cash on Hand (No Expansion)	NA		
16				
17	**Calculations (Expansion)**	**2014**	**2015**	**2016**
18	Total Sales Dollars	$350,000		
19	Cost of Goods Sold	$245,000		
20	Cost of Goods Sold (as a percent of Sales)	70%		
21	Interest Rate for Business Loan		NA	NA
22				
23	**Calculations (No Expansion)**	**2014**	**2015**	**2016**
24	Total Sales Dollars	$350,000		
25	Cost of Goods Sold	$245,000		
26	Cost of Goods Sold (as a percent of Sales)	70%		

Source: Used with permission from Microsoft Corporation

FIGURE C-1   Tutorial skeleton 1

	A	B	C	D
28	**Income and Cash Flow Statements (Expansion)**	**2014**	**2015**	**2016**
29	Beginning-of-year Cash on Hand	NA		
30	Sales (Revenue)	NA		
31	Cost of Goods Sold	NA		
32	*Business Loan Payment*	NA		
33	Income before Taxes	NA		
34	Income Tax Expense	NA		
35	Net Income after Taxes	NA		
36	End-of-year Cash on Hand	$15,000		
37				
38	**Income and Cash Flow Statements (No Expansion)**	**2014**	**2015**	**2016**
39	Beginning-of-year Cash on Hand	NA		
40	Sales (Revenue)	NA		
41	Cost of Goods Sold	NA		
42	Income before Taxes	NA		
43	Income Tax Expense	NA		
44	Net Income after Taxes	NA		
45	End-of-year Cash on Hand	$15,000		

Source: Used with permission from Microsoft Corporation

FIGURE C-2   Tutorial skeleton 2

Each spreadsheet section is discussed in detail next.

The Constants Section

This section holds values that are needed for the spreadsheet calculations. These values are usually given to you, and generally do not change for the exercise. However, you can change these values later if necessary; for example, you might need to borrow more or less money for your business expansion (cell C5). For this tutorial, the constants are the Tax Rate and the Loan Amount.

The Inputs Section

The Inputs section in Figure C-1 provides a place to designate the two possible economic outlooks and the two possible inflation outlooks. If you wanted to make these outlooks change by business year, you could leave blanks under both business years. However, as you will see later when you use Scenario Manager, this approach would greatly increase the complexity of interpreting the results. For simplicity's sake, assume that the same outlooks will apply to both 2015 and 2016.

The Summary of Key Results Section

This section summarizes the Year 2 and 3 Net Income after Taxes (profit) and the End-of-year Cash on Hand both for expanding the business and for not expanding. These cells are copied from the Income and Cash Flow Statements section at the bottom of the sheet. Summary sections are frequently placed near the top of a spreadsheet to allow managers to see a quick "bottom line" summary without having to scroll down the spreadsheet to see the final result. Summary sections can also make it easier to select cells for charting.

The Calculations Sections (Expansion and No Expansion)

The following areas are used to compute the following necessary results:

- The Total Sales Dollars, which is a function of the Year 2014 value and the Economic Outlook input
- The Cost of Goods Sold, which is the Total Sales Dollars multiplied by the Cost of Goods Sold (as a percent of Sales)
- The Cost of Goods Sold (as a percent of Sales), which is a function of the Year 2014 value and the Inflation Outlook input
- In addition, the Calculations section for the expansion includes the interest rate, which is also a function of the Economic Outlook input. This interest rate will be used to determine the Business Loan Payment in the Income and Cash Flow Statements section.

You could make these formulas part of the Income and Cash Flow Statements section. However, it makes more sense to use the approach shown here because it makes the formulas in the Income and Cash Flow Statements less complicated. In addition, when you create other DSS models that include unit costing and pricing calculations, you can enter the formulas in this section to facilitate managerial accounting cost analysis.

The Income and Cash Flow Statements Sections (Expansion and No Expansion)

These sections are the financial or accounting "body" of the spreadsheet. They contain the following values:

- Beginning-of-year Cash on Hand, which equals the *prior* year's End-of-year Cash on Hand.
- Sales (Revenue), which in this tutorial is simply the results of the Total Sales Dollars copied from the Calculations section.
- Cost of Goods Sold, which also is copied from the Calculations section.
- Business Loan Payment, which is calculated using the PMT (Payment) function and the inputs for loan amount and interest rate from the Constants and Calculations sections. Note that only the Income and Cash Flow Statement for Expansion includes a value for Business Loan Payment. If you do not expand, you do not need to borrow the money.
- Income before Taxes, which is Sales minus the Cost of Goods Sold; for the expansion scenarios, you also subtract the Business Loan Payment.
- Income Tax Expense, which is zero when there is no income or the income is negative; otherwise, this value is the Income before Taxes multiplied by the Tax Rate from the Constants section.
- Net Income after Taxes, which is Income before Taxes minus Income Tax Expense.
- End-of-year Cash on Hand, which is Beginning-of-year Cash on Hand plus Net Income after Taxes.

Note that this Income and Cash Flow Statement is greatly simplified. It does not address the issues of changes in Inventories, Accounts Payable, and Accounts Receivable, nor any period expenses such as Selling and General Administrative expenses, utilities, salaries, real estate taxes, insurance, or depreciation.

Construction of the Spreadsheet Model

Next, you will work through three steps to build the spreadsheet model:

1. Make a skeleton or "shell" of the spreadsheet. Save it with a name you can easily recognize, such as TUTC.xlsx or Tutorial C *YourName*.xlsx. When submitting electronic work to an instructor or supervisor, include your last name and first initial in the filename.
2. Fill in the "easy" cell formulas.
3. Then enter the "hard" spreadsheet formulas.

Again, you can use the spreadsheet skeleton if you prefer; it will save you time. Select Tutorial C from your data files, and then select **Thrift Shop Expansion Skeleton.xlsx**.

Making a Skeleton or "Shell"

The first step is to set up the skeleton worksheet. The skeleton should have headings, text labels, and constants. Do not enter any formulas yet.

Before you start entering data, you should first try to visualize a sensible structure for your worksheet. In Figures C-1 and C-2, the seven sections are arranged vertically down the page; the item descriptions are in the first column (A), and the time periods (years) are in the next three columns (B, C, and D). This is a widely accepted business practice, and is commonly called a "horizontal analysis." It is used to visually compare financial data side by side through successive time periods.

Because your key results depend on the Income and Cash Flow Statements, you usually set up that section first, and then work upward to the top of the sheet. In other words, you set up the Income and Cash Flow Statements section, then the Calculations section, and then the Summary of Key Results, Inputs, and Constants sections. Some might argue that the Income and Cash Flow Statements should be at the top of the sheet, but when you want to change values in the Constants or Inputs section or examine the Summary of Key Results, it does not make sense to have to scroll to the bottom of the worksheet. When you run the model, you do not enter anything in the Income and Cash Flow Statements—they are all calculations. So, it makes sense to put them last.

Here are some other general guidelines for designing effective DSS spreadsheets:

- Decide which items belong in the Calculations section. A good rule of thumb is that if your items have formulas but do not belong in the Income and Cash Flow Statements, put them in the Calculations section. Good examples are intermediate calculations such as unit volumes, costs and prices, markups, or changing interest rates.
- The Summary of Key Results section should be just that—*key* results. These outputs help you make good business decisions. Key results frequently include net income before taxes (profit) and end-of-year cash on hand (how much cash your business has). However, if you are creating a DSS model on alternative capital projects, your key results can also include cost savings, net present value of a project, or rate of return for an investment.
- The Constants section holds known values needed to perform other calculations. You use a Constants section rather than just including the values in formulas so that you can input new values if they change. This approach makes your DSS model more flexible.

AT THE KEYBOARD

Enter the Excel skeleton shown in Figures C-1 and C-2.

NOTE

When you see NA (Not Applicable) in a cell, do not enter any values or formulas in the cell. The cells that contain values in the 2014 column are used by other cells for calculations. In this example, you are mainly interested in what happens in 2015 and 2016. The rest of the cells are "Not Applicable."

Filling in the "Easy" Formulas

The next step in building a spreadsheet is to fill in the "easy" formulas. To begin, format all the cells that will contain monetary values as Currency with zero decimal places:

- Constants—C5
- Summary of Key Results—C12 to C15, D12 to D15
- Calculations (Expansion)—C18, C19, D18, D19
- Calculations (No Expansion)—C24, C25, D24, D25
- Income and Cash Flow Statements (Expansion)—B36, C29 to C36, D29 to D36
- Income and Cash Flow Statements (No Expansion)—B45, C39 to C45, D39 to D45

NOTE

With the insertion point in cell C12 (where the $0 appears), note the editing window—the white space at the top of the spreadsheet to the right of the f_x symbol. The cell's contents, whether it is a formula or value, should appear in the editing window. In this case, the window shows =C35.

The Summary of Key Results section (see Figure C-3) will contain the values you calculate in the Income and Cash Flow Statements sections. To copy the cell contents for this section, move your cursor to cell C12, click the cell, type =C35, and press Enter. If you formatted your money cells properly, a $0 should appear in cell C12.

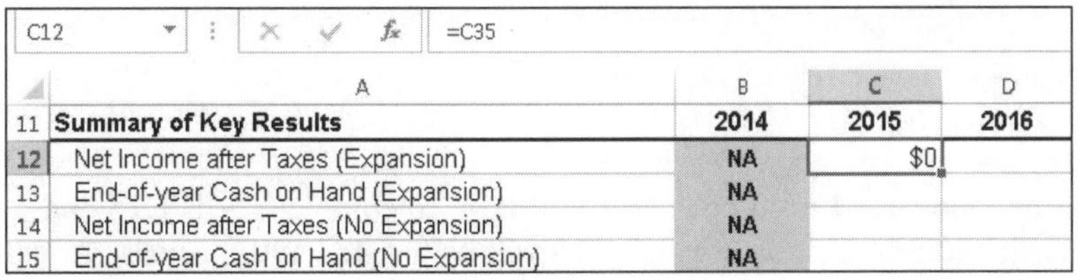

C12			f_x	=C35			
	A				B	C	D
11	**Summary of Key Results**				**2014**	**2015**	**2016**
12	Net Income after Taxes (Expansion)				NA	$0	
13	End-of-year Cash on Hand (Expansion)				NA		
14	Net Income after Taxes (No Expansion)				NA		
15	End-of-year Cash on Hand (No Expansion)				NA		

Source: Used with permission from Microsoft Corporation

FIGURE C-3 Value from cell C35 (Net Income after Taxes) copied to cell C12

Because cell C35 does not contain a value yet, Excel assumes that the empty cell has a numerical value of 0. When you put a formula in cell C35 later, cell C12 will echo the resulting answer. Because Net Income after Taxes (Expansion) for 2016 (cell D35) and its corresponding cell in Summary of Key Results (cell D12) are both directly to the right of the values for 2015, you can either type =D35 into cell D12 or copy cell C12 to D12. To perform the copy operation:

1. Click in a cell or click and drag to select the range of cells you want to copy.
2. Hold down the Control key and press C (Ctrl+C).
3. A moving dashed box called a *marquee* should now be animated over the cell(s) selected for copying.
4. Select the cell(s) where you want to copy the data.
5. Hold down the Control key and press V (Ctrl+V). Cell D12 should now contain $0, but actually it has a reference to cell D35. Click cell D12 and look again at the editing window; it should display =D35.

Cells C14, C15, D14, and D15 represent Net Income after Taxes and End-of-year Cash on Hand for both years of No Expansion; these cells are mirrors of cells C44, C45, D44, and D45 in the last section. Select cell C14, type =C44, and press Enter. Select cell C14 again, use the Copy command, and paste the contents into cell D14 (see Figure C-4).

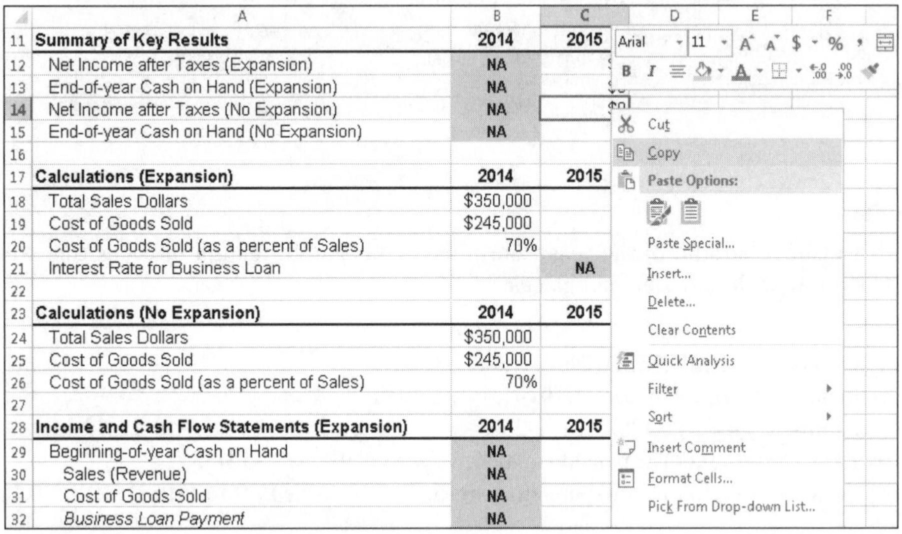

Source: Used with permission from Microsoft Corporation

FIGURE C-4   Copying the formula from cell C14 to cell D14

Because Excel uses *relative* cell references by default, copying cell C14 into cell D14 will copy and paste the contents of cell D44 (the cell adjacent to C44) into cell D14. See Figure C-5.

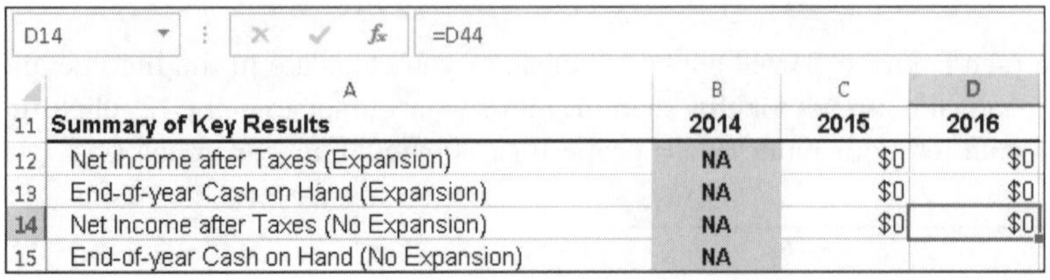

Source: Used with permission from Microsoft Corporation

FIGURE C-5   Formula from cell D44 pasted into cell D14

Use the Copy command again, this time downward from cells C14 and D14, to complete cells C15 and D15. If you are successful, the formula in the editing window for cell C15 will be "=C45" and for cell D15 will display "=D45."

You will create the formulas for the two Calculations sections last because they are the hardest formulas. Next, you will create the formulas for the two Income and Cash Flow Statements sections; all the cells in these two sections should be formatted as Currency with zero decimal places.

As shown in Figure C-6, the Beginning-of-year Cash on Hand for 2015 is the End-of-year Cash on Hand for 2014. In cell C29, type =B36. A handy shortcut is to type the "=" sign, immediately move your mouse pointer to the cell you want to designate, and then click the left mouse button. Excel will enter the cell location into the formula for you. This shortcut is especially useful if you want to avoid making a typing error.

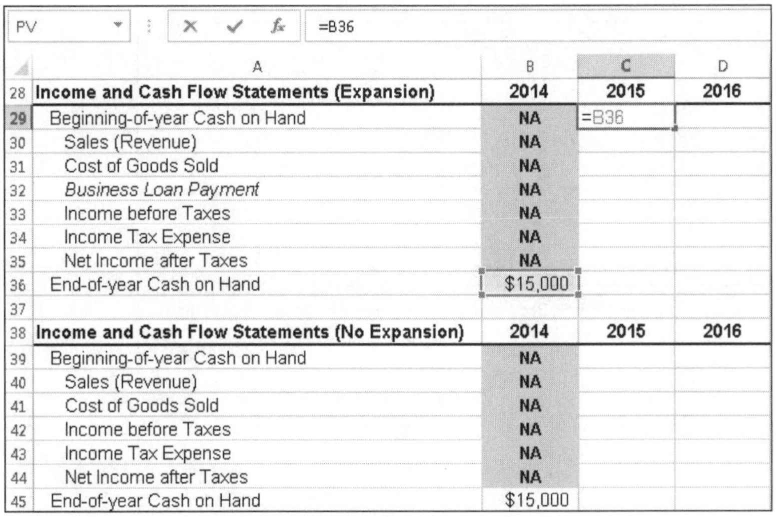

PV	:	X ✓ fx	=B36		
⊿	A		B	C	D
28	**Income and Cash Flow Statements (Expansion)**		**2014**	**2015**	**2016**
29	Beginning-of-year Cash on Hand		NA	=B36	
30	Sales (Revenue)		NA		
31	Cost of Goods Sold		NA		
32	*Business Loan Payment*		NA		
33	Income before Taxes		NA		
34	Income Tax Expense		NA		
35	Net Income after Taxes		NA		
36	End-of-year Cash on Hand		$15,000		
37					
38	**Income and Cash Flow Statements (No Expansion)**		**2014**	**2015**	**2016**
39	Beginning-of-year Cash on Hand		NA		
40	Sales (Revenue)		NA		
41	Cost of Goods Sold		NA		
42	Income before Taxes		NA		
43	Income Tax Expense		NA		
44	Net Income after Taxes		NA		
45	End-of-year Cash on Hand		$15,000		

Source: Used with permission from Microsoft Corporation

FIGURE C-6   End-of-year Cash on Hand for 2014 copied to Beginning-of-year Cash on Hand for 2015

Likewise, copy the other three End-of-year Cash on Hand cells to the Beginning-of-year Cash on Hand cells for both Income and Cash Flow Statements (cells D29, C39, and D39).

The Sales (Revenue) cells C30, D30, C40, and D40 are simply copies of cells C18, D18, C24, and D24, respectively, from the Calculations sections (both Expansion and No Expansion). Use the shortcut method to copy these cells. Note that all four cells will display $0 until you enter the formulas in the Calculations sections (see Figure C-7).

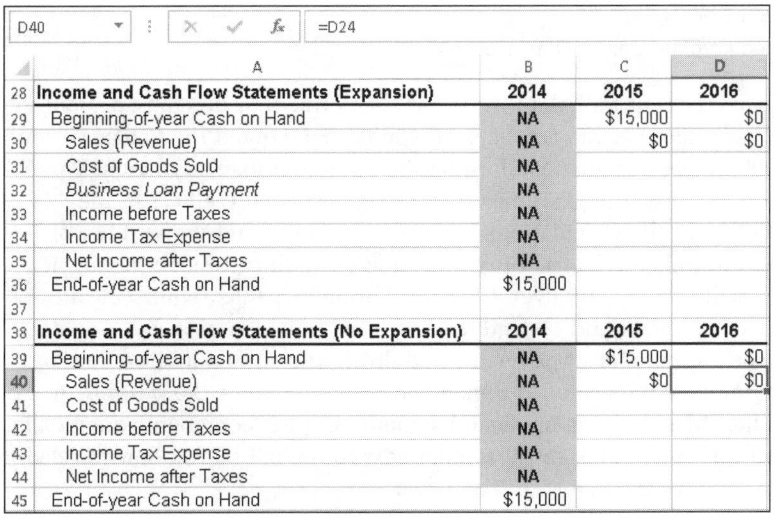

D40	:	X ✓ fx	=D24		
⊿	A		B	C	D
28	**Income and Cash Flow Statements (Expansion)**		**2014**	**2015**	**2016**
29	Beginning-of-year Cash on Hand		NA	$15,000	$0
30	Sales (Revenue)		NA	$0	$0
31	Cost of Goods Sold		NA		
32	*Business Loan Payment*		NA		
33	Income before Taxes		NA		
34	Income Tax Expense		NA		
35	Net Income after Taxes		NA		
36	End-of-year Cash on Hand		$15,000		
37					
38	**Income and Cash Flow Statements (No Expansion)**		**2014**	**2015**	**2016**
39	Beginning-of-year Cash on Hand		NA	$15,000	$0
40	Sales (Revenue)		NA	$0	$0
41	Cost of Goods Sold		NA		
42	Income before Taxes		NA		
43	Income Tax Expense		NA		
44	Net Income after Taxes		NA		
45	End-of-year Cash on Hand		$15,000		

Source: Used with permission from Microsoft Corporation

FIGURE C-7   Sales Revenue cells copied from the Calculations sections

The Cost of Goods Sold cells C31, D31, C41, and D41 are simply copies of the contents of cells C19, D19, C25, and D25, respectively, from the Calculations sections. Because the cells in both locations are directly below the Sales cells in the four locations, you can use the Copy command to fill those cells easily. As you can see in Figure C-8, you can drag your mouse pointer over both cells C40 and D40, right-click to see the shortcut menu, and select Copy. Move your mouse pointer to select cells C41 and D41, right-click the mouse, and select Paste. If you are uncomfortable copying and pasting with the mouse, you can type =C19, =D19, =C25, and =D25 in cells C31, D31, C41, and D41.

	A	B	C	D	E	F	G	H
28	**Income and Cash Flow Statements (Expansion)**	**2014**	**2015**	**2016**				
29	Beginning-of-year Cash on Hand	NA	$15,000	$0				
30	Sales (Revenue)	NA	$0	$0				
31	Cost of Goods Sold	NA						
32	*Business Loan Payment*	NA						
33	Income before Taxes	NA						
34	Income Tax Expense	NA						
35	Net Income after Taxes	NA						
36	End-of-year Cash on Hand	$15,000						
37								
38	**Income and Cash Flow Statements (No Expansion)**	**2014**	**2015**	**2016**				
39	Beginning-of-year Cash on Hand	NA	$15,000	$				
40	Sales (Revenue)	NA	$0	$0				
41	Cost of Goods Sold	NA						
42	Income before Taxes	NA						
43	Income Tax Expense	NA						
44	Net Income after Taxes	NA						
45	End-of-year Cash on Hand	$15,000						
46								
47								
48								
49								
50								
51								
52								
53								
54								
55								
56								
57								
58								
59								
60								
61								

Shortcut menu options: Cut, Copy, Paste Options:, Paste Special..., Insert..., Delete..., Clear Contents, Quick Analysis, Filter, Sort, Insert Comment, Format Cells..., Pick From Drop-down List..., Define Name..., Hyperlink...

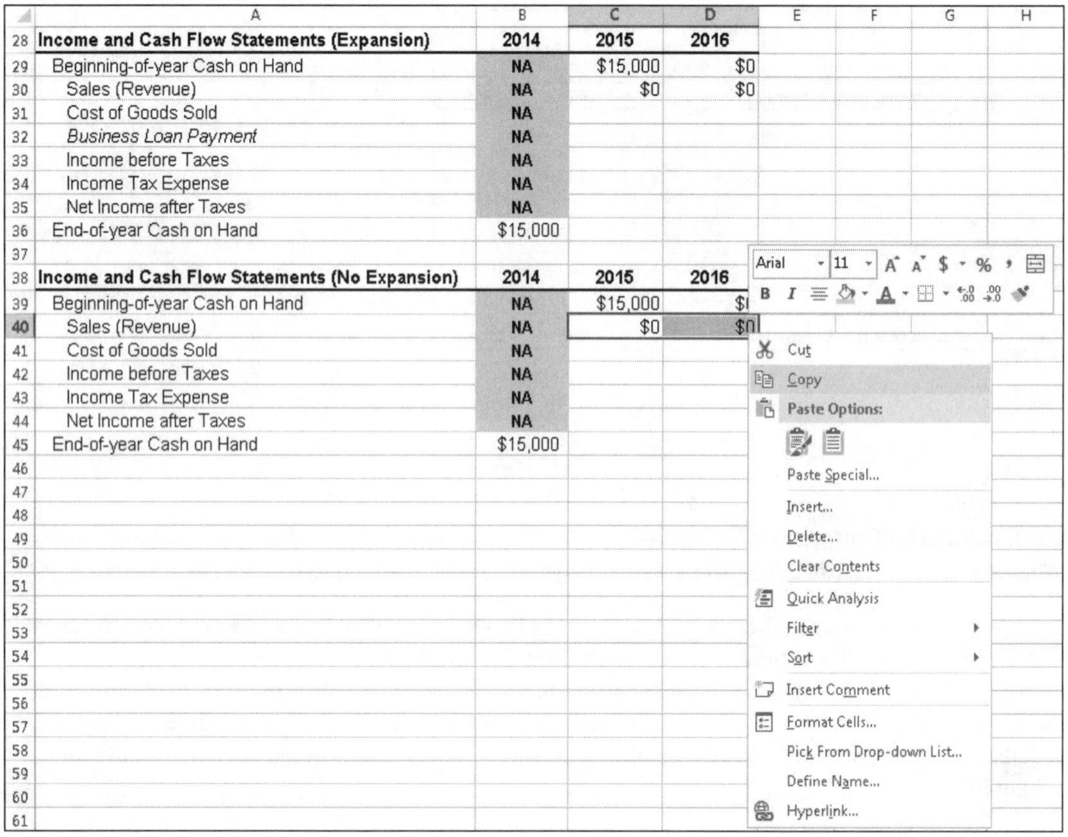

Source: Used with permission from Microsoft Corporation

FIGURE C-8   Cost of Goods Sold cells copied from the Calculations sections

Next you determine the Business Loan Payment for cells C32 and D32—notice that it is only present in the Income and Cash Flow Statements (Expansion) section, because if you do not expand the business, you do not need the business loan of $100,000. Excel has financial formulas to figure out loan payments. To determine a loan payment, you need to know three things: the amount being borrowed (cell C5 in the Constants section), the interest rate (cell B21 in the Calculations-Expansion section), and the number of payment periods. At the beginning of the tutorial, you learned that the bank was willing to loan money at either 5% or 4% interest compounded annually, to be paid over 10 years. Normally, banks require businesses to make monthly payments on their loans and compound the interest monthly, in which case you would enter 120 (12 months/year × 10 years) for the number of payments and divide the annual interest rate by 12 to get the period interest rate. This formula is important to remember when you enter the business world, but for now you will simplify the calculation by specifying one loan payment per year compounded annually. To put in the payment formula, click cell C32, then click the f_x symbol next to the editing window (circled in

Figure C-9). The Payment function is called PMT, so type PMT in the Insert Function window and click Go—you will see a short description of the function with its arguments, as shown in Figure C-9.

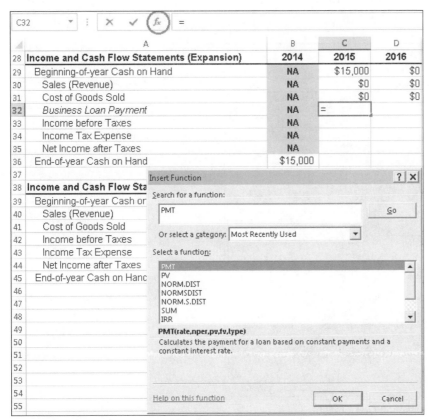

Source: Used with permission from Microsoft Corporation

FIGURE C-9   Accessing the PMT function in Excel for cell C32

NOTE

Rate is the interest rate per period of the loan, Nper is an abbreviation for the number of loan periods, and Pv is an abbreviation for Present Value, the amount of money you are borrowing "today." The PMT function can determine a series of equal loan payments necessary to pay back the amount borrowed, plus the accumulated compound interest over the life of the loan.

When you click OK, the resulting window allows you to enter the cells or values needed in the function arguments (see Figure C-10). In the Rate text box, enter B21, which is the cell that will contain the calculated interest rate. In the Nper text box, enter 10 (for 10 years). In the Pv text box, enter C5, which is the cell that contains the loan amount.

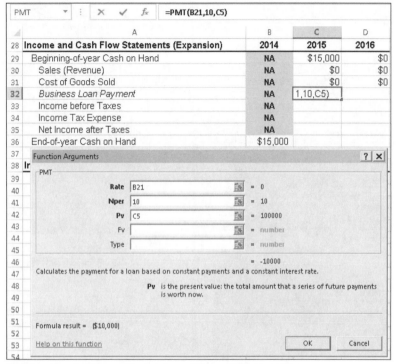

Source: Used with permission from Microsoft Corporation

FIGURE C-10   The Function Arguments window for the PMT function with the values filled in

> **NOTE**
>
> Be careful if you decide to copy the PMT formula from cell C32 into cell D32, because the Copy command will change the cells in the formula arguments to the next adjacent cells. To make the Copy command work correctly, you have two options. First, you can change the Rate and Pv cells in the cell C32 formula from *relative reference* (B21, C5) to *absolute reference* ($B$21, $C$5). Your other option is to re-insert the PMT function into cell D32 and type the same arguments as before in the boxes. Absolute referencing of a cell (using $ signs in front of the Column and Row designators) "anchors" the cell so that when the Copy command is used, the destination cell will refer back to the same cells that the source cell used. If necessary, consult the Excel online Help for an explanation of relative and absolute cell references.

When you click OK ($10,000) should appear in cell C32. Payments in Excel always appear as negative numbers, which is why the number has parentheses around it. (Depending on your cell formatting, the number may also appear in red.) Next, you need to have the same payment amount in cell D32 (for 2016). Because the PMT function creates equal payments over the life of the loan, you can simply type =C32 into cell D32.

The next line in the Income and Cash Flow Statements is Income before Taxes, which is an easy calculation. It is the Sales minus the Cost of Goods Sold, minus the Business Loan Payment. However, because the PMT function shows the loan payment as a negative number, you will instead add the Business Loan Payment. In cell C33, enter =C30-C31+C32. Again, a negative $10,000 should be displayed, as the cells other than the loan payment currently have zero in them. Copy cell C33 to cell D33. In cell C42 of the next section below (No Expansion), enter =C40-C41. (There is no loan payment in this section to put in the calculation.) Next, copy cell C42 to cell D42. At this point, your Income and Cash Flow Statements should look like Figure C-11.

D42	▾ : × ✓ fx	=D40-D41		

⊿	A	B	C	D
28	**Income and Cash Flow Statements (Expansion)**	**2014**	**2015**	**2016**
29	Beginning-of-year Cash on Hand	NA	$15,000	$0
30	Sales (Revenue)	NA	$0	$0
31	Cost of Goods Sold	NA	$0	$0
32	*Business Loan Payment*	NA	($10,000)	($10,000)
33	Income before Taxes	NA	-$10,000	-$10,000
34	Income Tax Expense	NA		
35	Net Income after Taxes	NA		
36	End-of-year Cash on Hand	$15,000		
37				
38	**Income and Cash Flow Statements (No Expansion)**	**2014**	**2015**	**2016**
39	Beginning-of-year Cash on Hand	NA	$15,000	$0
40	Sales (Revenue)	NA	$0	$0
41	Cost of Goods Sold	NA	$0	$0
42	Income before Taxes	NA	$0	$0
43	Income Tax Expense	NA		
44	Net Income after Taxes	NA		
45	End-of-year Cash on Hand	$15,000		

Source: Used with permission from Microsoft Corporation

FIGURE C-11   The Income and Cash Flow Statements completed up to Income before Taxes

Income Tax Expense is the most complex formula for these sections. Because you do not pay income tax when you have no income or a loss, you must use a formula that allows you to enter 0 if there is no income or a loss, or to calculate the tax rate on a positive income. You can use the IF function in Excel to enter one of two results in a cell, depending on whether a defined logical statement is true or false. To create an IF function, select cell C34, then click the f_x symbol next to the cell editing window (circled in Figure C-12). When the Insert Function window appears, type IF in the "Search for a function" text box, and click the Go button if necessary. The IF function should appear. When you click OK, the Function Arguments window appears (see Figure C-13).

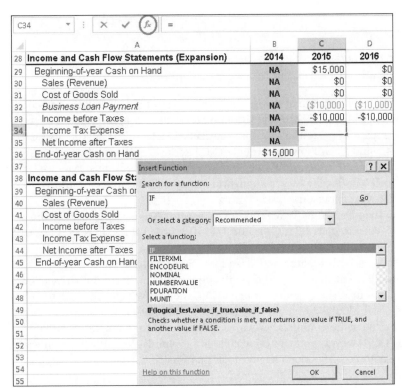

Source: Used with permission from Microsoft Corporation

FIGURE C-12   The IF function

Type the following in the Function Arguments window:

- Next to Logical_test, type C33<=0.
- Next to Value_if_true, type 0.
- Next to Value_if_false, type C33*C4 (the Income before Taxes multiplied by the Tax Rate for 2015).

As you fill in the arguments, Excel writes the formula for you in the formula editing window (circled in Figure C-13).

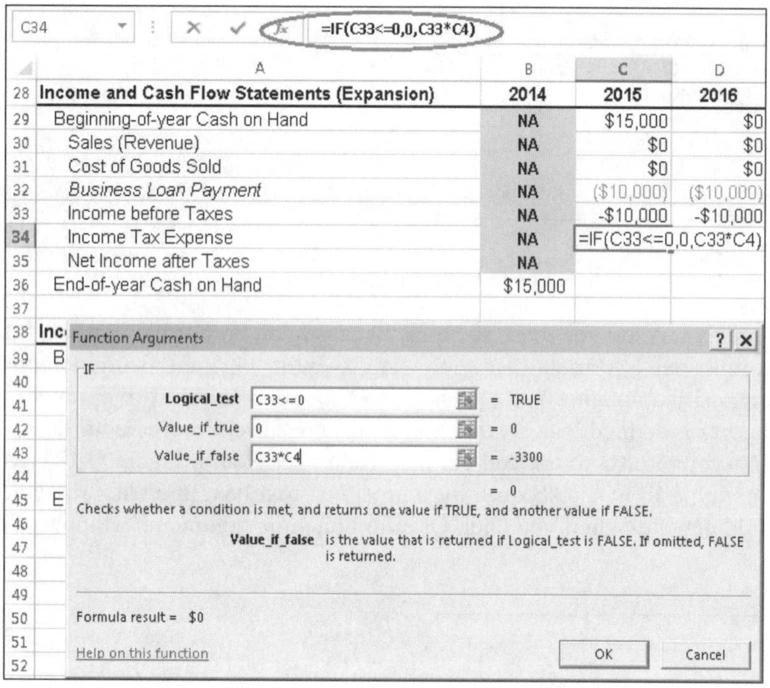

Source: Used with permission from Microsoft Corporation

FIGURE C-13   The Function Arguments window with the arguments filled in

Once you have entered the arguments, click OK; Excel enters the formula into the cell. Because you had negative income, the cell should display a zero for now. Because the same formula will be used in 2016 (but with the 2016 tax rate), you can simply copy and paste the formula from cell C34 to cell D34. You also have to calculate the income tax for the Income and Cash Flow Statements (No Expansion). In cell C43, use the same IF function, but in the Logical_test, Value_if_true, and Value_if_false arguments, you must type C42<=0, 0, and C42*C4, respectively. Again, the cell will display $0 for an answer. Copy cell C43 to cell D43 to complete the Income Tax Expense line for No Expansion.

Net Income after Taxes is simply the Income before Taxes minus the Income Tax Expense. Enter the formula into cell C35, then copy cell C35 over to cells D35, C44, and D44. If you did this correctly, cells C35 and D35 will display a negative $10,000, and cells C44 and D44 will display $0.

End-of-year Cash on Hand, the last line in both Income and Cash Flow Statements sections, is not difficult either. Conceptually, the cash you have at the end of the year is equal to your Beginning-of-year Cash on Hand plus your Net Income after Taxes. Enter the formula into cell C36, then copy cell C36 over to cell D36. Note that because the Income and Cash Flow Statements (No Expansion) do not have a line item for Business Loan Payment, you cannot copy the same command down to it. You have to enter the formula manually for cell C45, which is =C39+C44. However, you can copy cell C45 to cell D45 to finish the Income and Cash Flow Statements sections. The completed sections should look like Figure C-14.

D45	▼	:	×	✓	ƒx	=D39+D44		

	A	B	C	D
28	**Income and Cash Flow Statements (Expansion)**	**2014**	**2015**	**2016**
29	Beginning-of-year Cash on Hand	NA	$15,000	$5,000
30	Sales (Revenue)	NA	$0	$0
31	Cost of Goods Sold	NA	$0	$0
32	*Business Loan Payment*	NA	($10,000)	($10,000)
33	Income before Taxes	NA	-$10,000	-$10,000
34	Income Tax Expense	NA	$0	$0
35	Net Income after Taxes	NA	-$10,000	-$10,000
36	End-of-year Cash on Hand	$15,000	$5,000	-$5,000
37				
38	**Income and Cash Flow Statements (No Expansion)**	**2014**	**2015**	**2016**
39	Beginning-of-year Cash on Hand	NA	$15,000	$15,000
40	Sales (Revenue)	NA	$0	$0
41	Cost of Goods Sold	NA	$0	$0
42	Income before Taxes	NA	$0	$0
43	Income Tax Expense	NA	$0	$0
44	Net Income after Taxes	NA	$0	$0
45	End-of-year Cash on Hand	$15,000	$15,000	$15,000

Source: Used with permission from Microsoft Corporation

FIGURE C-14   The completed Income and Cash Flow Statements sections

Filling in the "Hard" Formulas

To finish the spreadsheet, you will enter values in the Inputs section and write the formulas in both Calculations sections.

AT THE KEYBOARD

In cell C8, enter an R for Recession, and in cell C9, enter H for High Inflation. You could enter any values here, but these two values will work with the IF functions you will write later. Recall that you did not use separate inputs for 2015 and 2016. You are assuming that the economic outlook or inflation rate that exists for 2015 will extend into 2016. However, because you are using the same inputs from these two locations, you must remember to use *absolute* cell references to both cells C8 and C9 in the various IF statements if you want to use a Copy command for adjacent cells. Your Inputs section should look like the one in Figure C-15.

	A	B	C	D
7	**Inputs**	**2014**	**2015**	**2016**
8	Economic Outlook (R=Recession, B=Boom)	NA	R	NA
9	Inflation Outlook (H=High, L=Low)	NA	H	NA

Source: Used with permission from Microsoft Corporation

FIGURE C-15   The Inputs section with values entered in cells C8 and C9

Remember that you referred to cell addresses in both Calculations sections in your formulas in the Income and Cash Flow Statements sections. Now you will enter formulas for these calculations. If necessary, format the four Total Sales Dollars cells and the four Cost of Goods Sold cells in the Calculations sections as Currency with no decimal places.

As described at the beginning of the tutorial, the forecast for Total Sales Dollars is a function of both the Economic Outlook and whether you expand the business. The following table lists the predicted sales growth percentages:

Sales Growth Forecast—Collegetown Thrift Shop

	Business Expansion	No Business Expansion
Recession-R	30%	20%
Boom-B	15%	5%

You will use IF formulas to forecast Total Sales Dollars. Click cell C18, then bring up the IF function and type the following in the text boxes:

Logical_test: $C$8="R" (Note that you must use absolute cell referencing for cell B8 and quotation marks for Excel to recognize a text string.)

Value_if_true: B18*1.3 (the 2014 sales multiplied by 1.3 for 30% sales growth)

Value_if_false: B18*1.15 (the 2014 sales multiplied by 1.15 for 15% sales growth)

Compare your entries to Figure C-16.

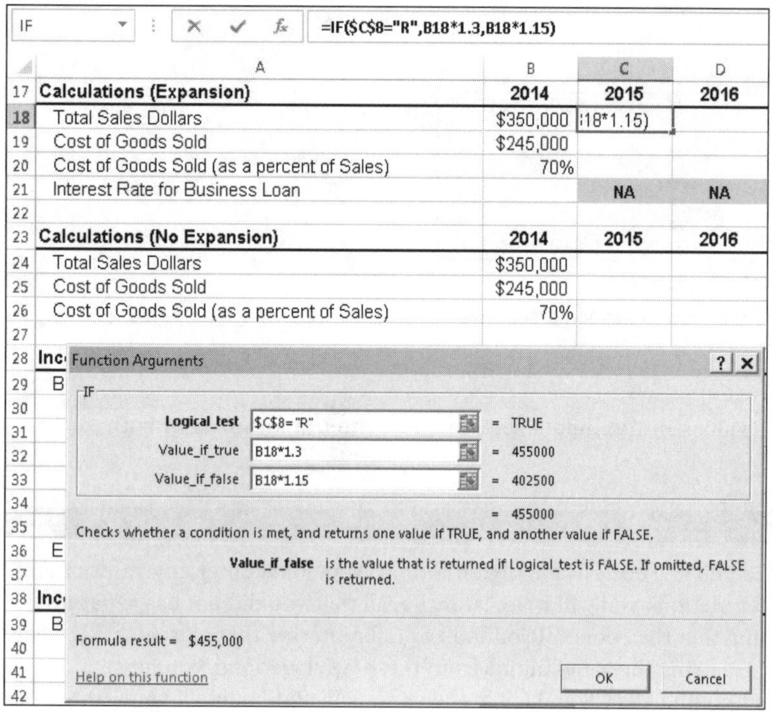

Source: Used with permission from Microsoft Corporation

FIGURE C-16 Using the IF function to enter the Total Sales Dollars forecast for 2015

When you click OK, cell C18 should display $455,000, because 30% of $350,000 is $105,000, and $350,000 plus $105,000 equals $455,000. So, it appears that the formula returned a "true" value with an R inserted in cell C8. Because you "anchored" cell C8 by entering $C$8, copy this formula over to cell D18 for the year 2016.

Once you complete the Total Sales Dollars cells for the Expansion scenario, go down to the Calculations (No Expansion) section and use IF statements to enter formulas for the Total Sales Dollars. Use a 20% sales growth factor for Recession and 5% for Boom. You can copy the formula from cell C18 into cell C24, but you then will have to use the editing window to change the values in the true and false arguments from 1.3 and 1.15 to 1.2 and 1.05, respectively, to reflect the fact that you did not expand the business. See Figure C-17.

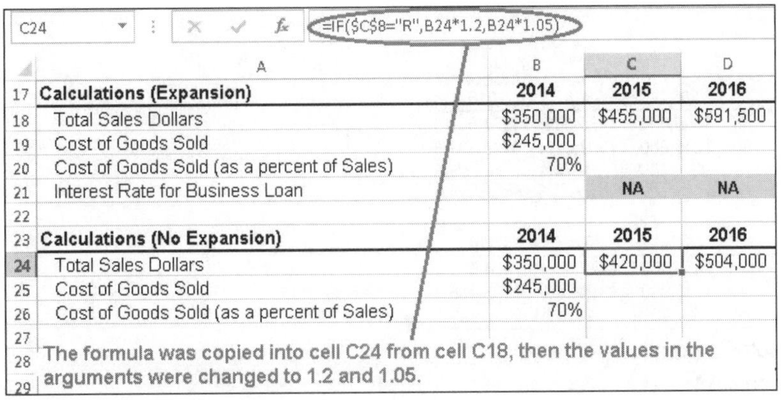

C24	▼	:	×	✓	fx	=IF($C$8="R",B24*1.2,B24*1.05)		

	A	B	C	D
		2014	**2015**	**2016**
17	**Calculations (Expansion)**			
18	Total Sales Dollars	$350,000	$455,000	$591,500
19	Cost of Goods Sold	$245,000		
20	Cost of Goods Sold (as a percent of Sales)	70%		
21	Interest Rate for Business Loan		NA	NA
22				
23	**Calculations (No Expansion)**	**2014**	**2015**	**2016**
24	Total Sales Dollars	$350,000	$420,000	$504,000
25	Cost of Goods Sold	$245,000		
26	Cost of Goods Sold (as a percent of Sales)	70%		
27				
28	The formula was copied into cell C24 from cell C18, then the values in the			
29	arguments were changed to 1.2 and 1.05.			

Source: Used with permission from Microsoft Corporation

FIGURE C-17   Copying cell C18 into cell C24 and then editing the IF function arguments to change the sales growth percentages

As before, you can now copy cell C24 to cell D24. You have completed the Total Sales Dollars calculations.

The Cost of Goods Sold (cells C19, D19, C25, and D25) is the Total Sales Dollars multiplied by the Cost of Goods Sold as a percent of Sales. In cell C19, type =C18*C20 and press Enter. Copy cell C19 and paste the contents into cells D19, C25, and D25. Your answers will be $0 until you enter the formulas for the Cost of Goods Sold as a percent of Sales.

The Cost of Goods Sold as a percent of Sales (cells C20, D20, C26, and D26) was 70% in 2014. In variety merchandising for resold items, it is easier to use an aggregate measure such as Cost of Goods Sold as a percent of Sales rather than trying to capture an individual Cost of Goods Sold for each item. From the 2014 data, you determined that for every dollar of sales you collected in 2014, you spent 70 cents purchasing the item and preparing it for resale. You will use that percentage as a basis for forecasting Cost of Goods Sold as a percent of Sales, applying an appropriate inflation factor for the cost of acquiring the stock for sale. The following table lists the predicted inflation percentages for Cost of Goods Sold.

Cost of Goods Sold Forecast—Collegetown Thrift Shop

	Business Expansion	**No Business Expansion**
High Inflation	6%	6%
Low Inflation	2%	2%

As with Total Sales Dollars previously, you will again use the IF function to calculate the Cost of Goods Sold as a percent of Sales. Now that you are familiar with the IF function, you can probably enter the function without using the windows. In cell C20, type the following:

=IF($C$9="H",B20*1.06,B20*1.02)

This expression means that if the text string in cell C9 is the letter H, you multiply the value in cell B20 by 1.06 (6% inflation). If the value in cell C9 is not an H, multiply the value in cell B20 by 1.02 (2% inflation). The value in cell B20 was the baseline Cost of Goods Sold as a percent of Sales in 2014, which was 70%. You can now copy cell C20 and paste the contents into cell D20.

Because the inflation percentages were exactly the same for both the Expansion and No Expansion calculations, you can also copy cell C20 and paste the contents into cells C26 and D26. Your Calculations sections should now look like Figure C-18.

| D26 | ▼ | : | × | ✓ | *fx* | =IF($C$9="H",C26*1.06,C26*1.02) |

	A	B	C	D
17	**Calculations (Expansion)**	**2014**	**2015**	**2016**
18	Total Sales Dollars	$350,000	$455,000	$591,500
19	Cost of Goods Sold	$245,000	$337,610	$465,227
20	Cost of Goods Sold (as a percent of Sales)	70%	74%	79%
21	Interest Rate for Business Loan		**NA**	**NA**
22				
23	**Calculations (No Expansion)**	**2014**	**2015**	**2016**
24	Total Sales Dollars	$350,000	$420,000	$504,000
25	Cost of Goods Sold	$245,000	$311,640	$396,406
26	Cost of Goods Sold (as a percent of Sales)	70%	74%	79%

Source: Used with permission from Microsoft Corporation

FIGURE C-18   Calculations sections nearly complete

The last item in the Calculations section is the Interest Rate for Business Loan (cell B21). Remember the bank's statement that if the economy recovers, it could lower the interest rate from 5% to 4%. So, you will need one more IF function to insert into cell B21 based on the economic outlook. If the economic outlook is for a Recession (R), then the interest rate will be 5% annually; if the outlook is for a Boom (B), then the interest rate will be 4% annually. Now that you are familiar with the IF function, you can simply type the expression into the cell yourself. Click cell B21, type =IF(C8="R",5%,4%), and press Enter.

You will immediately notice that 5% appears in the cell because you have R in the input cell for Economic Outlook. You may also notice that you now have a negative $12,950 in the Business Loan Payment cells (C32 and D32). See Figure C-19 to compare your results.

| B21 | ▼ | : | × | ✓ | *fx* | =IF(C8="R",5%,4%) |

	A	B	C	D
7	**Inputs**	**2014**	**2015**	**2016**
8	Economic Outlook (R=Recession, B=Boom)	**NA**	R	**NA**
9	Inflation Outlook (H=High, L=Low)	**NA**	H	**NA**
10				
11	**Summary of Key Results**	**2014**	**2015**	**2016**
12	Net Income after Taxes (Expansion)	**NA**	$69,974	$73,660
13	End-of-year Cash on Hand (Expansion)	**NA**	$84,974	$158,634
14	Net Income after Taxes (No Expansion)	**NA**	$72,601	$69,936
15	End-of-year Cash on Hand (No Expansion)	**NA**	$87,601	$157,537
16				
17	**Calculations (Expansion)**	**2014**	**2015**	**2016**
18	Total Sales Dollars	$350,000	$455,000	$591,500
19	Cost of Goods Sold	$245,000	$337,610	$465,227
20	Cost of Goods Sold (as a percent of Sales)	70%	74%	79%
21	Interest Rate for Business Loan	5%	**NA**	**NA**
22				
23	**Calculations (No Expansion)**	**2014**	**2015**	**2016**
24	Total Sales Dollars	$350,000	$420,000	$504,000
25	Cost of Goods Sold	$245,000	$311,640	$396,406
26	Cost of Goods Sold (as a percent of Sales)	70%	74%	79%
27				
28	**Income and Cash Flow Statements (Expansion)**	**2014**	**2015**	**2016**
29	Beginning-of-year Cash on Hand	**NA**	$15,000	$84,974
30	Sales (Revenue)	**NA**	$455,000	$591,500
31	Cost of Goods Sold	**NA**	$337,610	$465,227
32	*Business Loan Payment*	**NA**	($12,950)	($12,950)
33	Income before Taxes	**NA**	$104,440	$113,323
34	Income Tax Expense	**NA**	$34,465	$39,663
35	Net Income after Taxes	**NA**	$69,974	$73,660
36	End-of-year Cash on Hand	$15,000	$84,974	$158,634
37				
38	**Income and Cash Flow Statements (No Expansion)**	**2014**	**2015**	**2016**
39	Beginning-of-year Cash on Hand	**NA**	$15,000	$87,601
40	Sales (Revenue)	**NA**	$420,000	$504,000
41	Cost of Goods Sold	**NA**	$311,640	$396,406
42	Income before Taxes	**NA**	$108,360	$107,594
43	Income Tax Expense	**NA**	$35,759	$37,658
44	Net Income after Taxes	**NA**	$72,601	$69,936
45	End-of-year Cash on Hand	$15,000	$87,601	$157,537

Source: Used with permission from Microsoft Corporation

FIGURE C-19   The finished spreadsheet

You can change the economic inputs in four different combinations: R-H, R-L, B-H, B-L. This allows you to see the impact on your net income and cash on hand both for expanding and not expanding. However, you have another more powerful way to do this. In the next section, you will learn how to tabulate the financial results of the four possible combinations using an Excel tool called Scenario Manager.

SCENARIO MANAGER

You are now ready to evaluate the four possible outcomes for your DSS model. Because this is a simple, four-outcome model, you could have created four different spreadsheets, one for each set of outcomes, and then transferred the financial information from each spreadsheet to a Summary Report.

In essence, Scenario Manager performs the same task. It runs the model for all the requested outcomes and presents a tabular summary of the results. This summary is especially useful for reports and presentations needed by upper managers, financial investors, or in this case, the bank.

To review, the four possible combinations of input values are: R-H (Recession and High Inflation), R-L (Recession and Low Inflation), B-H (Boom and High Inflation), and B-L (Boom and Low Inflation). You could consider each combination of inputs a separate scenario. For each of these scenarios, you are interested in four outputs: Net Income after Taxes for Expansion and No Expansion, and End-of-year Cash on Hand for Expansion and No Expansion.

Scenario Manager runs each set of combinations and then records the specified outputs as a summary into a separate worksheet. You can use these summary values as a table of numbers and print it, or you can copy them into a Microsoft Word document or a PowerPoint presentation. You can also use the data table to build a chart or graph, which you can put into a report or presentation.

When you define a scenario in Scenario Manager, you name it and identify the input cells and input values. Then you identify the output cells so Scenario Manager can capture the outputs in a summary sheet.

AT THE KEYBOARD

To start, click the Data tab on the Ribbon. In the Data Tools group, click the What-If Analysis button, then click Scenario Manager from the menu that appears (see Figure C-20).

Source: Used with permission from Microsoft Corporation

FIGURE C-20   Scenario Manager option in the What-If Analysis menu

Scenario Manager appears (see Figure C-21), but no scenarios are defined. Use the window to add, delete, or edit scenarios.

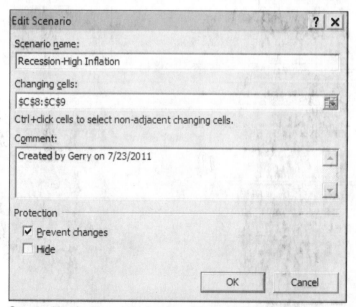

Source: Used with permission from Microsoft Corporation

FIGURE C-21   Initial Scenario Manager window

NOTE

When working with the Scenario Manager window and any following windows, do not use the Enter key to navigate. Use mouse clicks to move from one step to the next.

To define a scenario, click the Add button. The Edit Scenario window appears. (The title bar displays Add Scenario until you enter data in the fields.) Enter Recession-High Inflation in the field under Scenario name. Then type the input cells in the Changing cells field (in this case, C8:C9). Better yet, you can use the button next to the field to select the cells in your spreadsheet. If you do, Scenario Manager changes the cell references to absolute cell references, which is acceptable (see Figure C-22).

Source: Used with permission from Microsoft Corporation

FIGURE C-22   Defining a scenario name and input cells

Click OK to open the Scenario Values window. Enter the input values for the scenario. In the case of Recession and High Inflation, the values will be R and H for cells C8 and C9, respectively (see Figure C-23). Note that if you already have entered values in the spreadsheet, the window will display the current values. Make sure to enter the correct values.

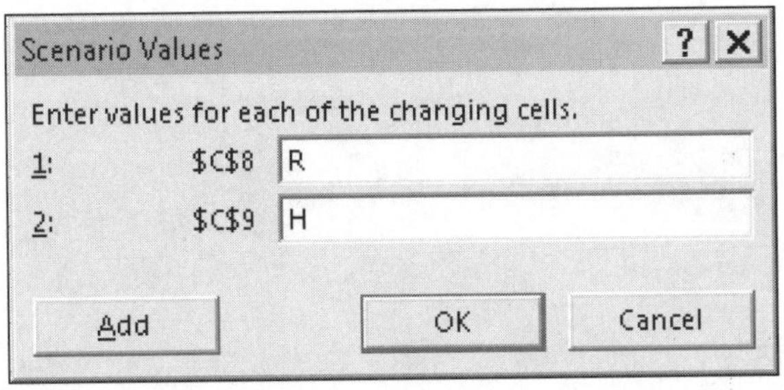

Source: Used with permission from Microsoft Corporation

FIGURE C-23   Entering values for the input cells

Click OK to return to the Scenario Manager window. Enter the other three scenarios: Recession-Low Inflation, Boom-High Inflation, and Boom-Low Inflation (R-L, B-H, and B-L), and their related input values. When you finish, you should see the names and changing cells for the four scenarios (see Figure C-24).

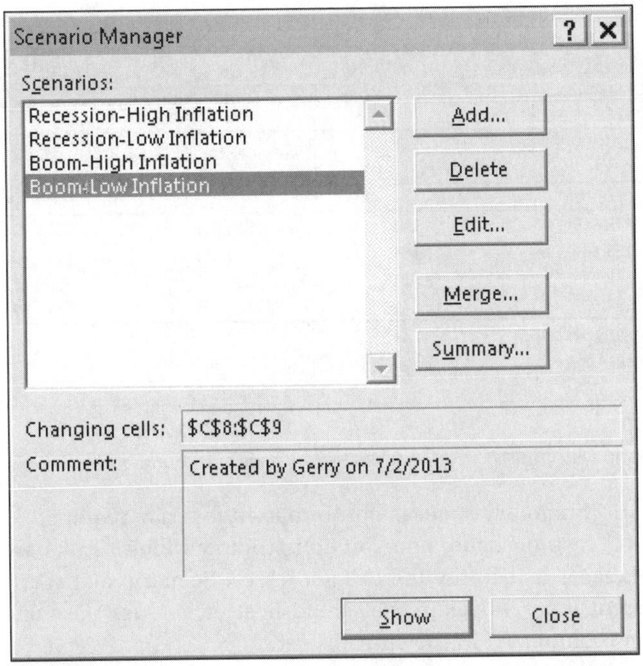

Source: Used with permission from Microsoft Corporation

FIGURE C-24   Scenario Manager window with all four scenarios entered

You can now create a summary sheet that displays the results of running the four scenarios. Click the Summary button to open the Scenario Summary window, as shown in Figure C-25. You must now enter the output cell addresses in Excel—they will be the same for all four scenarios. Recall that you created a section in your spreadsheet called Summary of Key Results. You are primarily interested in the results at the end of 2016, so you will choose the four cells that represent the Net Income after Taxes and End-of-year Cash on Hand, and then use them for both the expansion scenario and the non-expansion scenario. These cells are D12 to D15 in your spreadsheet. Either type D12:D15 or use the button next to the Result cells field and select those cells in the spreadsheet.

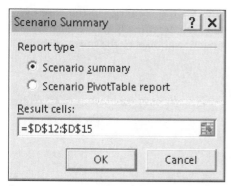

Source: Used with permission from Microsoft Corporation

FIGURE C-25   Scenario Summary window with Result cells entered

Another good reason for having a Summary of Key Results section is that it provides a contiguous range of cells to define for summary output. However, if you want to add output from other cells in the spreadsheet, simply separate each cell or range of cells in the window with a comma. Next, click OK. Excel runs each set of inputs in the background, collects the results from the result cells, and then creates a new sheet called Scenario Summary (the name on the sheet's lower tab), as shown in Figure C-26.

			Current Values:	Recession-High Inflation	Recession-Low Inflation	Boom-High Inflation	Boom-Low Inflation
Scenario Summary							
Changing Cells:							
$C$8	R		R		R	B	B
$C$9	H		H		L	H	L
Result Cells:							
$D$12			$73,660	$73,660	$96,052	$56,216	$73,738
$D$13			$158,634	$158,634	$189,562	$132,531	$157,605
$D$14			$69,936	$69,936	$89,015	$53,545	$68,152
$D$15			$157,537	$157,537	$184,496	$132,071	$153,573

Notes: Current Values column represents values of changing cells at time Scenario Summary Report was created. Changing cells for each scenario are highlighted in gray.

Source: Used with permission from Microsoft Corporation

FIGURE C-26   Scenario Summary sheet created by Scenario Manager

As you can see, the output created by the Scenario Summary sheet is not formatted for easy reading. You do not know which results are the net income and cash on hand, and you do not know which results are for Expansion vs. No Expansion, because Scenario Manager listed only the cell addresses. Scenario Manager also listed a separate column (column D) for the current input values in the spreadsheet, which are the same as the values in column E. It also left a blank column (column A) in the spreadsheet.

Fortunately, it is fairly easy to format the output. Delete columns D and A, put in the labels for cell addresses in the new column A, and then retitle the Scenario Summary as Collegetown Thrift Shop Financial Forecast, End of Year 2016 (because you are looking only at Year 2016 results). You can also make the results columns narrower by breaking the column headings into two lines; place your cursor in the editing window where you want to break the words, and then press Alt+Enter. Add a heading for column B (Cell Address). Finally, merge and center the title, and center the column headings and the input cell values (R, B, H, and L). Leave your financial data right-justified to keep the numbers lined up correctly. Add some border boxes around each column of results. Finally, delete the notes in Rows 13 through 15—they are no longer needed. Figure C-27 shows a formatted Scenario Summary worksheet.

	A	B	C	D	E	F
1						
2			Scenario Summary--Collegetown Thrift Shop Financial Forecast--End of Year 2016			
3		Cell Address	Recession-High Inflation	Recession-Low Inflation	Boom-High Inflation	Boom-Low Inflation
5	Changing Cells:					
6	Economic Outlook: R-Recession, B-Boom	$C$8	R	R	B	B
7	Inflation: H-High, L-Low	$C$9	H	L	H	L
8	Result Cells:					
9	Net Income after Taxes (Expansion)	$D$12	$73,660	$96,052	$56,216	$73,738
10	End-of-year Cash on Hand (Expansion)	$D$13	$158,634	$189,562	$132,531	$157,605
11	Net Income after Taxes (No Expansion)	$D$14	$69,936	$89,015	$53,545	$68,152
12	End-of-year Cash on Hand (No Expansion)	$D$15	$157,537	$184,496	$132,071	$153,573

Source: Used with permission from Microsoft Corporation

FIGURE C-27   Scenario Summary worksheet after formatting and adding labels

Interpreting the Results

Now that you have good data, what do you do with it? Remember, you wanted to see if taking a $100,000 business loan to expand the thrift shop was a good financial decision. This is a relatively simple business case, and the shop's success so far ($350,000 of sales in 2014) would seem to make expansion a good risk. But how good a risk is the expansion?

After building the spreadsheet and doing the analysis, you can make comparisons and interpret the results. Regardless of the economic outlook or inflation, all four scenarios indicate that expanding the business should provide greater Net Income After Taxes and End-of-year Cash on Hand (after two years) than not expanding. So, the DSS model not only provides a quantitative basis for expanding, it provides an analysis that you can present to prospective lenders.

What decision would you make about expansion if you looked only at the 2015 forecast? You could go back to the original spreadsheet and look at the figures for 2015, or you can go to Scenario Manager and create a new summary, specifying the 2015 cells C12 through C15. See Figure C-28.

	A	B	C	D	E	F	G	H
1	Tutorial Exercise--Collegetown Thrift Shop							
2								
3	Constants	2014	2015	2016				
4	Tax Rate	NA	33%	35%				
5	Loan Amount for Store Expansion	NA	$100,000	NA				
6								
7	Inputs	2014	2015	2016				
8	Economic Outlook (R=Recession, B=Boom)	NA	R	NA				
9	Inflation Outlook (H=High, L=Low)	NA	H	NA				
10								
11	Summary of Key Results	2014	2015	2016				
12	Net Income after Taxes (Expansion)	NA	$69,974	$73,660	Scenario Summary		? X	
13	End-of-year Cash on Hand (Expansion)	NA	$84,974	$158,634	Report type			
14	Net Income after Taxes (No Expansion)	NA	$72,601	$69,936	⊙ Scenario summary			
15	End-of-year Cash on Hand (No Expansion)	NA	$87,601	$157,537	○ Scenario PivotTable report			
16								
17	Calculations (Expansion)	2014	2015	2016	Result cells:			
18	Total Sales Dollars	$350,000	$455,000	$591,500	=$C$12:$C$15			
19	Cost of Goods Sold	$245,000	$337,610	$465,227	OK		Cancel	
20	Cost of Goods Sold (as a percent of Sales)	70%	74%	79%				

Source: Used with permission from Microsoft Corporation

FIGURE C-28   Creating a new Scenario Summary for 2015 instead of 2016

When you click OK, Excel creates a second Scenario Summary (appropriately named Scenario Summary 2), but this time the output values come from 2015, not 2016. After editing and formatting, the 2015 Scenario Summary should look like Figure C-29.

⬚	A	B	C	D	E	F
1						
2			Scenario Summary 2--Collegetown Thrift Shop Financial Forecast--End of Year 2015			
3		Cell Address	Recession-High Inflation	Recession-Low Inflation	Boom-High Inflation	Boom-Low Inflation
5	**Changing Cells:**					
6	Economic Outlook: R-Recession, B-Boom	$C$8	R	R	B	B
7	Inflation: H-High, L-Low	$C$9	H	L	H	L
8	**Result Cells:**					
9	Net Income after Taxes (Expansion)	$C$12	$69,974	$78,510	$61,316	$68,867
10	End-of-year Cash on Hand (Expansion)	$C$13	$84,974	$93,510	$76,316	$83,867
11	Net Income after Taxes (No Expansion)	$C$14	$72,601	$80,480	$63,526	$70,420
12	End-of-year Cash on Hand (No Expansion)	$C$15	$87,601	$95,480	$78,526	$85,420

Source: Used with permission from Microsoft Corporation

FIGURE C-29   Scenario Summary for End of Year 2015

As you can see, *not* expanding the business yields slightly better financial results at the end of 2015. As the original Scenario Summary points out, it will take two years for the business expansion to start making more money when compared with not expanding. You can also revise the original spreadsheet to copy the columns out to 2017, 2018, and beyond to forecast future income and cash flows. However, note that the accuracy of a forecast gets worse as you extend it in time.

Managers must also maintain a healthy skepticism about the validity of their assumptions when formulating a DSS model. Most assumptions about economic outlooks, inflation, and interest rates are really educated guesses. For example, who could have predicted the economic meltdown in 2007? Business DSS models for investments, new product launches, business expansion, or major capital projects commonly look at three possible outcomes: best case, most likely, and worst case. The most likely outcome is based on previous years' data already collected by the firm. The best-case and worst-case outcomes are formulated based on some percentage of performance that falls above or below the most likely scenario. At least these are data-driven forecasts, or what people in the business world call "guessing—with data."

So, how do you reduce risk when making financial decisions based on DSS model results? It helps to formulate the model based on valid data and to use conservative estimates for success. More importantly, collecting pertinent data and tracking the business results *after* deciding to invest or expand can help reduce the risk of failure for the enterprise.

Summary Sheets

When you start working on the Scenario Manager spreadsheet cases later in this book, you will need to know how to manipulate summary sheets and their data. Some of these operations are explained in the following sections.

Rerunning Scenario Manager

The Scenario Summary sheet does not update itself when you change formulas or inputs in the spreadsheet. To get an updated Scenario Summary, you must rerun Scenario Manager, as you did when changing the outputs from 2016 to 2015. Click the Summary button in the Scenario Manager window, verify the results, and then click OK. Another summary sheet is created; Excel numbers them sequentially (Scenario Summary, Scenario Summary2, etc.), so you do not have to worry about Excel overwriting any of your older summaries. That is why you should rename each summary with a description of the changes.

Deleting Unwanted Scenario Manager Summary Sheets

When working with Scenario Manager, you might produce summary sheets you do not want. To delete an unwanted sheet, move your mouse pointer to the group of sheet tabs at the bottom of the screen and *right*-click the tab of the sheet you want to delete. Click Delete from the menu that appears (see Figure C-30).

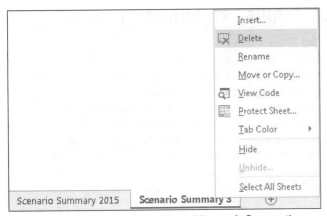

FIGURE C-30   Deleting unwanted worksheets

Charting Summary Sheet Data

You can easily chart Summary Sheet results using the Charts group in the Insert tab, as discussed in Tutorial F. Figure C-31 shows a 3D clustered column chart prepared from the data in the Scenario Summary for 2016. Charts are useful because they provide a visual comparison of results. As the chart shows, the best economic climate for the thrift shop is a Recession with Low Inflation.

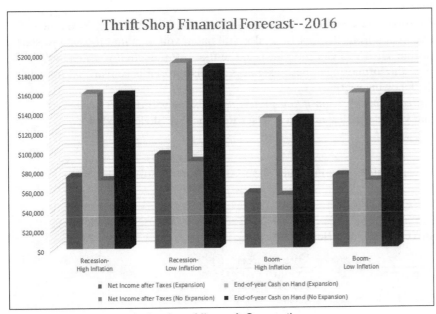

FIGURE C-31   3D Clustered column chart displaying data from the Summary Sheet

Copying Summary Sheet Data to the Clipboard

As you can with almost everything else in Microsoft Office, you can copy summary sheet data to other Office applications (a Word document or PowerPoint slide, for example) by using the Clipboard. Follow these steps:

1. Select the data range you want to copy.
2. Right-click the mouse and select Copy from the resulting menu.
3. Open the Word document or PowerPoint presentation into which you want to copy.
4. Click your cursor where you want the upper-left corner of the copied data to be displayed.
5. Right-click the mouse and select Paste from the resulting menu. The data should now appear on your document.

PRACTICE EXERCISE—TED AND ALICE'S HOUSE PURCHASE DECISION

Ted and Alice are a young couple who have been living in an apartment for the first two years of their marriage. They would like to buy their first house, but do not know whether they can afford it. Ted works as a carpenter's apprentice, and Alice is a customer service specialist at a local bank. In 2014, Ted's "take home" wages were $24,000 after taxes and deductions, and Alice's take-home salary was $30,000. Ted gets a 2% raise every year, and Alice gets a 3% raise. Their apartment rent is $1,200 per month ($14,400 per year), but the lease is up for renewal and the landlord said he needs to increase the rent for the next lease.

Ted and Alice have been looking at houses and have found one they can buy, but they will need to borrow $200,000 for a mortgage. Their parents are helping them with the down payment and closing costs. After talking to several lenders, Ted and Alice have learned that the state legislature is voting on a first-time home buyers' mortgage bond. If the bill passes, they will be able to get a 30-year fixed mortgage at 3% interest. Otherwise, they will have to pay 6% interest on the mortgage.

Because of the depressed housing market, Ted and Alice are not figuring equity value into their calculations. In addition, although the mortgage interest and real estate taxes will be deductible on their income taxes, these deductions will not be higher than the standard allowable tax deduction, so they are not figuring on any savings there either. Ted and Alice's other living expenses (such as car payments, food, and medical bills), the utilities expenses for either renting or buying, and estimated house maintenance expenses are listed in the Constants section (see Figure C-32).

Ted and Alice's primary concern is their cash on hand at the end of years 2015 and 2016. They are thinking of starting a family, but they know it will be difficult without adequate savings.

Getting Started on the Practice Exercise

If you closed Excel after the first tutorial exercise, start Excel again—it should automatically open a new workbook for you. If your Excel workbook from the first tutorial is still open, you may find it useful to start a new worksheet in the same workbook. Then you can refer back to the first tutorial when you need to structure or format the spreadsheet; the formatting of both exercises in this tutorial is similar. Set up your new worksheet as explained in the following sections.

You can also download the spreadsheet skeleton if you prefer; it will save you time. To access this skeleton file, select Tutorial C from your data files, and then select **Rent or Buy Skeleton.xlsx**.

Constants Section

Your spreadsheet should have the constants shown in Figure C-32. An explanation of the line items follows the figure.

	A	B	C	D
1	**Tutorial Exercise Skeleton--Ted and Alice's House Decision**			
2				
3	**Constants**	**2014**	**2015**	**2016**
4	Non-Housing Living Expenses (Cars, Food, Medical, etc)	NA	$36,000	$39,000
5	Mortgage Amount for Home Purchase	NA	$200,000	NA
6	Real Estate Taxes and Insurance on Home	NA	$3,000	$3,150
7	Utilities Expense (Heat & Electric)--Apartment	NA	$2,000	$2,200
8	Utilities Expense(Heat, Electric, Water, Trash)--House	NA	$2,500	$2,600
9	House Repair and Maintenance Expenses	NA	$1,200	$1,400

Source: Used with permission from Microsoft Corporation

FIGURE C-32   Constants section

- Non-Housing Living Expenses—This value represents Ted and Alice's estimate of all their other living expenses for 2015 and 2016.
- Mortgage Amount for Home Purchase
- Real Estate Taxes and Insurance on Home—A lender has given Ted and Alice estimates for these values; they are usually paid monthly with the house mortgage payment. The money is placed in an escrow account and then paid by the mortgage company to the state or county and insurance company.
- Utilities Expense—Apartment—This value is Ted and Alice's estimate for 2015 and 2016 based on their 2014 bills.

- Utilities Expense—House—Currently the apartment rent includes fees for water, sewer, and trash disposal. If they get a house, Ted and Alice expect the utilities to be higher.
- House Repair and Maintenance Expenses—In an apartment, the landlord is responsible for repair and maintenance. Ted and Alice will have to budget for repair and maintenance on the house.

Inputs Section

Your spreadsheet should have the inputs shown in Figure C-33. An explanation of line items follows the figure.

	A	B	C	D
	Inputs	2014	2015	2016
11				
12	Rental Occupancy (H=High, L=Low)	NA		NA
13	First Time Buyer Bond Loans Available (Y=Yes, N=No)	NA		NA

Source: Used with permission from Microsoft Corporation

FIGURE C-33 Inputs section

- Rental Occupancy (H=High, L=Low)—When the housing market is depressed (in other words, people are not buying homes), rental housing occupancy percentages are high, which allows landlords to charge higher rents when leases are renewed. Ted and Alice think their rent will increase in 2015. The amount of the increase depends on the Rental Occupancy. If the occupancy is high, Ted and Alice expect to see a 10% increase in rent in both 2015 and 2016. If occupancy is low, they only expect a 3% increase for each year.
- First Time Buyer Bond Loans Available (Y=Yes, N=No)—As described earlier, when housing markets are depressed, local governments will frequently pass a bond bill to provide low-interest mortgage money to first-time home buyers. If the bond loans are available, Ted and Alice can obtain a 30-year fixed mortgage at only 3%, which is half the interest rate they would otherwise pay for a conventional mortgage.

Summary of Key Results Section

Figure C-34 shows what key results Ted and Alice are looking for. They want to know their End-of-year Cash on Hand for both 2015 and 2016 if they decide to stay in the apartment and if they decide to purchase the house.

	A	B	C	D
15	Summary of Key Results	2014	2015	2016
16	End-of-year Cash on Hand (Rent)	NA		
17	End-of-year Cash on Hand (Buy)	NA		

Source: Used with permission from Microsoft Corporation

FIGURE C-34   Summary of Key Results section

These results are copied from the End-of-year Cash on Hand sections of the Income and Cash Flow Statements sections (for both renting and buying).

Calculations Section

Your spreadsheet will need formulas to calculate the apartment rent, house payments, and interest rate for the mortgage (see Figure C-35). You will use the rent and house payments later in the Income and Cash Flow Statements for both renting and buying.

	A	B	C	D
19	Calculations	2014	2015	2016
20	Apartment Rent	$14,400		
21	House Payments	NA		
22	Interest Rate for House Mortgage		NA	NA

Source: Used with permission from Microsoft Corporation

FIGURE C-35   Calculations section

- Apartment Rent—The 2014 amount is given. Use IF formulas to increase the rent by 10% if occupancy rates are high, or by 3% if occupancy rates are low.
- House Payments—This value is the total of the 12 monthly payments made on the mortgage. An important point to note is that house mortgage interest is always compounded *monthly*, not annually, as in the thrift shop tutorial. To properly calculate the house payments for the year, you divide the annual interest rate by 12 to determine the monthly interest. You also have to multiply a 30-year mortgage by 12 to get 360 payments, and then multiply the PMT formula by 12 to get the total amount for your annual house payments. Also, you will precede the PMT function with a negative sign to make the payment amount a positive number. Your formula should look like the following:

  =–PMT(B22/12,360,C5)*12

- Interest Rate for House Mortgage—Use the IF formula to enter a 3% interest rate if the bond money is available, and a 6% interest rate if no bond money is available.

Income and Cash Flow Statements Sections

As with the thrift shop tutorial, you want to see the Income and Cash Flow Statements for two scenarios—in this case, for continuing to rent and for purchasing a house. Each section begins with cash on hand at the end of 2014. As you can see in Figure C-36, Ted and Alice have only $4,000 in their savings.

	A	B	C	D
24	**Income and Cash Flow Statement (Continue to Rent)**	**2014**	**2015**	**2016**
25	Beginning-of-year Cash on Hand	NA		
26	Ted's Take Home Wages	$24,000		
27	Alice's Take Home Salary	$30,000		
28	Total Take Home Income	$54,000		
29	*Apartment Rent*	NA		
30	Utilities (Apartment)	NA		
31	Non-Housing Living Expenses	NA		
32	Total Expenses	NA		
33	End-of-year Cash on Hand	$4,000		
34				
35	**Income and Cash Flow Statement (Purchase House)**	**2014**	**2015**	**2016**
36	Beginning-of-year Cash on Hand	NA		
37	Ted's Take Home Wages	$24,000		
38	Alice's Take Home Salary	$30,000		
39	Total Take Home Income	$54,000		
40	*House Payments*	NA		
41	*Real Estate Taxes and Insurance*	NA		
42	Utilities (House)	NA		
43	*House Repair and Maintenance Expense*	NA		
44	Non-Housing Living Expenses	NA		
45	Total Expenses	NA		
46	End-of-year Cash on Hand	$4,000		

Source: Used with permission from Microsoft Corporation

FIGURE C-36   Income and Cash Flow Statements sections (for both rent and purchase)

- Beginning-of-year Cash on Hand—This value is the End-of-year Cash on Hand from the previous year.
- Ted's Take Home Wages—This value is given for 2014. To get values for 2015 and 2016, increase Ted's wages by 2% each year.
- Alice's Take Home Salary—This value is given for 2014. To get values for 2015 and 2016, increase Alice's salary by 3% each year.
- Total Take Home Income—The sum of Ted and Alice's pay.
- Apartment Rent—The rent is copied from the Calculations section.
- House Payments—The house payments are also copied from the Calculations section.

- Real Estate Taxes and Insurance, Utilities (Apartment or House), House Repair and Maintenance Expense, and Non-Housing Living Expenses—These values all are copied from the Constants section.
- Total Expenses—This value is the sum of all the expenses listed above. Note that the house payment is now a positive number, so you can sum it normally with the other expenses.
- End-of-year Cash on Hand—This value is the Beginning-of-year Cash on Hand plus the Total Take Home Income minus the Total Expenses.

Scenario Manager Analysis

When you have completed the spreadsheet, set up Scenario Manager and create a Scenario Summary sheet. Ted and Alice want to look at their End-of-year Cash on Hand in 2016 for renting or buying under the following four scenarios:

- High occupancy and bond money available
- High occupancy and no bond money available
- Low occupancy and bond money available
- Low occupancy and no bond money available

If you have done your spreadsheet and Scenario Manager correctly, you should get the results shown in Figure C-37.

	A	B	C	D	E	F
1	Scenario Summary--Ted & Alice's House Purchase Decision--2016					
2			Hi Occ- Bond $	Hi Occ- No Bond $	Lo Occ- Bond $	Lo Occ- No Bond $
4	**Changing Cells:**					
5	**Rental Occupancy (H-High, L-Low)**	$C$12	H	H	L	L
6	**Bond Mortgage Available (Y or N)**	$C$13	Y	N	Y	N
7	**Result Cells:**					
8	**End of Year Cash on Hand (Rent)**	$D$16	$3,713	$3,713	$6,868	$6,868
9	**End of Year Cash on Hand (Buy)**	$D$17	$7,090	($1,452)	$7,090	($1,452)

Source: Used with permission from Microsoft Corporation

FIGURE C-37   Scenario Summary results

Interpreting the Results

Based on the Scenario Summary results, what should Ted and Alice do? At first glance, it looks like the safe decision is to stay in the apartment. Actually, their decision hinges on whether they can get the lower-interest mortgage from the first-time buyers' bond issue. If they can, and if occupancy levels in apartments stay high, purchasing a house will give them about $3,300 more in savings at the end of 2016 than if they continued renting. Some other intangible factors are that home owners do not need permission to have pets, detached houses are quieter than apartments, and homes usually have a yard for pets and children to play in. Also, for the purposes of this exercise, you did not consider the tax benefits of home ownership. Depending on the amount of mortgage interest and real estate taxes Ted and Alice have to pay, they may be able to itemize their deductions and pay less income tax. If the income tax savings are more than $1,500, they can purchase the house even at the higher interest rate. In any case, because you did the DSS model for them, Ted and Alice now have a quantitative basis to help them make a good decision.

Visual Impact: Charting the Results

Charts and graphs often add visual impact to a Scenario Summary. Using the data from the Scenario Summary output table, try to create a chart similar to the one in Figure C-38 to illustrate the financial impact of each outcome.

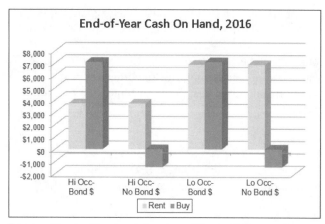

Source: Used with permission from Microsoft Corporation

FIGURE C-38 A 3D clustered column chart created from Scenario Summary data

Printing and Submitting Your Work

Ask your instructor which worksheets need to be printed for submission. Make sure your printouts of the spreadsheet, the Scenario Manager Summary table, and the graph (if you created one) fit on one printed page apiece. Click the File tab on the Ribbon, click the Print button, and then click Page Setup at the bottom of the Print Navigation pane. When the Page Setup window opens, click the Page tab if it is not already open, then click the Fit to radio button and click 1 page wide by 1 page tall. Your spreadsheet, table, and graph will be fitted to print on one page apiece.

REVIEW OF EXCEL BASICS

This section reviews some basic operations in Excel and provides some tips for good work practices. Then you will work through some cash flow calculations. Working through this section will help you complete the spreadsheet cases in the following chapters.

Save Your Work Often—and in More Than One Place

To guard against data loss in case of power outages, computer crashes, and hard drive failure, it is always a good idea to save your work to a separate storage device. Copying a file into two separate folders on the same hard disk is *not* an adequate safeguard. If you are working on your college's computer network and you have been assigned network storage, the network storage is usually "mirrored"; in other words, it has duplicate drives recording data to prevent data loss if the system goes down. However, most laptops and home computers lack this feature. An excellent way to protect your work from accidental deletion is to purchase a USB "thumb" drive and copy all of your files to it.

When you save your Excel files, Windows will usually store them in the My Documents folder unless you specify the storage location. Instead of just clicking the Save icon, a good idea is to click the File group in the Ribbon, and then click Save As in the left column menu. A window will appear with icons on the left side, as shown in Figure C-39. If you have previously saved your file to a particular location, it will appear in the Save in text box at the top of the window. To save the file in the same location, click Save. If your work is stored elsewhere, you can find the location using the icons on the left side of the window. If you are saving to a USB thumb drive, it will appear as a storage device when you click the Computer icon. Click the folder where you want to save your file.

> **N O T E**
>
> If you are trying an operation that might damage your spreadsheet and you do not want to use the Undo command, you can use the Save As command, and then add a number or letter to the filename to save an additional copy to "play with." Your original work will be preserved.

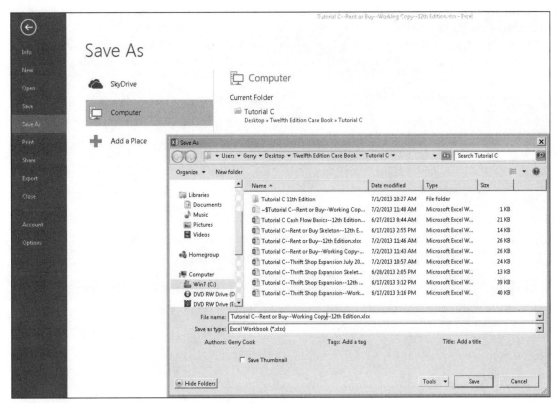

Source: Used with permission from Microsoft Corporation

FIGURE C-39   The Save As window in Excel

Basic Operations

To begin, you will review the following topics: formatting cells, displaying the spreadsheet cell formulas, circular reference errors, using the AND and OR logical operators in IF statements, and using nested IF statements to produce more than two outcomes.

Formatting Cells

Cell Alignment

Headings for columns are usually centered, while numbers in cells are usually aligned to the right. To set the alignment of cell data:

1. Highlight the cell or cell range to format.
2. Select the Home tab.
3. In the Alignment group, click the button representing the horizontal alignment you want for the cell (Align Left, Center, or Align Right).
4. Also in the Alignment group, above the horizontal alignment buttons, click the vertical alignment you want (Top Align, Middle Align, or Bottom Align). Middle Align is the most common vertical alignment for cells.

Cell Borders

Bottom borders are common for headings, and accountants include borders and double borders to indicate subtotals and grand totals on spreadsheets. Sometimes it is also useful to put a "box" border around a table of values or a section of a spreadsheet. To create borders:

1. Highlight the cell or cell range that needs a border.
2. Select the Home tab.
3. In the Font group, click the drop-down arrow of the Border icon. A menu of border selections appears (see Figure C-40).
4. Choose the desired border for the cell or group of cells. Note that All Borders creates a box border around each cell, while Outside Borders draws a box around a group of cells.

Source: Used with permission from Microsoft Corporation
FIGURE C-40   Selections in the Borders menu

Number Formats

For financial numbers, you usually use the Currency format. (Do not use the Accounting format, as it places the $ sign to the far left side of the cell). To apply the appropriate Currency format:

1. Highlight the cell or cell range to be formatted.
2. Select the Home tab.
3. In the Number group, select Currency in the Number Format drop-down list.
4. To set the desired number of decimal places, click the Increase Decimal or Decrease Decimal button in the bottom-right corner of the group (see Figure C-41).

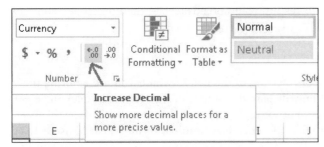

Source: Used with permission from Microsoft Corporation

FIGURE C-41   Increase Decimal and Decrease Decimal buttons

If you do not know what a button does in Office, hover your mouse pointer over the button to see a description.

Format "Painting"

If you want to copy *all* the format properties of a certain cell to other cells, use the Format Painter. First, select the cell whose format you want to copy. Then click the Format Painter button (the paintbrush icon) in the Clipboard group under the Home tab (see Figure C-42). When you click the button, the mouse pointer turns into a paintbrush. Click the cell you want to reformat. To format multiple cells, select the cell whose format you want to copy, and then click *twice* on the Format Painter button. The mouse cursor will become a paintbrush, and the paint function will stay on so you can reformat as many cells as you want. To turn off the Format Painter, click its button again or press the Esc key.

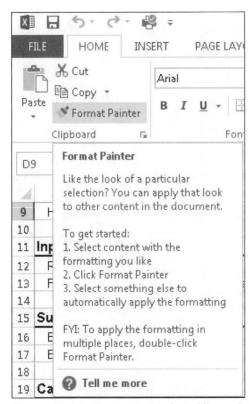

Source: Used with permission from Microsoft Corporation

FIGURE C-42   The Format Painter button

Showing the Excel Formulas in the Cells

Sometimes your instructor might want you to display or print the formulas in the spreadsheet cells. If you want the spreadsheet cells to display the actual cell formulas, follow these steps:

1. While holding down the Ctrl key, press the key in the upper-left corner of the keyboard that contains the back quote (`) and tilde (~). The spreadsheet will display the formulas in the cells. The columns may also become quite wide—if so, do not resize them.

2. The Ctrl+`~ key combination is a toggle; to restore your spreadsheet to the normal cell contents, press Ctrl+`~ again.

Understanding Circular Reference Errors

When entering formulas, you might make the mistake of referring to the cell in which you are entering the formula as part of the formula, even though it should only display the output of that formula. Referring a cell back to itself in a formula is called a *circular reference*. For example, suppose that in cell B2 of a worksheet, you enter =B2-B1. A terrible but apt analogy for a circular reference is a cannibal trying to eat himself! Excel 2010 would inform you when you tried to enter a circular reference into a formula, and would warn you if you tried to open an existing spreadsheet that has one or more circular references.

Excel 2013 has changed its treatment of circular reference errors. It allows you to enter them without warning, displaying a 0 in the cell, unless you do the following:

1. Click the File tab.
2. Click Options in the left column menu.
3. Click Formulas in the left column of the Excel Options window.
4. Uncheck the "Enable iterative calculation" check box (see Figure C-43), and then click OK. Now, when you enter a circular reference in the spreadsheet, you receive a warning message (see Figure C-44).

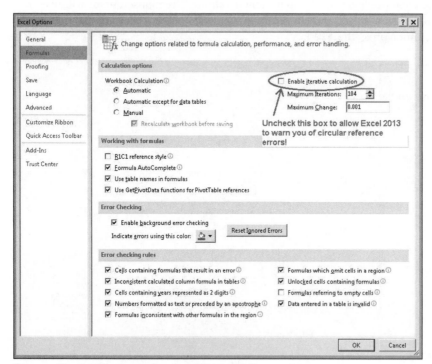

Source: Used with permission from Microsoft Corporation

FIGURE C-43   Setting formula options in Excel

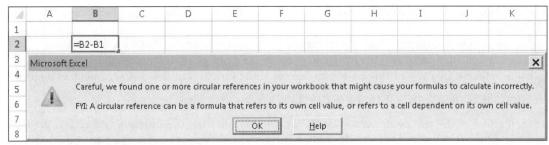

Source: Used with permission from Microsoft Corporation

FIGURE C-44   Excel circular reference warning

Using AND and OR Functions in IF Statements

Recall that the IF function has the following syntax:

IF(test condition, result if test is True, result if test is False)

The test conditions in the previous example IF statements tested only one cell's value, but a test condition can test more than one value of a cell.

For example, look at the thrift shop tutorial again. The Total Sales Dollars for 2015 depended on the economic outlook (recession or boom). The original IF statement was =IF($C$8="R",B18*1.3,B18*1.15), as shown in Figure C-45. This function increased the baseline 2014 Total Sales Dollars by 30% if there was continued recession ("R" entered in cell C8), but only increased the total by 15% if there was a boom.

C18   ▾   :   ×   ✓   *fx*   =IF($C$8="R",B18*1.3,B18*1.15)

	A	B	C	D
7	**Inputs**	**2014**	**2015**	**2016**
8	Economic Outlook (R=Recession, B=Boom)	NA	R	NA
9	Inflation Outlook (H=High, L=Low)	NA	H	NA
10				
11	**Summary of Key Results**	**2014**	**2015**	**2016**
12	Net Income after Taxes (Expansion)	NA	$69,974	$73,660
13	End-of-year Cash on Hand (Expansion)	NA	$84,974	$158,634
14	Net Income after Taxes (No Expansion)	NA	$72,601	$69,936
15	End-of-year Cash on Hand (No Expansion)	NA	$87,601	$157,537
16				
17	**Calculations (Expansion)**	**2014**	**2015**	**2016**
18	Total Sales Dollars	$350,000	$455,000	$591,500
19	Cost of Goods Sold	$245,000	$337,610	$465,227
20	Cost of Goods Sold (as a percent of Sales)	70%	74%	79%
21	Interest Rate for Business Loan	5%	NA	NA

Source: Used with permission from Microsoft Corporation

FIGURE C-45   The original IF statement used to calculate the Total Sales Dollars for 2015

To take the IF argument one step further, assume that the Total Sales Dollars for 2015 depended not only on the Economic Outlook, but on the Inflation Outlook (High or Low). Suppose there are two possibilities:

- Possibility 1: If the economic outlook is for a Recession and the inflation outlook is High, the Total Sales Dollars for 2015 will be 30% higher than in 2014.
- Possibility 2: For the other three cases (Recession and Low Inflation, Boom and High Inflation, and Boom and Low Inflation), assume that the 2015 Total Sales Dollars will only be 15% higher than in 2014.

The first possibility requires two conditions to be true at the same time: C8="R" and C9="H". You can include an AND( ) function inside the IF statement to reflect the additional condition as follows:

=IF(AND($C$8="R",$C$9="H"), B18*130%,B18*115%)

When the test argument uses the AND() function, conditions "R" *and* "H" both must be present at the same time for the statement to use the true result (multiplying last year's sales by 130%). Any of the other three outcome combinations will cause the statement to use the false result (multiplying last year's sales by 115%).

You can also use an OR() function in an IF statement. For example, assume that instead of both conditions (Recession and High Inflation) having to be present, only one of the two conditions needs to be present for sales to increase by 30%. In this case, you use the OR() function in the test argument as follows:

=IF(OR($C$8="R",$C$9="H"), B18*130%,B18*115%)

In this case, if *either* of the two conditions (C8="R" or C9="H"), is true, the function will return the true argument, multiplying the 2014 sales by 130%. If *neither* of the two conditions is true, then the function will return the false argument, multiplying the 2014 sales by 115% instead.

Using IF Statements Inside IF Statements (Also Called "Nesting IFs")

By now you should be familiar with IF statements, but here is a quick review of the syntax:

=IF(test condition, result if test is True, result if test is False)

In the preceding examples, only two courses of action were possible for each of the inputs: Recession or Boom, High Inflation or Low Inflation, Rental Occupancy High or Low, Bond Money Available or No Bond Money Available. The tutorial used only two possible outcomes to keep them simple.

However, in the business world, decision support models are frequently based on three or more possible outcomes. For capital projects and new product launches, you will frequently project financial outcomes based on three possible scenarios: Most Likely, Worst Case, and Best Case. You can modify the IF statement by placing another IF statement inside the result argument if the first test is false, creating the ability to launch two more alternatives from the second IF statement. This is called "nesting" your IF statements.

Try a simple nested IF statement: In your thrift shop example, assume that three economic outlooks are possible: Recession (R), Boom (B), or Stable (S). As before, the 2015 Total Sales Dollars (cell C18) will be the 2014 Total Sales Dollars increased by some fixed percentage. In a Recession, sales will increase by 30%, in a Boom they will increase by 15%, and for a Stable Economic Outlook, sales will increase by 22%, which is roughly midway between the other two percentages. You can "nest" the IF statement in cell C18 to reflect the third outcome as follows:

=IF($C$8= "R",B18*130%,IF($C$8="B",B18*115%,B18*122%))

Note the added IF statement inside the False value argument. You can break down this statement:

- If the value in cell C8 is "R", multiply the value in cell B18 by 130%, and enter the result in cell C18.
- If the value in cell C8 is not "R", check whether the value in cell C8 is "B". If it is, multiply the value in cell B18 by 115%, and enter the result in cell C18.
- If the value in cell C8 is not "B", multiply the value in cell B18 by 122%, and enter the result in cell C18.

If you have four or more alternatives, you can keep nesting IF statements inside the false argument for the outer IF statements. (Excel 2007 and later versions have a limit of 64 levels of nesting in the IF function, which should take care of every conceivable situation.)

NOTE

The "embedded IFs" in a nested IF statement are not preceded by an equals sign. Only the first IF gets the equals sign.

Cash Flow Calculations: Borrowing and Repayments

The Scenario Manager cases that follow in this book require accounting for money that the fictional company will have to borrow or repay. This money is not like the long-term loan that the Collegetown Thrift Shop is considering for its expansion. Instead, this money is short-term borrowing that companies use to pay current obligations, such as purchasing inventory or raw materials. Such short-term borrowing is called a line of credit, and is extended to businesses by banks, much like consumers have credit cards. Lines of credit usually involve interest payments, but for simplicity's sake, focus instead on how to do short-term borrowing and repayment calculations.

To work through cash flow calculations, you must make two assumptions about a company's borrowing and repayment of short-term debt. First, you assume that the company has a desired *minimum* cash level at the end of a fiscal year (which is also its cash level at the start of the next fiscal year), to ensure that the company can cover short-term expenses and purchases. Second, assume the bank that serves the company will provide short-term loans (a line of credit) to make up the shortfall if the end-of-year cash falls below the desired minimum level.

NCP stands for Net Cash Position, which equals beginning-of-year cash plus net income after taxes for the year. NCP represents the available cash at the end of the year, *before* any borrowing or repayment. For the three examples shown in Figure C-46, set up a simple spreadsheet in Excel and determine how much the company needs to borrow to reach its minimum year-end cash level. Use the IF function to enter 0 under Amount to Borrow if the company does not need to borrow any money.

	A	B	C	D
1	Example	NCP	Minimum Cash Required	Amount to Borrow
2	1	$25,000	$10,000	?
3	2	$9,000	$10,000	?
4	3	($12,000)	$10,000	?

Source: Used with permission from Microsoft Corporation

FIGURE C-46 Examples of borrowing

You can also assume that the company will use some of its cash on hand at the end of the year to pay off as much of its outstanding debt as possible without going below its minimum cash on hand required. The "excess" cash is the company's NCP *less* the minimum cash on hand required—any cash above the minimum is available to repay any debt. In the examples shown in Figure C-47, compute the excess cash and then compute the amount to repay. In addition, compute the ending cash on hand after the debt repayment.

	A	B	C	D	E	F
9	Example	NCP	Minimum Cash Required	Beginning-of-Year Debt	Repay?	Ending Cash
10	1	$12,000	$10,000	$5,000	?	?
11	2	$13,000	$10,000	$8,000	?	?
12	3	$20,000	$10,000	$0	?	?
13	4	$60,000	$10,000	$40,000	?	?
14	5	($20,000)	$10,000	$10,000	?	?

Source: Used with permission from Microsoft Corporation

FIGURE C-47 Examples of debt repayment

In the Scenario Manager cases of the following chapters, your spreadsheets may need two bank financing sections beneath the Income and Cash Flow Statements sections. You will build the first section to calculate any needed borrowing or repayment at year's end to compute year-end cash on hand. The second section will calculate the amount of debt owed at the end of the year after any borrowing or repayment.

Return to the Collegetown Thrift Shop tutorial and assume that it includes a line of credit at a local bank for short-term cash management. The first new section extends the end-of-year cash calculation, which was shown for the thrift shop in Figure C-19. Figure C-48 shows the structure of the new section highlighted in boldface.

	A	B	C	D
29	**Income and Cash Flow Statements (Expansion)**	**2014**	**2015**	**2016**
30	Beginning-of-year Cash on Hand	NA	$15,000	$0
31	Sales (Revenue)	NA	$455,000	$591,500
32	Cost of Goods Sold	NA	$337,610	$465,227
33	*Business Loan Payment*	NA	($12,950)	($12,950)
34	Income before Taxes	NA	$104,440	$113,323
35	Income Tax Expense	NA	$34,465	$39,663
36	Net Income after Taxes	NA	$69,974	$73,660
37	**Net Cash Position NCP** **Beginning-of-year Cash on Hand** **plus Net Income after Taxes**	**NA**	**$84,974**	**$73,660**
38	**Line of credit borrowing from bank**	**NA**		
39	**Line of credit repayments to bank**	**NA**		
40	**End-of Year Cash on Hand**	**$15,000**		

Source: Used with permission from Microsoft Corporation

FIGURE C-48   Calculation section for End-of-Year Cash on Hand with borrowing and repayments added

The heading in cell A36 was originally End-of-year Cash on Hand in Figure C-19, but you will add line-of-credit borrowing and repayment to the end-of-year totals. You must add the line-of-credit borrowing from the bank to the NCP and subtract the line-of-credit repayments to the bank from the NCP to obtain the End-of-Year Cash on Hand.

The second new section you add will compute the End-of year debt owed. This section is called Debt Owed, as shown in Figure C-49.

	A	B	C	D
42	**Debt Owed**	**2014**	**2015**	**2016**
43	**Beginning-of-year debt owed**	**NA**		
44	**Borrowing from bank line of credit**	**NA**		
45	**Repayment to bank line of credit**	**NA**		
46	**End-of-year debt owed**	**$47,000**		

Source: Used with permission from Microsoft Corporation

FIGURE C-49   Debt Owed section

As you can see, the thrift shop currently owes $47,000 on its line of credit at the end of 2014. The End-of-year debt owed equals the Beginning-of-year debt owed plus any new borrowing from the bank's line of credit, minus any repayment to the bank's line of credit. Therefore, the formula in cell C46 would be:

=C43+C44-C45

Assume that the amounts for borrowing and repayment (cells C44 and C45) were calculated in the first new section (for the year 2015, the amounts would be in cells C38 and C39), and then copied into the second section. The formula for cell C44 would be =C38, and for cell C45 would be =C39. The formula for cell C43, Beginning-of-year debt owed in 2015, would simply be the End-of-year debt owed in 2014, or =B46.

Now that you have added the spreadsheet entries for borrowing and repayment, consider the logic for the borrowing and repayment formulas.

Calculation of Borrowing from the Bank Line of Credit

When using logical statements, it is sometimes easier to state the logic in plain language and then turn it into an Excel formula. For borrowing, the logic in plain language is:

> If (cash on hand before financing transactions is greater than the minimum cash required,
>> then borrowing is not needed; else,
>> borrow enough to get to the minimum)

You can restate this logic as the following:

> If (NCP is greater than minimum cash required,
>> then borrowing from bank=0;
>> else, borrow enough to get to the minimum)

You have not added minimum cash at the end of the year as a requirement, but you could add it to the Constants section at the top of the spreadsheet (in this case the new entry would be cell C6). Assume that you want $50,000 as the minimum cash on hand at the end of both 2015 and 2016. Assuming that the NCP is shown in cell C37, you could restate the formula for borrowing (cell C38) as the following:

IF(NCP>Minimum Cash, 0; otherwise, borrow enough to get to the minimum cash)

You have cell addresses for NCP (cell C37) and for Minimum Cash (cell $C$6). To develop the formula for cell C38, substitute the cell address for the test argument; the true argument is simply zero (0), and the false argument is the minimum cash minus the current NCP. The formula stated in Excel for cell C38 would be:

=IF(C37>=$C$6, 0, $C$6-C37)

Calculation of Repayment to the Bank Line of Credit

Simplify the statements first in plain language:

IF(beginning of year debt=0, repay 0 because nothing is owed, but

   IF(NCP is less than the minimum, repay 0, because you must borrow, but

      IF(extra cash equals or exceeds the debt, repay the whole debt,

         ELSE (to stay above the minimum cash, repay the extra cash above the minimum)

Look at the following formula. If you assume that the repayment amount will be in cell C39, the beginning-of-year debt is in cell C43, and the minimum cash target is still in cell $C$6, the repayment formula for cell C39 with the nested IFs should look like the following:

=IF(C43=0,0,IF(C37<=$C$6,0,IF(C37-$C$6>=C43,C43,C37-$C$6)))

The new sections of the thrift shop spreadsheet would look like those in Figure C-50.

C39	▼ : × ✓ fx	=IF(C43=0,0,IF(C37<=$C$6,0,IF(C37-$C$6>=C43,C43,C37-$C$6)))		
	A	B	C	D
29	**Income and Cash Flow Statements (Expansion)**	**2014**	**2015**	**2016**
30	Beginning-of-year Cash on Hand	NA	$15,000	$50,000
31	Sales (Revenue)	NA	$455,000	$591,500
32	Cost of Goods Sold	NA	$337,610	$465,227
33	*Business Loan Payment*	NA	($12,950)	($12,950)
34	Income before Taxes	NA	$104,440	$113,323
35	Income Tax Expense	NA	$34,465	$39,663
36	Net Income after Taxes	NA	$69,974	$73,660
37	**Net Cash Position NCP** **Beginning-of-year Cash on Hand** **plus Net Income after Taxes**	NA	$84,974	$123,660
38	**Line of credit borrowing from bank**	NA	$0	$0
39	**Line of credit repayments to bank**	NA	$34,974	$12,026
40	**End-of Year Cash on Hand**	$15,000	$50,000	$111,634
41				
42	**Debt Owed**	**2014**	**2015**	**2016**
43	**Beginning-of-year debt owed**	NA	$47,000	$12,026
44	**Borrowing from bank line of credit**	NA	$0	$0
45	**Repayment to bank line of credit**	NA	$34,974	$12,026
46	**End-of-year debt owed**	$47,000	$12,026	$0

Source: Used with permission from Microsoft Corporation

FIGURE C-50 Thrift shop spreadsheet with line-of-credit borrowing, repayments, and Debt Owed added

Answers to the Questions about Borrowing and Repayment

Figures C-51 and C-52 display solutions for the borrowing and repayment calculations.

	A	B	C	D
1	Example	NCP	Minimum Cash Required	Amount to Borrow
2	1	$25,000	$10,000	$0
3	2	$9,000	$10,000	$1,000
4	3	($12,000)	$10,000	$22,000

Source: Used with permission from Microsoft Corporation

FIGURE C-51   Answers to examples of borrowing

In Figure C-51, the formula in cell D2 for the amount to borrow is =IF(B2>=C2,0,C2-B2).

	A	B	C	D	E	F
9	Example	NCP	Minimum Cash Required	Beginning-of-Year Debt	Repay?	Ending Cash
10	1	$12,000	$10,000	$5,000	$2,000	$10,000
11	2	$13,000	$10,000	$8,000	$3,000	$10,000
12	3	$20,000	$10,000	$0	$0	$20,000
13	4	$60,000	$10,000	$40,000	$40,000	$20,000
14	5	($20,000)	$10,000	$10,000	$0	NA

Source: Used with permission from Microsoft Corporation

FIGURE C-52   Answers to examples of repayment

In Figure C-52, the formula in cell E10 for the amount to repay is

=IF(B10>=C10,IF(D10>0,MIN(B10-C10,D10),0),0).

Note the following points about the repayment calculations shown in Figure C-52.

- In Example 1, only $2,000 is available for debt repayment ($12,000 – $10,000) to avoid dropping below the Minimum Cash Required.
- In Example 2, only $3,000 is available for debt repayment.
- In Example 3, the Beginning-of-Year Debt was zero, so the Ending Cash is the same as the Net Cash Position.
- In Example 4, there was enough cash to repay the entire $40,000 debt, leaving $20,000 in Ending Cash.
- In Example 5, the company has cash problems—it cannot repay any of the Beginning-of-Year Debt of $10,000, and it will have to borrow an additional $30,000 to reach the Minimum Cash Required target of $10,000.

You should now have all the basic tools you need to tackle Scenario Manager in Cases 6 and 7. Good luck!

NEW ENGLAND ENERGY INVESTMENT STRATEGY DECISION

Decision Support Using Microsoft Excel

PREVIEW

New England Energy (NEE) is a regional electricity provider that operates its own generating plants and contracts for the construction and operation of government-owned utilities throughout New England and eastern Canada. As part of its investment strategy, NEE faces a major decision in 2014: the Canadian government has approached the company with a proposal to build a hydroelectric dam in Quebec. Prior to this proposal, NEE was considering whether to build the world's first large methane (natural gas) fuel cell plant in upstate New York. By implementing either alternative, NEE would make an additional 1000 megawatts of electrical power available for sale on the New England power grid. However, variations in business and economic factors would affect the profitability of either alternative.

Each option for power generation has its advantages and drawbacks. Hydroelectric power generation has been a proven technology for more than 120 years, and it generates no appreciable carbon footprint—in fact, it was the original renewable energy source. The operating cost of a hydroelectric plant is relatively low as well, at approximately $20 per megawatt-hour (MWh). Building a renewable energy-generating station would also enable NEE to sell "carbon credits" on the world market. However, the construction costs would be significantly higher than those for a methane fuel cell plant, and the hydroelectric plant would take twice as long to build and implement as a fuel cell plant. Hydroelectric stations also have to be offline for about two weeks per year for dam and turbine maintenance.

Besides requiring less time and money to build, the methane fuel cell plant would be offline for only one week per year because the fuel cell has no moving parts other than auxiliary equipment. However, large fuel cells are relatively new technology, so there is some risk that their actual capacity for power generation might not be as great as their theoretical capacity. The operating cost of the methane fuel cell plant would be considerably higher than that of the hydroelectric plant because the cost of natural gas can be as much as $45 per megawatt-hour. Also, fuel cells generate electricity using methane gas, which is a nonrenewable fossil fuel that emits carbon dioxide and water vapor. However, the total greenhouse gas emissions from the fuel cell plant would be extremely low when compared with other methods of generating power from fossil fuels.

The Finance and Engineering departments at NEE have created some initial inputs for you to develop an investment decision model using a Microsoft Excel workbook. Your finished decision model will be instrumental in helping NEE choose the better investment alternative.

PREPARATION

- Review the spreadsheet concepts discussed in class and in your textbook.
- Your instructor may assign Excel exercises to help prepare you for this case.
- Tutorial C has an excellent review of IF and nested IF statements to help you with this case.
- Review the file-saving instructions in Tutorial C—it is always a good idea to save an extra copy of your work on a USB thumb drive.
- Review Tutorial F if you need to brush up on your presentation skills.
- Because the NEE case uses a strategic investment decision model, you will calculate the internal rate of return (IRR) in the model. If you are unfamiliar with the IRR function in Excel, the case includes a section that explains how to set it up.

BACKGROUND

You are an information analyst working for NEE. The company president has asked you to prepare a quantitative analysis of financial, sales, and operations data to help determine which power plant investment offers the better strategic opportunity for the company. The department managers have been asked to provide the following data from their functional areas:

- Financial and Accounting—The current cash position of the company, the cash outlay for the two investment choices, data for operating costs and period costs for each alternative, and the corporate income tax rates
- Sales—Forecasts for regional electricity demand and sales price, a formula for calculating sales demand based on weather, and forecasts of carbon credit sales for the hydroelectric alternative
- Engineering—Plant power capacities in megawatts, projected operating hours per year, and timelines for construction of either alternative

The departments have given you the following data:

- Carbon credits available for sale from the hydroelectric plant, 2016 through 2019
- Operating costs per megawatt-hour (MWh), 2015 through 2019
- Estimated market price per megawatt-hour, 2015 through 2019
- Hours of scheduled operation per year for both alternatives, 2015 through 2019
- Capital investment for hydroelectric and fuel cell plants, 2014
- Design generation capacity in megawatts for either alternative
- Projected corporate income tax rates, 2015 through 2019
- Forecasted additional market demand base for energy, 2014
- Period cost allocation base, 2014
- End-of-year cash on hand for 2014; this money is available for investment

Assignment 1 contains information you need to write formulas for the Calculations section, Income and Cash Flow Statements section, IRR Calculation section, and Canadian Bond Payment Calculation section. Because construction of the hydroelectric plant costs more than twice as much as the fuel cell plant, and because NEE has enough cash to build only the fuel cell plant, the Canadian government has agreed to issue a tax-exempt bond for $1 billion that NEE can repay over 10 years at 4 percent interest. This financial assistance and the higher generating capacity of the plant will be factored into the analysis.

You will use Excel to see how much profit and positive cash flow each alternative will generate for NEE from 2015 through 2019, and you will use Excel to calculate an internal rate of return for each alternative. You will also examine the effects of the weather (mild or severe) and the market price of methane (low or high) on projected electricity sales and profits for each alternative. In summary, your DSS will include the following inputs:

- Your decision to invest in the hydroelectric plant or methane fuel cell plant
- Whether the long-term weather outlook is mild or severe
- Whether the price of methane is low or high

Your DSS model must account for the effects of the preceding three inputs on costs, selling prices, sales demand, and other variables. If you design the model well, it will let you develop "what-if" scenarios with all the inputs, see the results, and show a preferred alternative for NEE to adopt.

ASSIGNMENT 1: CREATING A SPREADSHEET FOR DECISION SUPPORT

In this assignment, you create a spreadsheet that models the business decision NEE is seeking. In Assignment 2, you use the spreadsheet to gather data needed to determine the best investment decision. In Assignment 2A, you use Scenario Manager to summarize the financial outcomes for different combinations of inputs. In Assignment 2B, you modify the model to account for a financing change for the hydroelectric plant. In Assignment 2C, you modify the model to examine the impacts of futures contracts for purchasing methane and of improving the efficiency of the fuel cell plant. In Assignment 3, you write a report to NEE's chief executive officer that summarizes your analysis and recommendations. In Assignment 4, you prepare and give a presentation of your analysis and recommendations.

To begin, you create the spreadsheet model of the company's financial and marketing data. The model will cover sales from 2015 through 2019 for the type of generating plant selected. Assume that the preliminary research and development has been completed for the fuel cell alternative, and that the Canadian government has completed the hydroelectric dam and requires NEE only to install the turbines and generators to put the plant into operation. Under these assumptions, the fuel cell plant can start operating in 2015 and the hydroelectric plant can begin operations in 2016.

This section helps you set up each of the following spreadsheet components before entering the cell formulas:

- Constants
- Inputs
- Summary of Key Results
- Calculations
- Income and Cash Flow Statements
- Internal Rate of Return Calculation
- Bond Payment Calculation

The Internal Rate of Return Calculation section is needed because Excel financial formulas such as IRR work better if the cash outflow and inflow data are arranged in a vertical column with the years in ascending order, as opposed to taking the cash flows from across the page or from nonadjacent cells.

Constants Section

First, build the skeleton of your spreadsheet. *You can also download the spreadsheet skeleton if you prefer.* To access this skeleton file, select Case 6 from your data files, and then select **New England Energy Skeleton.xlsx**. Set up your Constants section as shown in Figure 6-1. An explanation of the line items follows the figure.

	A	B	C	D	E	F	G
1	**New England Energy Strategic Investment Decision**						
2							
3	**Constants**	**2014**	**2015**	**2016**	**2017**	**2018**	**2019**
4	Carbon Credits available for sale Hydroelectric Plant	NA	NA	200,000	200,000	200,000	200,000
5	Operating Cost per Megawatt-Hour Hydroelectric Plant	NA	NA	$20.00	$20.60	$21.22	$21.85
6	Operating Cost per Megawatt-Hour Methane Fuel Cell Plant (base)	NA	$45.00	$46.35	$47.74	$49.17	$50.65
7	Estimated Market Price per Megawatt-Hour	NA	$70.00	$73.50	$77.18	$81.03	$85.09
8	Estimated Sales Price for Carbon Credit (base)	NA	$80.00	$82.00	$84.00	$86.00	$88.00
9	Hours of Scheduled Operation in a Year, Hydroelectric Plant	8424	NA	NA	NA	NA	NA
10	Hours of Scheduled Operation in a Year, Fuel Cell Power Plant	8592	NA	NA	NA	NA	NA
11	Capital Investment for Hydroelectric Plant	$2,000,000,000	NA	NA	NA	NA	NA
12	Capital Investment for Fuel Cell Plant	$900,000,000	NA	NA	NA	NA	NA
13	Generation Capacity of Hydroelectric Plant, Megawatts	3,000	NA	NA	NA	NA	NA
14	Generation Capacity of Fuel Cell Plant, Megawatts	2,200	NA	NA	NA	NA	NA
15	Corporate Income Tax Rate	25%	25%	26%	26%	27%	27%

Source: Used with permission from Microsoft Corporation

FIGURE 6-1   Constants section

- Carbon Credits available for sale, Hydroelectric Plant—This value is the number of metric tons of carbon saved each year by operating the hydroelectric plant. Each metric ton of carbon equals one carbon credit. Carbon credits can be sold on the international market.
- Operating Cost per Megawatt-Hour, Hydroelectric Plant—This value is the direct cost of generating one megawatt-hour of power needed to operate the hydroelectric plant. Most of this cost would be spent on station operators' salaries, monitoring technology, parts, and maintenance of the water turbines and generators. This cost increases by 3 percent per year.
- Operating Cost per Megawatt-Hour, Methane Fuel Cell Plant (base)—This value is the direct cost of generating one megawatt-hour of power needed to operate the fuel cell plant. Unlike the hydroelectric plant, the fuel cell stacks have no moving parts, so the major component of the direct cost is the methane purchased to make the conversion to electricity through the fuel cells. The base cost increases by 3 percent per year, but it could vary considerably due to fluctuations in the market price of methane.
- Estimated Market Price per Megawatt-Hour—This value is the projected market price for one megawatt-hour of electricity for the next five years. In practice, the market price can vary considerably over the seasons according to supply and demand, but these seasonal prices have

been averaged to keep the value simple. In the spreadsheet, the average market price increases by 5 percent per year.

- Estimated Sales Price for Carbon Credit (base)—This value is the projected sales price for a carbon credit over the next five years. The value increases by two dollars each year. The actual sales price for carbon credits will depend on whether enough nations require industrial polluters to purchase carbon credits.
- Hours of Scheduled Operation in a Year, Hydroelectric Plant—This value is the number of hours in 50 weeks of continuous plant operation. Hydroelectric plants typically require two weeks offline each year for preventive maintenance.
- Hours of Scheduled Operation in a Year, Fuel Cell Power Plant—This value is the number of hours in 51 weeks of continuous plant operation. Because the fuel cell stacks contain no moving parts or rotating machinery, the Engineering department estimates that the new technology would require only one week offline per year for preventive maintenance, mostly for the cooling and auxiliary equipment.
- Capital Investment for Hydroelectric Plant—This value is the total amount of investment needed in equipment, materials, and construction costs to build the hydroelectric plant.
- Capital Investment for Fuel Cell Plant—This value is the total amount of investment needed in equipment, materials, and construction costs to build the methane fuel cell plant.
- Generation Capacity of Hydroelectric Plant, Megawatts—This value is the designed output capacity for the hydroelectric plant, measured in megawatts.
- Generation Capacity of Fuel Cell Plant, Megawatts—This value is the designed output capacity for the methane fuel cell plant, measured in megawatts.
- Corporate Income Tax Rate—These values are the projected corporate income tax rates for NEE. These rates are lower than usual corporate income tax rates because NEE would receive research and development tax credits for any new technologies it can incorporate into either plant.

Inputs Section

Your spreadsheet model must include the following inputs that will apply from 2015 through 2019, as shown in Figure 6-2.

	A	B
17	**Inputs**	
18	Weather Outlook (M=Mild, S=Severe)	
19	Market Price of Methane (L=Low, H=High)	
20	Investment Selection (W=Hydroelectric, F=Methane Fuel Cell)	

Source: Used with permission from Microsoft Corporation

FIGURE 6-2   Inputs section

- Weather Outlook—This value is either Mild (M) or Severe (S). Weather is a major determinant of electricity demand. Higher demand allows generating plants to operate closer to design capacity, which increases sales revenues.
- Market Price of Methane—This value is either Low (L) or High (H). The market price of natural gas would have a significant effect on the operating cost of the methane fuel cell plant. In Assignment 2C later in this case, you explore how NEE can minimize the impact of fluctuations in methane's market price by executing futures contracts for purchases.
- Investment Selection—This value is the basic input for the strategic decision to build either the proven hydroelectric plant (W) or the innovative methane fuel cell plant (F).

Summary of Key Results Section

This section (see Figure 6-3) contains the results data, which is of primary interest to the management team at NEE. This data includes income and end-of-year cash on hand information, as well as the annualized internal rate of return for a particular set of business inputs. This section summarizes the values from the Calculations, Income and Cash Flow Statements, and Internal Rate of Return Calculation sections.

	A	B	C	D	E	F	G
22	**Summary of Key Results**	**2014**	**2015**	**2016**	**2017**	**2018**	**2019**
23	Net Income After Taxes	NA					
24	End-of-year Cash on hand	NA					
25	Internal Rate of Return for Investment	NA	NA	NA	NA	NA	

Source: Used with permission from Microsoft Corporation

FIGURE 6-3   Summary of Key Results section

For each year from 2015 to 2019, your spreadsheet should show net income after taxes and end-of-year cash on hand. The net income after taxes is also the cash inflow for the IRR calculation.

> **N O T E**
>
> When writing formulas in the Calculations and Income and Cash Flow Statements sections, be careful if you copy a formula from one cell to another and the formula references a value from a particular cell in the Constants or Inputs section. For instance, if you copy the formula for Forecasted Additional Market Demand for Energy from cell C28 to cells D28 through G28, all destination cells must reference the value in cell B18 in order to complete the IF statement successfully. Therefore, you must use an *absolute* cell reference for cell B18 ($B$18) in the formula for cell C28 so the destination cells will retain cell B18 in the copied formulas. If you do not use an absolute cell reference, the Copy command will incorrectly use cells C18 through F18 in the copied formula values. To use absolute cell referencing, add $ signs before the column and row designations; these signs "anchor" the cell and ensure that destination cells refer to the correct source cell when information is copied. If necessary, consult the Excel online help for an explanation of relative and absolute cell references.

Calculations Section

The Calculations section includes the calculations you must perform to determine the market demand for energy, the projected annual electricity sales, the projected annual carbon credit sales for the hydroelectric alternative, the annual operating costs for both alternatives, and the period costs allocated to both alternatives. See Figure 6-4.

	A	B	C	D	E	F	G
27	**Calculations**	**2014**	**2015**	**2016**	**2017**	**2018**	**2019**
28	Forecasted Additional Market Demand for Energy--MW	2,000					
29	Available Additional Generation Capacity for Sale	NA					
30	Annual Electricity Sales	NA					
31	Annual Carbon Credit Sales (Hydroelectric Only)	NA	NA				
32	Annual Operating Cost for Hydroelectric Plant	NA	NA				
33	Annual Operating Cost for Methane Fuel Cell Plant	NA					
34	Period Cost Allocation (Sales, General, & Administrative)	$150,000,000					

Source: Used with permission from Microsoft Corporation

FIGURE 6-4   Calculations section

- Forecasted Additional Market Demand for Energy--MW—This value is the amount of additional market demand for energy, measured in megawatts. Cell B28 contains the demand in 2014 (2,000 megawatts). The forecasted demand for 2015 through 2019 depends on the weather outlook selected in the Inputs section. If the weather outlook is Mild (M), the demand for each succeeding year will be 3 percent higher than the previous year's (in other words, the previous year's demand multiplied by 1.03). For example, the value in cell C28 is the value in cell B28 multiplied by 1.03. If the weather outlook is Severe (S), the demand for each succeeding year will be 10 percent higher than the previous year's. Note that when you see the word *if* in the text, you need to write a formula that uses the IF function for the target cell.
- Available Additional Generation Capacity for Sale—This value is the amount of additional megawatts that the selected power station can provide for sale. The additional capacity depends on the investment selection from the Inputs section. In 2015 (cell C29), only the methane fuel cell alternative can provide additional capacity—the hydroelectric plant would be under construction and unavailable until 2016. For this cell, you must construct an IF formula that provides a zero value for the Hydroelectric option (W); for the methane fuel cell (F) selection, the formula must provide the fuel cell plant capacity from cell B14 of the Constants section. For the years 2016 through 2019 (cells D29 through G29), both plants would be producing electricity, so the appropriate generation capacity would be selected from cells B13 and B14, depending on the investment selection.

- Annual Electricity Sales—The formula for these cells depends on the investment selection from the Inputs section and whether the demand value in Row 28 is smaller or greater than the supply value in Row 29. You cannot sell more electricity than you have the capacity to produce or than the market demands. In 2015 (cell C30), only the methane fuel cell plant can provide additional capacity, so the formula for that cell must take the lesser of the two values in cells C28 and C29. You can use the MIN function to determine this value. Next, multiply that value by the Estimated Market Price per Megawatt-Hour (cell C7) and by the hours of scheduled operation per year for the fuel cell power plant (cell B10). The formula for cell C30 does not require an IF formula because cell C29 will return a zero for the 2015 available capacity of the hydroelectric plant. However, years 2016 through 2019 (cells D30 through G30) require an IF formula to select the correct hours of scheduled operation in a year for inclusion in the electricity sales calculation. Multiply the Estimated Market Price per Megawatt-Hour in cells D7 through G7 by either the hydroelectric plant or fuel cell plant's scheduled hours of operation in a year (cells B9 or B10).

- Annual Carbon Credit Sales (Hydroelectric Only)—Only the hydroelectric investment selection will allow NEE to sell carbon credits. Build an IF formula that returns a zero in cells D31 through G31 for the fuel cell alternative. For the hydroelectric alternative, the IF formula multiplies the carbon credits available for sale (cells D4 through G4) by the Estimated Sales Price for Carbon Credit (cells D8 through G8), and then enters the answer in cells D31 through G31.

- Annual Operating Cost for Hydroelectric Plant—This value is the operating cost per megawatt-hour for the hydroelectric plant (cells D5 through G5) multiplied by the hours of scheduled operation per year for the hydroelectric plant (cell B9), multiplied by the generation capacity of the hydroelectric plant in megawatts (cell B13). All three values are from the Constants section. Note that the hydroelectric plant only incurs operating costs in 2016 through 2019 (cells D32 through G32).

- Annual Operating Cost for Methane Fuel Cell Plant—The base calculation for this value is the operating cost per megawatt-hour for the methane fuel cell plant (cells C6 through G6) multiplied by the hours of scheduled operation per year for the methane fuel cell plant (cell B10), multiplied by the generation capacity of the fuel cell plant in megawatts (cell B14). All three values are from the Constants section. This formula must also include a factor for the market price of methane from the Inputs section. If the market price is Low (L), the base calculation is reduced by 10 percent; in other words, multiply the base calculation by .9. If the market price of methane is High (H), the base calculation is increased by 10 percent; in other words, multiply the base calculation by 1.1. The formulas for 2015 through 2019 are written for cells C33 through G33.

- Period Cost Allocation (Sales, General, & Administrative)—The value for 2014 (cell B34) is a projected cost allocation base to be assigned to the selected generating plant. Period costs are corporate costs that are not directly traceable to an operating plant, but that must be absorbed by the company's operations. The formula for the 2015 period cost (cell C34) depends on the investment selection. If the hydroelectric plant is selected, no period cost can be allocated because the plant will not be in operation yet. Therefore, the value is zero. If the fuel cell plant is selected, the period cost allocation for 2015 is the 2014 period cost increased by 3 percent. In other words, multiply the 2014 base value by 1.03. The 2016 period cost (cell D34) needs a specific IF formula as well. If the hydroelectric plant is selected, the period cost allocation for 2016 is the 2014 period cost increased by 6 percent. In other words, multiply the 2014 base value by 1.06. If the fuel cell plant is selected, the period cost allocation for 2016 is the 2015 period cost increased by 3 percent. For each of the years 2017 through 2019 (cells E34, F34, and G34), the period cost calculation is the same for either investment alternative—increase the preceding year's period cost by 3 percent.

Income and Cash Flow Statements Section

The statements for income and cash flow start with the cash on hand at the beginning of the year. Because NEE is funding the capital investments *internally*—that is, with its own cash on hand—you must deduct the invested funds from the cash on hand at the end of 2014. Figure 6-5 and the following list show how you should structure the Income and Cash Flow Statements section.

36	Income and Cash Flow Statements	2014	2015	2016	2017	2018	2019
		A	B	C	D	E	F	G
37	Beginning-of-year Cash on Hand (deduct Investment for 2014)	NA					
38	Electricity Sales	NA					
39	Carbon Credit Sales	NA	NA				
40	Total Revenues	NA					
41	less: Operating Costs	NA					
42	Gross Profit	NA					
43	less: Bond Payment (Hydroelectric Plant only)	NA					
44	less: Period Costs Allocated	NA					
45	Net Profit before Income Tax	NA					
46	less: Income Tax Expense	NA					
47	Net Income after Taxes (Cash Inflow)	NA					
48	End-of-year Cash on Hand	$1,200,000,000					

Source: Used with permission from Microsoft Corporation

FIGURE 6-5   Income and Cash Flow Statements section

- Beginning-of-year Cash on Hand—For 2015, this value is the end-of-year cash on hand from 2014 minus the capital investment, depending on the investment selection. If you choose the hydroelectric plant, the capital investment will be $1 billion—in other words, cell B11 in the Constants section minus the amount of the Canadian bond, which is cell F51 in the Bond Payment Calculation section. If you choose the methane fuel cell plant, the capital investment will be $900 million (cell B12 in the Constants section). For the years 2016 through 2019, the beginning-of-year cash on hand is the end-of-year cash on hand from the previous year.
- Electricity Sales—This value is the Annual Electricity Sales from the Calculations section (cells C30 through G30). Note that the hydroelectric plant investment will have no electricity sales in 2015 because the plant will not be completed yet.
- Carbon Credit Sales—This value is the Annual Carbon Credit Sales from the Calculations section (cells D31 through G31). Note that the investment in the methane fuel cell plant will be zero for carbon credit sales.
- Total Revenues—This value is the sum of the Electricity Sales and the Carbon Credit Sales.
- Less: Operating Costs—If the hydroelectric plant is selected, the operating costs are zero in 2015 and are copied from cells D32 through G32 for the years 2016 through 2019. If the methane fuel cell plant is selected, the operating costs for 2015 are copied from cell C33 in the Calculations section. For the years 2016 through 2019, the operating costs are copied from cells D33 through G33.
- Gross Profit—This value is the Total Revenues minus the Operating Costs.
- Less: Bond Payment—If the hydroelectric plant is selected, the bond payment is copied from the Annual Payment calculated in the Bond Payment Calculation section (cell F54). This value must be converted to a positive number, so you need to insert a minus sign in front of the cell in the IF statement argument. If the fuel cell plant is selected, the bond payment is zero.
- Less: Period Costs Allocated—This value is copied directly from the Period Cost Allocation row in the Calculations section (cells C34 through G34).
- Net Profit before Income Tax—This value is the Gross Profit minus the Bond Payment and the Period Costs Allocated.
- Less: Income Tax Expense—If you make a profit (in other words, if the Net Profit before Income Tax is greater than zero), this value is the Net Profit before Income Tax multiplied by the Corporate Income Tax Rate from the Constants section. If you make nothing or have a net loss, the Income Tax Expense is zero.
- Net Income after Taxes (Cash Inflow)—This value is the Net Profit before Income Tax minus the Income Tax Expense. From a strict accounting standpoint, the net income after taxes is not the cash inflow; you would have to add back all noncash expenses such as depreciation or depletion to determine the true cash inflow. However, for the purposes of this case, assume that net income after taxes is equal to cash inflow.
- End-of-year Cash on Hand—This value is the Beginning-of-year Cash on Hand plus the Net Income after Taxes.

Internal Rate of Return Calculation Section

This section, as shown in Figure 6-6, is set up to help you use Excel's built-in IRR function.

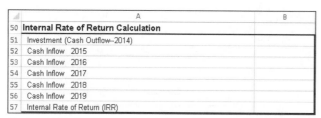

	A	B
50	**Internal Rate of Return Calculation**	
51	Investment (Cash Outflow--2014)	
52	Cash Inflow 2015	
53	Cash Inflow 2016	
54	Cash Inflow 2017	
55	Cash Inflow 2018	
56	Cash Inflow 2019	
57	Internal Rate of Return (IRR)	

Source: Used with permission from Microsoft Corporation

FIGURE 6-6   Internal Rate of Return Calculation section

- Investment (Cash Outflow)—This value depends on the investment selection. If the hydroelectric plant is selected, the total cost is $2 billion. However, because the company is getting a $1 billion loan from the Canadian government, the company's investment is actually the capital investment for the hydroelectric plant from the Constants section (cell B11) minus the bond amount from the Bond Payment Calculation section (cell F51). You might think it would be easier to enter $1 billion into the IF formula, but this approach would not work if the bond amount changed in cell F51. The fuel cell plant would be financed entirely from internal funds, so the company's investment would be the capital investment for the fuel cell plant from the Constants section (cell B12). The selected investment amount must be converted to a *negative* number to represent it as a cash outflow. (Think of it as money out of your pocket.)
- Cash Inflow 2015—This value is the net income after tax for 2015.
- Cash Inflow 2016—This value is the net income after tax for 2016.
- Cash Inflow 2017—This value is the net income after tax for 2017.
- Cash Inflow 2018—This value is the net income after tax for 2018.
- Cash Inflow 2019—This value is the net income after tax for 2019.
- Internal Rate of Return (IRR)—This value is the annual rate of return that the project generates for the company. Many companies set a minimum required IRR for a project or investment before it can be selected for implementation. To calculate the IRR, click cell B57, which is where you want to record the IRR result. Next, click the f_x symbol next to the cell-editing window. The Insert Function window appears (see Figure 6-7). Type IRR in the "Search for a function" text box, and then click Go.

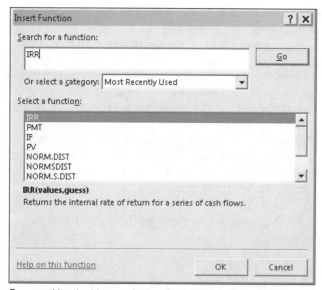

Source: Used with permission from Microsoft Corporation

FIGURE 6-7   The Insert Function window

When you click OK, the Function Arguments window appears to help you build the formula (see Figure 6-8). In the Values text box, enter the cells that contain all your cash outflows and inflows (B51:B56), or click and drag your mouse to select cells B51 through B56. Notice that Excel enters the formula for you in cell B57: =IRR(B51:B56). You do not have to enter a value in the Guess text box. When you click OK, Excel calculates the IRR and places the result in cell B57.

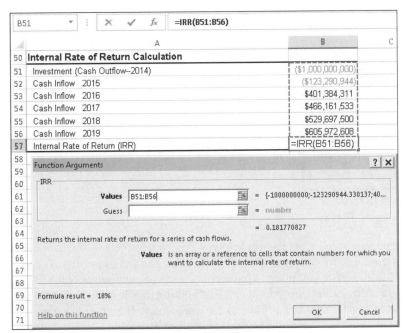

Source: Used with permission from Microsoft Corporation

FIGURE 6-8   The Function Arguments window for the IRR function

Bond Payment Calculation Section for the Hydroelectric Plant

The investment required for the Canadian hydroelectric plant is more than twice that of the methane fuel cell plant: $2 billion versus $900 million. NEE has only $1.2 billion of cash to invest in a new plant, so the Canadian government has agreed to issue tax-exempt bonds to raise $1 billion of the capital investment required to build the hydroelectric plant. If NEE chooses to build the hydroelectric plant, it must repay the Canadian government by making annual payments for 10 years at 4 percent compounded interest. The Bond Payment Calculation section (see Figure 6-9) is set up to help you use Excel's built-in PMT (Payment) function.

	D	E	F
50	**Bond Payment Calculation (Hydroelectric Plant)**		
51	Amount of Bond (Principal)		$1,000,000,000
52	Term of Bond (Yrs)		10
53	Interest Rate		4%
54	Annual Payment		

Source: Used with permission from Microsoft Corporation

FIGURE 6-9   Bond Payment Calculation section

- Amount of Bond (Principal)—Enter $1,000,000,000 in cell F51.
- Term of Bond (yrs)—Enter 10 in cell F52.
- Interest Rate—Enter 4% in cell F53.
- Annual Payment—This value is the annual payment required for the term of the bond. To calculate the annual payment, click cell F54, which is where you want to record the payment result. Next, click the f_x symbol next to the cell-editing window. The Insert Function window appears (see Figure 6-10). Type Payment or PMT in the "Search for a function" text box, and then click Go.

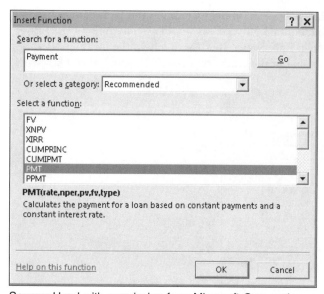

Source: Used with permission from Microsoft Corporation
FIGURE 6-10   The Insert Function window for the PMT function

When you click OK, the Function Arguments window appears to help you build the formula (see Figure 6-11). In the Rate text box, enter the cell that contains the interest rate (F53) or click and drag your mouse to select cell F53. Similarly, enter cell F52 in the Nper (number of payments) text box, and enter cell F51 in the Pv (present value) text box. Excel enters the formula for you in cell F54 and displays it in the cell-editing window at the top of the page: =PMT(F53,F52,F51). You do not have to enter a value in the Fv or Type text box. Before you click OK, note that the bottom of the Function Arguments window displays a preview of the results based on your input. When you click OK, Excel calculates the payment and places the result in cell F54.

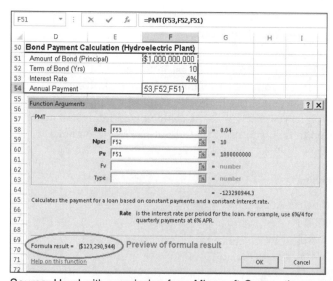

Source: Used with permission from Microsoft Corporation
FIGURE 6-11   The Function Arguments window for the PMT function

After you complete all the formulas, try testing your spreadsheet with various combinations of the three values from the Inputs section. There are eight possible combinations, as listed in the next section. If you receive any error messages or see strange values in the cells, go back and check your formulas.

The DSS spreadsheet contains some values that represent millions or billions of dollars. Accountants often simplify their spreadsheets by listing outputs in multiples of millions or billions. It is not hard to do—you simply divide the cell values by a million or billion, depending on the scale—but for the purposes of this case, you should keep the large numbers in the spreadsheets. If you see cell results listed as a group of "#" signs when working with large numbers (see Figure 6-12), the column is not wide enough to display the contents of that cell. Simply widen the column until the number is displayed.

A	B	G
36 **Income and Cash Flow Statements**	2014	2019
37 Beginning-of-year Cash on Hand (deduct Investment for 2014)	NA	###########
38 Electricity Sales	NA	###########
39 Carbon Credit Sales	NA	$17,600,000
40 Total Revenues	NA	###########
41 less: Operating Costs	NA	###########
42 Gross Profit	NA	###########
43 less: Bond Payment (Hydroelectric Plant only)	NA	###########
44 less: Period Costs Allocated	NA	###########
45 Net Profit before Income Tax	NA	###########
46 less: Income Tax Expense	NA	###########
47 Net Income after Taxes (Cash Inflow)	NA	###########
48 End-of-year Cash on Hand	$1,200,000,000	###########

Source: Used with permission from Microsoft Corporation

FIGURE 6-12   Column G is not wide enough to display the numbers

ASSIGNMENT 2: USING THE SPREADSHEET FOR DECISION SUPPORT

Next, you use the spreadsheet to gather data needed to determine the best investment decision and to document your recommendations in a report to NEE's management.

The DSS model has eight possible financial outcomes:

1. Hydroelectric plant (W)
   a. Mild weather and low methane price (M/L/W)
   b. Mild weather and high methane price (M/H/W)
   c. Severe weather and low methane price (S/L/W)
   d. Severe weather and high methane price (S/H/W)
2. Methane fuel cell plant (F)
   a. Mild weather and low methane price (M/L/F)
   b. Mild weather and high methane price (M/H/F)
   c. Severe weather and low methane price (S/L/F)
   d. Severe weather and high methane price (S/H/F)

You are primarily interested in the company's financial position based on each of these possible outcomes. The Summary of Key Results section lets you see the net income after taxes for each year of the new plant's operations, the end-of-year cash on hand for each year, and the internal rate of return by the end of 2019. The management team wants to make sure that the selected investment does not exhaust the company's cash on hand at the end of any year from 2015 through 2019. In addition, the team prefers the technology that gives the company the best chance of making its targeted internal rate of return of 18% or more on the investment.

Because there are eight (2^3) possible combinations of inputs for the weather outlook, methane price, and selected investment, you might want to run the spreadsheet model eight times, changing the inputs according to the preceding list. You should do this for two reasons:

- To ensure that no single year from 2015 through 2019 has negative end-of-year cash on hand (in other words, to make sure the company does not run out of money)
- To print each spreadsheet to meet the requirements of Assignment 1

You could then transcribe the results to a summary sheet. Next, you know that the management team is very interested in the financial data from the end of 2019. You can summarize that data easily using Scenario Manager.

Assignment 2A: Using Scenario Manager to Summarize Data

For each of the eight situations listed earlier, you want to know the net income after taxes and the end-of-year cash on hand for 2019, as well as the internal rate of return generated by the different inputs.

You will run "what-if" scenarios with the eight sets of input values using Excel Scenario Manager. If necessary, review Tutorial C for tips on using Scenario Manager. In this case, the input values are stored together in one vertical group of cells (B18 through B20) in the Inputs section, as are the three output cells (G23 through G25) in the Summary of Key Results section, so selecting the cells is easy. Run Scenario

Manager to gather your data in a report called the Scenario Summary. Format this summary to make it presentable, and then print it for your instructor.

Assignment 2B: Financing Change for the Hydroelectric Plant

In Assignment 1, the Canadian government was willing to help finance the hydroelectric plant by lending NEE $1 billion through a government bond issue, which is usually subject to the legislative process. If the Canadian Parliament did not vote for the bond, NEE would have to find its own financing. Assume that NEE can negotiate a commercial loan for $1.2 billion for five years at 7% interest. (NEE will need to borrow $200 million more for the commercial loan than for the bond; otherwise, it will have almost no cash on hand at the end of 2015.) Make a copy of your original worksheet by right-clicking the tab at the bottom of the worksheet, checking the Create a copy box in the Move or Copy window, and clicking OK, as shown in Figure 6-13. Rename the worksheet **New England Energy—Bank Loan**. In your new worksheet, change the values in the Bond Payment Calculation section to reflect the new loan terms and then rerun Scenario Manager, which will create a new summary report; name it **Summary Report 2**. How do the investment alternatives look now? Format this summary to make it presentable, and then print it for your instructor.

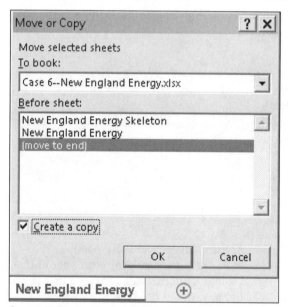

Source: Used with permission from Microsoft Corporation
FIGURE 6-13   The Move or Copy worksheet window

Assignment 2C: Methane Futures Contracts and Increasing Fuel Cell Efficiency

You have probably learned by now that the market price of methane is the most critical factor in determining the profitability of the methane fuel cell plant. When fuel prices are a critical cost for companies such as utilities and airlines, their supply-chain managers frequently negotiate long-term contracts to purchase fuel. Futures contracts frequently guarantee a maximum or "cap" price, with a small reduction for large price decreases. In this assignment, you modify your DSS model to examine the effect of a futures contract negotiated by NEE. Also, the case preview pointed out that the methane fuel cell plant was relatively new technology. Therefore, the original operating costs developed by the Engineering department were calculated conservatively. However, the Engineering department has developed lower operating cost estimates based on optimized fuel cell generation. To enter these new estimates, you must change the values for the operating cost of the methane fuel cell plant in the Constants section. Make a copy of your original worksheet and rename it **New England Energy—CH4 Futures**. Enter the new costs in cells C6 through G6, as shown in Figure 6-14.

▲	A	B	C	D	E	F	G
1	New England Energy Strategic Investment Decision--Methane Futures Contract and (Improved Fuel Cell Efficiency)						
2							
3	Constants	2014	2015	2016	2017	2018	2019
4	Carbon Credits available for sale Hydroelectric Plant	NA	NA	200,000	200,000	200,000	200,000
5	Operating Cost per Megawatt-Hour Hydroelectric Plant	NA	NA	$20.00	$20.60	$21.22	$21.85
6	Operating Cost per Megawatt-Hour Methane Fuel Cell Plant (base)	NA	$40.50	$41.72	$42.97	$44.26	$45.58
7	Estimated Market Price per Megawatt-Hour	NA	$70.00	$73.50	$77.18	$81.03	$85.09
8	Estimated Sales Price for Carbon Credit	NA	$80.00	$82.00	$84.00	$86.00	$88.00
9	Hours of Scheduled Operation in a Year, Hydroelectric Plant	8424	NA	NA	NA	NA	NA
10	Hours of Scheduled Operation in a Year, Fuel Cell Power Plant	8592	NA	NA	NA	NA	NA
11	Capital Investment for Hydroelectric Plant	$2,000,000,000	NA	NA	NA	NA	NA
12	Capital Investment for Fuel Cell Plant	$900,000,000	NA	NA	NA	NA	NA
13	Generation Capacity of Hydroelectric Plant, Megawatts	3,000	NA	NA	NA	NA	NA
14	Generation Capacity of Fuel Cell Plant, Megawatts	2,200	NA	NA	NA	NA	NA
15	Corporate Income Tax Rate	25%	25%	26%	26%	27%	27%

Source: Used with permission from Microsoft Corporation

FIGURE 6-14   The new values for operating cost per megawatt-hour in the fuel cell plant

A methane futures contract will reduce the variation in the price of methane. To reflect this change, you must modify the formula for the Annual Operating Cost for Methane Fuel Cell Plant in Row 33 of the Calculations section. Click cell C33. In the formula displayed in the cell-editing window (see Figure 6-15), change the values from .90 and 1.10 to .95 and 1, respectively. Copy the formula change to cells D33 through G33.

C33	▼	:	✕	✓	*fx*	=IF($B$19="L",C6*$B$14*$B$10*0.95,C6*$B$14*$B$10*1)	

▲	A Change .90 and 1.10 in this formula to .95 and 1 because of the futures contract	B	C
27	Calculations		2015
28	Forecasted Additional Market Demand for Energy--MW	2,000	2,060
29	Available Additional Generation Capacity for Sale	NA	0
30	Annual Electricity Sales	NA	$0
31	Annual Carbon Credit Sales (Hydroelectric Only)	NA	NA
32	Annual Operating Cost for Hydroelectric Plant	NA	NA
33	Annual Operating Cost for Methane Fuel Cell Plant	NA	$727,269,840
34	Period Cost Allocation (Sales, General, & Administrative)	$150,000,000	$0

Source: Used with permission from Microsoft Corporation

FIGURE 6-15   Formula change in Row 33 to reflect the methane futures contract

Rerun Scenario Manager, which will create a new summary report; name it **Summary Report 3**. How do the investment alternatives look now? The methane fuel cell plant should look more attractive than before, but is it a better option than the hydroelectric plant? Format the summary report to make it presentable, and then print it for your instructor.

ASSIGNMENT 3: DOCUMENTING YOUR RECOMMENDATIONS IN A REPORT

Use Microsoft Word to write a brief report to the chief executive officer of New England Energy, James Patterson. State the results of your analysis and recommend which investment choice to make (the hydroelectric generating plant or the methane fuel cell plant). Your report must meet the following requirements:

- The first paragraph must summarize the investment choices facing NEE and must state the purpose of the analysis.
- Summarize the results of your analysis and state your recommended action. Remember the environmental impacts of each decision, as mentioned in the case preview.
- Support your recommendation with a table outlining the Scenario Summary results. Figure 6-16 shows a recommended table format in Microsoft Word.
- If your report is well formatted, you might choose to embed an Excel object of the Scenario Summary in the body of your report. Tutorial C includes a brief description of how to copy and paste Excel objects.
- Your instructor might also ask you to provide a graph of the internal rates of return for the eight possible combinations of inputs.

Investment Choice	Weather Outlook	Methane Prices	2019 Net Income ($ millions)	2019 End-of-year Cash on Hand ($ millions)	Internal Rate of Return
Hydroelectric Plant	Mild	Low			
		High			
	Severe	Low			
		High			
Methane Fuel Cell Plant	Mild	Low			
		High			
	Severe	Low			
		High			

Source: © Cengage Learning 2015

FIGURE 6-16   Recommended format of a table to insert in your report

In Figure 6-16, divide each of your net income and cash-on-hand results by 1 million to display them in multiples of millions.

ASSIGNMENT 4: GIVING AN ORAL AND SLIDE PRESENTATION

Your instructor may ask you to summarize your analysis in an oral presentation. If so, assume that the management team at NEE wants you to explain your analysis and recommendations in 10 minutes or less. A well-designed PowerPoint presentation, with or without handouts, is considered appropriate in a business setting. Tutorial F provides excellent tips for preparing and delivering a presentation.

DELIVERABLES

Your completed case should include the following deliverables for the instructor:

- A printed copy of your report to management
- Printouts of your spreadsheets
- Electronic copies of all your work, including your report, PowerPoint presentation, and Excel DSS model. Ask your instructor for guidance on which items you should submit for grading.

THE CITY HOTEL RECOVERY ANALYSIS

Decision Support Using Excel

PREVIEW

A city-owned hotel cannot meet its debt service requirements. In this case, you will use Microsoft Excel to analyze options for improving the hotel's cash flow.

PREPARATION

- Review spreadsheet concepts discussed in class and in your textbook.
- Complete any exercises that your instructor assigns.
- Complete any part of Tutorial C that your instructor assigns. You may need to review the use of IF statements and the section called "Cash Flow Calculations: Borrowings and Repayments."
- Review file-saving procedures for Windows programs.
- Refer to Tutorials E and F as necessary.

BACKGROUND

Almost a decade ago, the administrators of a large American city approved the construction of a downtown hotel to complement its modern convention center, major-league baseball stadium, and professional football stadium. The administrators recognized that people who visited downtown for tourism, conventions, and sporting events needed more places to stay overnight. The city issued $300 million in bonds to finance construction, and the 750-room hotel opened in 2006 under city ownership and management.

Each year, the hotel's revenue has exceeded its operating expenses, but not by a margin great enough to fully cover the $2.2 million yearly construction loan payment. The hotel has had to borrow from the city's general fund every year to cover the construction loan payment, and now the hotel owes $15 million to the city. In effect, the city acts as the hotel's bank; the loans are called "working capital" loans because they address the hotel's working capital shortfall.

City administrators are under public pressure to end the hotel subsidy and get out of the hotel business. Some critics think that the hotel's management should find ways to improve the occupancy rate or simply charge more for rooms. However, the hotel's occupancy rate is 60 percent, which is above the average for city-owned hotels, and management does not think this rate can be improved. The hotel's $170 average daily room rate is also competitive in the industry; management thinks that raising rates would merely drive people to other hotels in the area.

The administration is examining two options for improving the hotel's cash flow. One possibility is to lease a block of 50 rooms to a local college that has a downtown campus nearby. Another option is to sell rooms to a time-share management company.

Lease Rooms to a Local College

The college has a thriving downtown campus that needs more classroom and administrative space. The hotel could offer a long-term lease of a block of 50 rooms on the side of the hotel that faces the campus. In the first year, the college would pay $3 million for the rooms; the rent would increase by a reasonable rate each year. As part of the leasing agreement, the hotel would pay for needed remodeling work. The lease would mean that the hotel would have only 700 rooms to offer, but hotel managers think the occupancy rate would remain at 60 percent.

Sell Rooms to a Time-share Company

A time-share company sells interests in its units to people who want guaranteed occupancy for part of a year. For example, a person who wants access to a ski lodge for one month each year could buy a one-twelfth ownership interest in a unit held by one of the company's ski resorts. The company would try to sell the remaining 11 months to other people.

Fortunately for the city, a large time-share company is looking for desirable downtown units. The city would like the company to commit to buying up to 10 hotel rooms for each of the next four years. The company offers to pay $100,000 per unit for the first year, and to increase the payment in succeeding years. As with the college, the hotel would agree to pay for needed remodeling work. The deal would mean that the hotel would have 10 fewer rooms to offer each year, but the occupancy rate would be expected to stay at 60 percent.

In addition to remodeling, the hotel would agree to maintain any rooms it rented to the college or sold to the time-share company. In other words, the hotel staff would continue to clean and service the rooms, and the hotel would continue to provide utilities such as heating and cooling. The hotel would ask to be paid $10,000 per room for cleaning and other maintenance each year.

Background Summary

The hotel earns more revenue than it spends on operations, but it needs to do better so that it can make the yearly loan payment without municipal assistance. Otherwise, the city might consider selling the hotel to a national chain that would be willing to assume the loan.

The city comptroller has asked you to create a DSS model of hotel operations. Your model needs to show the effects of the two options for increasing cash flow. You can use the model to develop "what-if" scenarios with the inputs, see the financial results, and then help you decide if the hotel can be made into a sufficiently profitable operation.

ASSIGNMENT 1: CREATING A SPREADSHEET FOR DECISION SUPPORT

In this assignment, you will produce a spreadsheet that models the hotel's business for the next four years. In Assignment 2, you will write a memorandum that documents your analysis and conclusions. In Assignment 3, you will prepare and give an oral presentation of your analysis and conclusions to hotel management and the city comptroller.

First, you create the spreadsheet model of the financial situation. The model covers the four years from 2015 to 2018. This section helps you set up each of the following spreadsheet components before entering cell formulas:

- Constants
- Inputs
- Summary of Key Results
- Calculations
- Income and Cash Flow Statements
- Working Capital Debt Owed

A discussion of each section follows. The spreadsheet skeleton is available for you to use; it will save you time. To access this skeleton, go to your data files, select Case 7, and then select **Hotel.xlsx**.

Constants Section

Your spreadsheet should include the constants shown in Figure 7-1. An explanation of the line items follows the figure.

	A	B	C	D	E	F
1	THE HOTEL RECOVERY ANALYSIS					
2						
3	Constants	2014	2015	2016	2017	2018
4	Minimum cash required	NA	$ 3,000,000.00	$ 3,000,000.00	$ 3,000,000.00	$ 3,000,000.00
5	Number of rooms in hotel	NA	750	750	750	750
6	Interest rate on working capital debt owed	NA	2%	2%	3%	3%
7	Income tax rate	NA	35%	35%	35%	35%
8	Expected occupancy rate	NA	60%	60%	60%	60%
9	Rent offered for 50 rooms by local college	NA	$ 3,000,000	$ 3,100,000	$ 3,200,000	$ 3,300,000
10	Expected sale price of a time-share unit	NA	$ 100,000	$ 103,000	$ 105,000	$ 107,000
11	Average daily amenity revenue per room rental	NA	$ 30	$ 31	$ 32	$ 33
12	Expected operating expense	NA	$ 33,500,000	$ 33,750,000	$ 34,000,000	$ 34,250,000
13	Interest on construction loan	NA	$ 773,534	$ 716,177	$ 656,526	$ 594,489
14	Construction loan principal payment	NA	$ 1,433,918	$ 1,491,275	$ 1,550,926	$ 1,612,963

Source: Used with permission from Microsoft Corporation

FIGURE 7-1 Constants section

- Minimum cash required—The hotel needs to have at least $3 million in cash at the beginning of each year. The city government will lend the amount needed at the end of a year to begin the new year with $3 million.
- Number of rooms in hotel—The hotel currently has 750 rooms to rent each day. The cash flow improvement options discussed in the Background section would reduce this number.
- Interest rate on working capital debt owed—The interest rate floats on this debt on a yearly basis. The rate is expected to be 2 percent in 2015 and 2016, and then is expected to rise to 3 percent in 2017 and 2018.
- Income tax rate—The hotel pays federal, state, and local income taxes. The combined tax rate is expected to be 35 percent each year.
- Expected occupancy rate—The hotel is completely booked on some days, almost empty on others, and more than half-full on most days. The average occupancy rate is expected to be 60 percent each year.
- Rent offered for 50 rooms by local college—The local college has offered to pay $3 million for a block of 50 rooms in 2015. The payment would increase by $100,000 each year.
- Expected sale price of a time-share unit—The time-share company has offered to pay $100,000 per unit in 2015, and would pay more in each succeeding year.
- Average daily amenity revenue per room rental—The hotel has two excellent restaurants and a gift shop. The average daily revenue from amenities per room rental is expected to be $30 in 2015, and is expected to increase in succeeding years.
- Expected operating expense—Hotel operating expenses are expected to increase each year, as shown.
- Interest on construction loan—The $300 million construction loan requires a payment each year. Part of the payment is interest and part is principal on the loan. The interest expense for each year is shown.
- Construction loan principal payment—The principal portion of each year's loan payment is shown.

Inputs Section

Your spreadsheet should include the following inputs for the years 2015 to 2018, as shown in Figure 7-2.

	A	B	C	D	E	F
16	Inputs	2014	2015	2016	2017	2018
17	Rent fifty rooms to college? (Y/N)	NA		NA	NA	NA
18	Number of rooms sold to time-share	NA				
19	Inflation rate	NA				

Source: Used with permission from Microsoft Corporation

FIGURE 7-2 Inputs section

- Rent fifty rooms to college? (Y/N)—Enter Y if the rooms will be rented to the college, and N if not. The entry applies to all four years.
- Number of rooms sold to time-share—Enter the number of rooms expected to be sold each year. Valid numbers range from 0 to 10.
- Inflation rate—Enter a value for the expected increase in the general level of prices each year. For example, if no increase is expected in any year, the entry would be .00, .00, .00, .00. If a

3 percent increase is expected each year, the entry would be .03, .03, .03, .03. Cells should be formatted for percentage.

Summary of Key Results Section

Your spreadsheet should include the results shown in Figure 7-3. An explanation of each item follows the figure.

A	B	C	D	E	F
21 **Summary of Key Results**	2014	2015	2016	2017	2018
22 Income before interest expense	NA				
23 Net Income (Loss) in Year	NA				
24 End-of-Year cash on hand	NA				
25 End-of-year debt owed to city government	NA				

Source: Used with permission from Microsoft Corporation

FIGURE 7-3   Summary of Key Results section

For each year, your spreadsheet should show income before interest expense, net income (or loss) in the year, cash on hand at the end of the year, and debt owed to the city at the end of the year. The net income, cash, and debt cells should be formatted as currency with no decimals. These values are computed elsewhere in the spreadsheet and should be echoed here.

Calculations Section

You should calculate intermediate results that will be used in the income and cash flow statements that follow. Calculations, as shown in Figure 7-4, may be based on expected year-end 2014 values. When called for, use absolute referencing properly. Values must be computed by cell formula; hard-code numbers in formulas only when you are told to do so. Cell formulas should not reference a cell with a value of "NA."

An explanation of each item in this section follows the figure.

A	B	C	D	E	F
27 **Calculations**	2014	2015	2016	2017	2018
28 Average daily room rate	$ 170.00				
29 Complete hotel room rental revenue in year	NA				
30 Hotel room rental lost to sold time-shares	NA				
31 Hotel room rental lost to college	NA				
32 Revenue from sale of time-share units	NA				
33 Number of hotel rooms available	NA				
34 Number of hotel room rentals in year	NA				
35 Amenity revenue in year	NA				
36 Maintenance revenue	NA				
37 College remodeling expense	NA				
38 Time-share remodeling expense	NA				

Source: Used with permission from Microsoft Corporation

FIGURE 7-4   Calculations section

- Average daily room rate—The average daily room rate will increase by the expected inflation rate for the year. This yearly rate is taken from the Inputs section. For example, assuming a 1 percent increase each year, the average daily room rate for 2015 would be $170 × (1.01) = $171.70. The 2016 average would then be $171.70 × 1.01. Format cells for currency and two decimals.
- Complete hotel room rental revenue in year—This amount is a function of the number of days in a year (365), the expected occupancy rate (from the Constants section), the expected average daily room rate (from the Calculations section), and the number of rooms in the hotel (from the Constants section). This amount assumes no college rentals or time-share sales. You can hard-code the number of days in a year. Format cells for currency with no decimals.
- Hotel room rental lost to sold time-shares—This amount is a function of the number of days in a year (365), the expected occupancy rate, the expected average daily room rate, and the number of rooms expected to be sold to the time-share company, which is taken from the Inputs section. Note that the number of units sold increases each year. You can hard-code the number of days in a year. Format cells for currency with no decimals.
- Hotel room rental lost to college—This amount is zero if no units are rented. If units are rented, this amount is a function of the number of days in a year (365), the number of units rented (50),

the expected occupancy rate, and the expected average daily room rate. You can hard-code the number of days in a year and the number of units to be rented. Format cells for currency with no decimals.

- Revenue from sale of time-share units—This amount is a function of the number of rooms expected to be sold in the year (from the Inputs section) and the expected sale price for the year (from the Constants section). Format cells for currency with no decimals.
- Number of hotel rooms available—This amount is a function of the total number of rooms in the hotel, the number of rooms rented to the college, and the number of rooms sold as time-shares. Note that the number of rooms rented to the college may be zero, and that the number of rooms available may decrease each year because of time-share sales. Format cells as numbers with no decimals.
- Number of hotel room rentals in year—This amount is a function of the number of hotel rooms available during the year (from the Calculations section), the number of days in a year (365), and the expected occupancy rate for the year (from the Constants section). You can hard-code the number of days in a year. Format cells for numbers with no decimals.
- Amenity revenue in year—This amount is a function of the number of hotel room rentals during the year (from the Calculations section) and the expected daily amenity revenue for the year (from the Constants section). Format cells for currency with no decimals.
- Maintenance revenue—This amount is a function of the number of units sold or rented and the yearly maintenance rate, which is $10,000. You can hard-code the maintenance rate. Format cells for currency with no decimals.
- College remodeling expense—If no units are rented to the college, this amount is zero. If the units are rented, the amount is $200,000. You may hard-code the amount of remodeling expenses. This expense would occur only in 2015. Format cells for currency with no decimals.
- Time-share remodeling expense—This amount is a function of the number of units sold as time-shares in a year and the remodeling expense per unit, which is $1,000. You may hard-code the amount of remodeling expenses. Format cells for currency with no decimals.

Income and Cash Flow Statements

The forecast for net income and cash flow starts with the cash on hand at the beginning of the year. This value is followed by the income statement and the calculation of cash on hand at year's end. For readability, format cells in this section as currency with no decimals. Values must be computed by cell formula; hard-code numbers in formulas only if you are told to do so. Cell formulas should not reference a cell with a value of "NA." Your spreadsheets should look like those in Figures 7-5 and 7-6. A discussion of each item in the section follows each figure.

	A	B	C	D	E	F
40	**Income and Cash Flow Statements**	**2014**	**2015**	**2016**	**2017**	**2018**
41	Beginning-of-year cash on hand	NA				
42						
43	Revenue:					
44	Hotel room rental	NA				
45	Amenity revenue	NA				
46	Maintenance revenue	NA				
47	Rent from college for 50 rooms	NA				
48	Sale of time-share units	NA				
49	Total Revenue	NA				
50	Operating expenses:					
51	Operating expense	NA				
52	College remodeling expense	NA				
53	Time-share remodeling expense	NA				
54	Total operating expense	NA				
55	Income before interest expense	NA				
56	Interest expense:	NA				
57	Construction loan	NA				
58	Working capital loan	NA				
59	Total interest expense	NA				
60	Income before taxes	NA				
61	Income tax expense	NA				
62	Net income (loss) after taxes	NA				

Source: Used with permission from Microsoft Corporation

FIGURE 7-5   Income and Cash Flow Statements section

- Beginning-of-year cash on hand—This value is the cash on hand at the end of the prior year.
- Hotel room rental—This amount is the hotel's total revenue minus revenue lost to time-share sales and rentals to the college. These amounts are shown in the Calculations section.
- Amenity revenue—This amount is from the Calculations section and can be echoed here.
- Maintenance revenue—This amount is from the Calculations section and can be echoed here.
- Rent from college for 50 rooms—This amount is zero if no rooms are rented to the college. Otherwise, the amount for the year is taken from the Constants section.
- Sale of time-share units—This amount is from the Calculations section and can be echoed here.
- Total Revenue—This amount is the sum of hotel room revenue, amenity revenue, maintenance revenue, revenue from college rentals, and revenue from time-share unit sales.
- Operating expense—This amount is from the Constants section and can be echoed here.
- College remodeling expense—This amount is from the Calculations section and can be echoed here.
- Time-share remodeling expense—This amount is from the Calculations section and can be echoed here.
- Total operating expense—This amount is the sum of operating expenses, college remodeling expenses, and time-share remodeling expenses.
- Income before interest expense—This amount is the difference between total revenue and total operating expense.
- Construction loan interest expense—This amount is from the Constants section and can be echoed here.
- Working capital loan interest expense—This amount is the product of the working capital debt owed to the city at the beginning of the year and the working capital interest rate for the year. The interest rate is a value from the Constants section.
- Total interest expense—This amount is the sum of the construction loan interest and working capital interest.
- Income before taxes—This amount is the difference between income before interest expense and total interest expense.
- Income tax expense—This amount is zero if income before taxes is zero or negative. Otherwise, income tax expense is the product of the year's tax rate and income before taxes. The tax rate is taken from the Constants section.
- Net income (loss) after taxes—This amount is the difference between income before taxes and income tax expense.

Line items for the year-end cash calculation are shown in Figure 7-6. In the figure, column B represents 2014, column C is for 2015, and so on. Year 2014 values are NA except for End-of-year cash on hand, which is $3 million. Values must be computed by cell formula; hard-code numbers in formulas only when you are told to do so. Cell formulas should not reference a cell with a value of "NA." An explanation of each item follows the figure.

	A	B	C	D	E	F
64	Construction loan principal payment	NA				
65						
66	Net cash position (NCP)	NA				
67	Borrowing from city government in year	NA				
68	Repayment to city government in year	NA				
69	End-of-year cash on hand	$ 3,000,000				

Source: Used with permission from Microsoft Corporation

FIGURE 7-6   End-of-year cash on hand section

- Construction loan principal payment—The outlay for the year is from the Constants section and can be echoed here.
- Net cash position (NCP)—The NCP at the end of a year equals the cash at the beginning of the year, plus the year's net income (or loss), minus the year's construction loan principal payment.
- Borrowing from city government in year—Assume that the city government will lend enough working capital at the end of the year to reach the minimum cash needed to start the next year. If the NCP is less than this minimum, the hotel must borrow enough to start the next year with the minimum. Borrowing increases the cash on hand, of course.

- Repayment to city government in year—If the NCP is more than the minimum cash needed and some working capital debt is owed to the city at the beginning of the year, the hotel must pay off as much working capital debt as possible (but not take cash below the minimum amount required to start the next year). Repayments reduce cash on hand, of course.
- End-of-year cash on hand—This amount is the NCP plus any borrowing and minus any repayments.

Working Capital Debt Owed Section

This section shows a calculation of working capital debt owed to the city at year's end (see Figure 7-7). Year 2014 values are NA except for End-of-year debt owed, which is $15 million. Values must be computed by cell formula; hard-code numbers in formulas only when you are told to do so. Cell formulas should not reference a cell with a value of "NA." An explanation of each item follows the figure.

	A	B	C	D	E	F
71	**Working Capital Debt Owed**	**2014**	**2015**	**2016**	**2017**	**2018**
72	Beginning-of-year debt owed	NA				
73	Borrowing from city government	NA				
74	Repayment to city government	NA				
75	End-of-year debt owed to city government	$ 15,000,000				

Source: Used with permission from Microsoft Corporation

FIGURE 7-7   Working Capital Debt Owed section

- Beginning-of-year debt owed—Debt owed at the beginning of a year equals the debt owed at the end of the prior year.
- Borrowing from city government—This amount has been calculated elsewhere and can be echoed to this section. Borrowing increases the amount of debt owed.
- Repayment to city government—This amount has been calculated elsewhere and can be echoed to this section. Repayments reduce the amount of debt owed.
- End-of-year debt owed to city government—In 2015 through 2018, this is the amount owed at the beginning of a year, plus borrowing during the year, and minus repayments during the year.

ASSIGNMENT 2: USING THE SPREADSHEET FOR DECISION SUPPORT

Complete the case by (1) using the spreadsheet to gather data about possible scenarios and (2) documenting your findings in a memo.

The hotel has been profitable, but not profitable enough to cover the yearly construction loan payment. Hotel and city management want to generate more cash to start paying down the working capital debt. Both management groups want to know the projected financial results for four different scenarios:

1. No change—No rooms are rented to the college and no rooms are sold to the time-share company. The rate of inflation is 1, 2, 3, and 4 percent for the years 2015 to 2018, respectively. What will happen to working capital debt if no changes are made?
2. Rent rooms—Fifty rooms are rented to the college, but no rooms are sold to the time-share company. Inflation is 1, 2, 3, and 4 percent for the years 2015 to 2018, respectively. What will happen to working capital debt?
3. Sell time-shares—No rooms are rented to the college, but 10 units are sold to the time-share company in each of the four years. Inflation is 1, 2, 3, and 4 percent for the years 2015 to 2018, respectively. What will happen to working capital debt?
4. Do both—Fifty rooms are rented to the college, and 10 units are sold to the time-share company in each of the four years. Will profitability be restored and debt eliminated? Inflation is 1, 2, 3, and 4 percent for the years 2015 to 2018, respectively. What will happen to working capital debt?

In each scenario, management wants to know the projected amount of working capital debt owed for each of the four years. Management wants to know which scenarios, if any, will lead to less working capital debt. If none of the scenarios work, management will have to consider other options, including selling the hotel to a chain.

Assignment 2A: Using the Spreadsheet to Gather Data

You have built the spreadsheet to model the business situation. For each of the four scenarios, you want to know how much working capital debt is owed to the city in each of the years from 2015 to 2018.

You will run "what-if" scenarios with the four sets of input values using Scenario Manager. (See Tutorial C for details on using Scenario Manager.) Set up the four scenarios. Your instructor may ask you to use conditional formatting to make sure that your input values are proper. Note that in Scenario Manager you can enter noncontiguous cell ranges, such as C19, D19, C20:F20.

The relevant output cells are the end-of-year debt owed cells for 2015 through 2018 in the Summary of Key Results section. Run Scenario Manager to gather the data in a report. When you finish, print the spreadsheet with the input for any of the scenarios, print the Scenario Manager summary sheet, and then save the spreadsheet file a final time.

Assignment 2B: Documenting Your Recommendations in a Memo

Use Microsoft Word to write a brief memo documenting your analysis and conclusions. You can address the memo to hotel management and the city comptroller. Observe the following requirements:

- Set up your memo as described in Tutorial E.
- In the first paragraph, briefly state the business situation and the purpose of your analysis.
- Next, provide the answer to management's question: What scenarios lead to reduced working capital debt?
- State your recommendation. Do any scenarios look promising, or should management consider selling the hotel?
- Support your statements graphically, as your instructor requires. Your instructor may ask you to return to Excel and copy the Scenario Manager summary sheet results into the memo. (See Tutorial C for details on this procedure.) Your instructor might also ask you to make a summary table in Word based on the Scenario Manager summary sheet results. (This procedure is described in Tutorial E.)

Your table should resemble the format shown in Figure 7-8.

	No Change Scenario	Rent Rooms Scenario	Sell Time-shares Scenario	Do Both Scenario
2015 Debt owed				
2016 Debt owed				
2017 Debt owed				
2018 Debt owed				

Source: © Cengage Learning 2015

FIGURE 7-8   Format of table to insert in memo

ASSIGNMENT 3: GIVING AN ORAL PRESENTATION

Your instructor may ask you to explain your analysis and recommendations in an oral presentation. If so, assume that hotel and city managers want the presentation to last 10 minutes or less. Use visual aids or handouts that you think are appropriate. See Tutorial F for tips on preparing and giving an oral presentation.

DELIVERABLES

Your completed case should include the following deliverables for your instructor:

1. A printed copy of your memo
2. Printouts of your spreadsheet and scenario summary
3. Electronic copies of your memo and Excel DSS model

PART 3

DECISION SUPPORT CASES USING MICROSOFT EXCEL SOLVER

BUILDING A DECISION SUPPORT SYSTEM USING MICROSOFT EXCEL SOLVER

In Tutorial C, you learned that Decision Support Systems (DSS) are programs used to help managers solve complex business problems. Cases 6 and 7 were DSS models that used Microsoft Excel Scenario Manager to calculate and display financial outcomes given certain inputs, such as economic outlooks and mortgage interest rates. You used the outputs from Scenario Manager to see how different combinations of inputs affected cash flows and income so that you could make the best decision for expanding your business or selecting a technology to develop and market.

Many business situations require models in which the inputs are not limited to two or three choices, but include large ranges of numbers in more than three variables. For such business problems, managers want to know the best or optimal solution to the model. An optimal solution can either maximize an objective variable, such as income or revenues, or minimize the objective variable, such as operating costs. The formula or equation that represents the target income or operating cost is called an objective function. Optimizing the objective function requires the use of constraints (also called constraint equations), which are rules or conditions you must observe when solving the problem. The field of applied mathematics that addresses problem solving with objective functions and constraint equations is called linear programming. Before the advent of digital computers, linear programming required the knowledge of complex mathematical techniques. Fortunately, Excel has a tool called Solver that can compute the answers to optimization problems.

This tutorial has five sections:

1. **Adding Solver to the Ribbon**—Solver is not installed by default with Excel 2013; you must add it to the application. You may need to use Excel Options to add Solver to the Ribbon.
2. **Using Solver**—This section explains how to use Solver. You will start by determining the best mix of vehicles for shipping exercise equipment to stores throughout the country.
3. **Extending the example**—This section tests your knowledge of Solver as you modify the transportation mix to accommodate changes: additional stores to supply and redesign of the product to reduce shipping volume.
4. **Using Solver on a new problem**—In this section, you will use Solver on a new problem: maximizing the profits for a mix of products.
5. **Troubleshooting Solver**—Because Solver is a complex tool, you will sometimes have problems using it. This section explains how to recognize and overcome such problems.

NOTE

If you need a refresher, Tutorial C offers guidance on basic Excel concepts such as formatting cells and using the =IF( ) and AND( ) functions.

ADDING SOLVER TO THE RIBBON

Before you can use Solver, you must determine whether it is installed in Excel. Start Excel; if necessary, open a file or select a template to gain access to the Ribbon. Next, click the Data tab on the Ribbon. If you see a group on the right side named Analysis that contains Solver, you do not need to install Solver (see Figure D-1).

Source: Used with permission from Microsoft Corporation

FIGURE D-1   Analysis group with Solver installed

If the Analysis group or Solver is not shown on the Data tab of the Ribbon, do the following:

1. Click the File tab.
2. Click Options (see Figure D-2).
3. Click Add-Ins (see Figure D-3) to display the available add-ins in the right pane.
4. Click Go at the bottom of the right pane. The window shown in Figure D-4 appears.
5. Click the Solver Add-in box as well as the Analysis ToolPak and Analysis ToolPak-VBA boxes. (You will need the latter options in a subsequent case, so install them now with Solver.)
6. Click OK to close the window and return to the Ribbon. If you click the Data tab again, you should see the Analysis group with Data Analysis and Solver on the right.

Source: Used with permission from Microsoft Corporation

FIGURE D-2   Excel Options selection

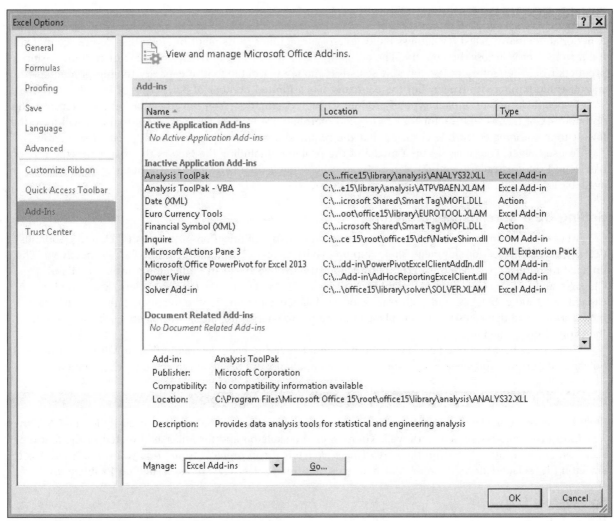

Source: Used with permission from Microsoft Corporation
FIGURE D-3 Add-Ins pane

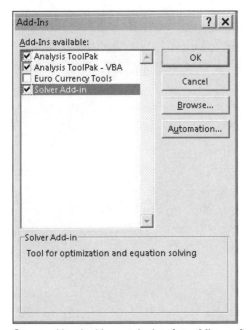

Source: Used with permission from Microsoft Corporation
FIGURE D-4 Add-Ins window with Solver, Analysis ToolPak, and Analysis ToolPak VBA selected

USING SOLVER

A fictional company called CV Fitness builds exercise machines in its plant in Memphis, Tennessee and ships them to its stores across the country. The company has a small fleet of trucks and tractor-trailers to ship its products from the factory to its stores. It costs less money per cubic foot of capacity to ship products with tractor-trailers than with trucks, but the company has a limited number of both types of vehicles and must ship a specified amount of each type of product to each destination. You have been asked to determine the optimal mix of trucks and tractor-trailers to send merchandise to each store. The optimal mix will have the lowest total shipping cost while ensuring that the required quantity of products is shipped to each store.

To use Solver, you must set up a model of the problem, including the factors that can vary (the mix of trucks and tractor-trailers) and the constraints on how much they can vary (the number of each vehicle available). Your goal is to minimize the shipping cost.

Setting Up a Spreadsheet Skeleton

CV Fitness makes three fitness machines: exercise bikes (EB), elliptical cross-trainers (CT), and treadmills (TM). When packaged for shipment, their shipping volumes are 12, 15, and 22 cubic feet, respectively. The finished machines are shipped via ground transportation to five stores in Philadelphia, Atlanta, Miami, Chicago, and Los Angeles. Your vehicle fleet consists of 12 trucks and six tractor-trailers. Each truck has a capacity of 1500 cubic feet, and each tractor-trailer has a capacity of 2350 cubic feet. The spreadsheet includes the road distances from your plant in Memphis to each store, along with each store's demand for the three fitness machines.

What is the best mix of trucks and tractor-trailers to send to each destination? You will learn how to use Solver to determine the answer. The spreadsheet components are discussed in the following sections.

AT THE KEYBOARD

Start by saving your blank spreadsheet. Use a descriptive filename so you can find it easily later—**CV Fitness Trucking Problem.xlsx** should work well. Then enter the skeleton and formulas as directed in the following sections. *You can also download the spreadsheet skeleton if you prefer; it will save you time.* To access this skeleton file, select Tutorial D from your data files, and then select **CV Fitness Trucking Problem.xlsx**.

Spreadsheet Title

Resize Column A, as illustrated in Figure D-5, to give your spreadsheet a small border on the left side. Enter the spreadsheet title in cell B1. Merge and center cells B1 through F1 using the Merge & Center button in the Alignment group of the Home tab.

Constants Section

Your spreadsheet should have a section for values that will not change. Figure D-5 shows a skeleton of the Constants section and the values you should enter. A discussion of the line items follows the figure.

	A	B	C	D	E	F
1		CV Fitness, Inc. Truck Load Management Problem				
2						
3		Constants Section:				
4			Volume Cu. Ft.	Operating Cost per mi.	Operating Cost per mi-cu. Ft.	Available Fleet
5		Truck	1500	$1.00	$0.000667	12
6		Tractor Trailer	2350	$1.30	$0.000553	6
7						
8		Exercise Bike (EB)	12			
9		Elliptical Crosstrainer (CT)	15			
10		Treadmill (TM)	22			

Source: Used with permission from Microsoft Corporation

FIGURE D-5   Spreadsheet title and Constants section

- In column C, enter the Volume Cu. Ft., which is the cubic-foot capacity of the vehicles as well as the shipping volume for each item of exercise equipment.
- In column D, enter the Operating Cost per mi., which is the cost per mile driven for each type of vehicle.
- In column E, enter the Operating Cost per mi.-cu.ft. This value is actually a formula: the operating cost per mile divided by the vehicle volume in cubic feet. Normally you do not put formulas in the Constants section, but in this case it lets you see the relative cost efficiencies of each vehicle. Assuming that both types of vehicles can be filled to capacity, the tractor-trailer is the preferred vehicle for shipping cost efficiency.
- In column F, enter the values for the Available Fleet, which is the number of each type of vehicle your company owns or leases.

You can update the Constants section as the company adds more products to its offerings or adds vehicles to its fleet.

NOTE

The column headings in the Constants section contain two or three lines to keep the columns from becoming too wide. To create a new line in a cell, hold down the Alt key and press Enter.

Now is a good time to save your workbook again. Keep the name you assigned earlier.

Calculations and Results Section

The structure and format of your Calculations and Results section will vary greatly depending on the nature of the problem you need to solve. In some Solver models, you might need to maximize income, which means you might also have an Income Statement section. In other Solver models, you may want to have a separate Changing Cells section that contains cells Solver will manipulate to obtain a solution. In this tutorial, you want to minimize shipping costs while meeting the product demand of your stores. You can accomplish this task by building a single unified table that includes the distances to the stores, the product demand for each store, and the shipping alternatives and costs.

A unified Calculations and Results section makes sense in this model for several reasons. First, it simplifies writing and copying the formulas for the needed shipping volumes, the vehicle capacity totals, and the shipping costs to each destination. Second, a well-organized table allows you to easily identify the changing cells, which Solver will manipulate to optimize the solution, as well as the total cost (or optimization cell). Finally, a unified table allows your management team to visualize both the problem and its solution.

When creating a complex table, it is often a good idea to sketch the table's structure first to see how you want to organize the data. Format the table structure, then enter the data you are given for the problem. Write the cells that contain the formulas last, starting with all the formulas in the first row. If you do a good job structuring your table, you will be able to copy the first-row formulas to the other rows.

Build the blank table shown in Figure D-6. A discussion of the rows and columns follows the figure.

NOTE

Leave rows 11 and 12 blank between the Constants section and the Calculations and Results section. You then will have room to add an extra product to your Constants section later.

	A	B	C	D	E	F	G	H	I	J	K	L	M	N
12														
13		Calculations and Results Section:												
14		Distance/Demand Table		Store Demand			Vehicle Loading							Cost
15		Distance Table (from Memphis Plant)	Miles	EB	CT	TM	Volume Required	Trucks	Volume for Trucks	Tractor-Trailers	Volume for Tractor-Trailers	Total Vehicle Capacity	% of Vehicle Capacity Utilized	Shipping Cost
16		Philadelphia Store	1010	140	96	86								
17		Atlanta Store	380	76	81	63								
18		Miami Store	1000	56	64	52								
19		Chicago Store	540	115	130	150								
20		Los Angeles Store	1810	150	135	180								
21						Totals:								
22		Fill Legend:		Changing Cells										Total Cost
23				Optimization Cell										

Source: Used with permission from Microsoft Corporation

FIGURE D-6 Blank table for Calculations and Results section

- In row 13, enter "Calculations and Results Section:" as the title of the table.
- In row 14, columns B and C, enter "Distance/Demand Table" as a column heading. Merge and center the heading in the two columns.
- In row 14, columns D, E, and F, enter "Store Demand" as a column heading. Merge and center the heading in the three columns.
- In row 14, columns G through M, enter "Vehicle Loading" as a column heading. Merge and center the heading across the columns.
- In row 14, column N, enter "Cost" as a centered column heading.
- In row 15, column B, enter "Distance Table (from Memphis Plant)" as a centered column heading.
- In row 15, column C, enter "Miles" as a centered column heading.
- In row 15, columns D, E, and F, enter "EB," "CT," and "TM," respectively, as equipment headings.
- In row 15, columns G through N, enter "Volume Required," "Trucks," "Volume for Trucks," "Tractor-Trailers," "Volume for Tractor-Trailers," "Total Vehicle Capacity," "% of Vehicle Capacity Utilized," and "Shipping Cost," respectively, as column headings.
- In rows 16 through 20, column B, enter the destination store locations.
- In rows 16 through 20, column C, enter the number of miles to the destination store locations.
- In rows 16 through 20, columns D through F, enter the number of exercise bikes (EB), cross-trainers (CT), and treadmills (TM) to be shipped to each store location.
- Rows 16 through 20, columns G through N, will contain formulas or "seed values" later. Leave them blank for now, but fill cells H16 through H20 and cells J16 through J20 with a light color to indicate that they are the changing cells for Solver. To fill a cell, use the Fill Color button in the Font group.
- In cell F21, enter "Totals:" to label the following cells in the row.
- Cells G21 through N21 will be used for column totals. Fill cell N21 with a slightly darker shade than you used for the changing cells. Cell N21 is your optimization cell.
- In cell B22, enter "Fill Legend:" as a label.
- Fill cell C22 with the fill color you selected for the changing cells.
- In cells D22 and E22, enter "Changing Cells" as the label for the fill color. Merge and center the label in the cells.
- In cell N22, enter "Total Cost" as the label for the value in cell N21.
- Fill cell C23 with the fill color you selected for the optimization cell.
- In cells D23 and E23, enter "Optimization Cell" as the label for the fill color. Merge and center the label in the cells.

Figure D-7 illustrates a magnified section of the Distance/Demand table in case the numbers in Figure D-6 are difficult to read.

	Distance Table (from Memphis Plant)	Miles	EB	CT	TM
	Calculations and Results Section:				
	Distance/Demand Table		**Store Demand**		
16	Philadelphia Store	1010	140	96	86
17	Atlanta Store	380	76	81	63
18	Miami Store	1000	56	64	52
19	Chicago Store	540	115	130	150
20	Los Angeles Store	1810	150	135	180
21					Totals:
22	Fill Legend:		Changing Cells		
23			Optimization Cell		

Source: Used with permission from Microsoft Corporation

FIGURE D-7   Magnified view of the Distance/Demand table

Use the Borders menu in the Font group to select and place appropriate borders around parts of the Calculations and Results section (see Figure D-8). The All Borders and Outside Borders selections are the most useful borders for your table.

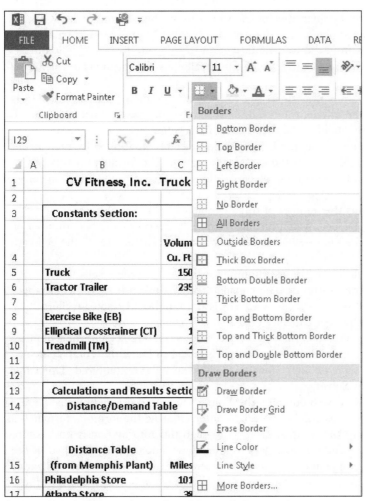

Source: Used with permission from Microsoft Corporation

FIGURE D-8   Borders menu

Next, you write the formulas for the volume and cost calculations. Figure D-9 shows a magnified view of the Vehicle Loading and Cost sections. A discussion of the formulas required for the cells follows the figure.

	G	H	I	J	K	L	M	N
13								
14			Vehicle Loading					Cost
15	Volume Required	Trucks	Volume for Trucks	Tractor-Trailers	Volume for Tractor-Trailers	Total Vehicle Capacity	% of Vehicle Capacity Utilized	Shipping Cost
16								
17								
18								
19								
20								
21								
22								Total Cost

Source: Used with permission from Microsoft Corporation

FIGURE D-9   Vehicle Loading and Cost sections

For illustration purposes, the cell numbers in the following list refer to values for the Philadelphia store.

- Volume Required—Cell G16 contains the total shipping volume of the three types of equipment shipped to the Philadelphia store. The formula for this cell is =D16*$C$8+E16*$C$9+F16*$C$10. Cells D16, E16, and F16 are the quantities of each item to be shipped, and cells $C$8, $C$9, and $C$10 are the shipping volumes for the exercise bike, cross-trainer, and treadmill, respectively. When taking values from the Constants section to calculate formulas, you almost always should use absolute cell references ($) because you will copy the formulas down the columns.
- Trucks—Cell H16 contains the number of trucks selected to ship the merchandise. Cell H16 is a changing cell, which means Solver will determine the best number of trucks to use and place the number in this cell. For now, you should "seed" the cell with a value of 1.
- Volume for Trucks—Cell I16 contains the number of trucks selected, multiplied by the capacity of a truck. The capacity value is taken from the Constants section. The formula for this cell is =H16*$C$5. Cell H16 is the number of trucks selected, and cell $C$5 is the volume capacity of the truck in cubic feet.
- Tractor-Trailers—Cell J16 contains the number of tractor-trailers selected to ship the merchandise. Cell J16 is a changing cell, which means Solver will determine the best number of tractor-trailers to use and place the number in this cell. For now, you should "seed" the cell with a value of 1.
- Volume for Tractor-Trailers—Cell K16 contains the number of tractor-trailers selected, multiplied by the capacity of a tractor-trailer. The capacity value is taken from the Constants section. The formula for this cell is =J16*$C$6. Cell J16 is the number of tractor-trailers selected, and cell $C$6 is the cubic feet capacity of the tractor-trailer.
- Total Vehicle Capacity—Cell L16 contains the sum of the Volume for Trucks and the Volume for Tractor-Trailers. The formula for this cell is =I16+K16. You need to know the Total Vehicle Capacity to make sure that you have enough capacity to ship the Volume Required. This value will be one of your constraints in Solver.
- % of Vehicle Capacity Utilized—Cell M16 contains the Volume Required divided by the Total Vehicle Capacity. The formula for this cell is =G16/L16; after entering the formula, format it as a percentage using the % button in the Number group. Although this information is not required to minimize shipping costs, it is useful for managers to know how much space was filled in the selected vehicles. Alternatively, you could run Solver to determine the highest space utilization on the vehicles rather than the lowest cost. Note that you cannot use more than 100% of the available space on the vehicles.

- Shipping Cost—Cell N16 contains the following calculation:

Mileage to destination store × Number of trucks selected × Cost per mile for trucks + Mileage to destination store × Number of tractor-trailers selected × Cost per mile for tractor-trailers

The formula for this cell is =H16*C16*$D$5+J16*C16*$D$6. Note that absolute cell references for the cost-per-mile values are taken from the Constants section.

If you entered the formulas correctly in row 16, your table should look like Figure D-10.

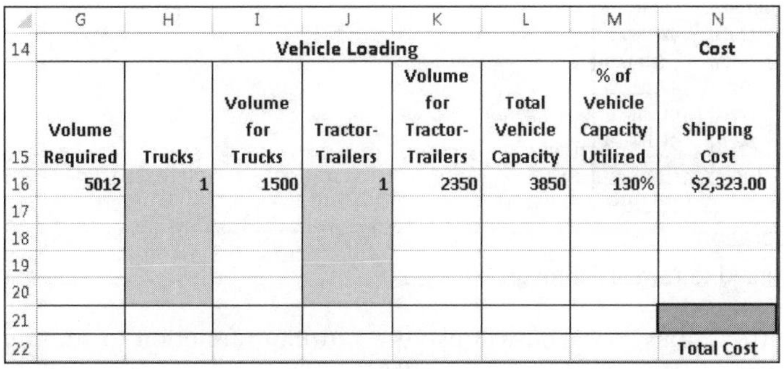

	G	H	I	J	K	L	M	N
14			Vehicle Loading					Cost
15	Volume Required	Trucks	Volume for Trucks	Tractor-Trailers	Volume for Tractor-Trailers	Total Vehicle Capacity	% of Vehicle Capacity Utilized	Shipping Cost
16	5012	1	1500	1	2350	3850	130%	$2,323.00
17								
18								
19								
20								
21								
22								Total Cost

Source: Used with permission from Microsoft Corporation

FIGURE D-10   Vehicle Loading and Cost sections with formulas entered in the first row

To complete the empty cells in rows 17 through 20, you can copy the formulas from cells G16 through N16 to the rest of the rows. Click and drag to select cells G16 through N16, then right-click and select Copy from the menu (see Figure D-11).

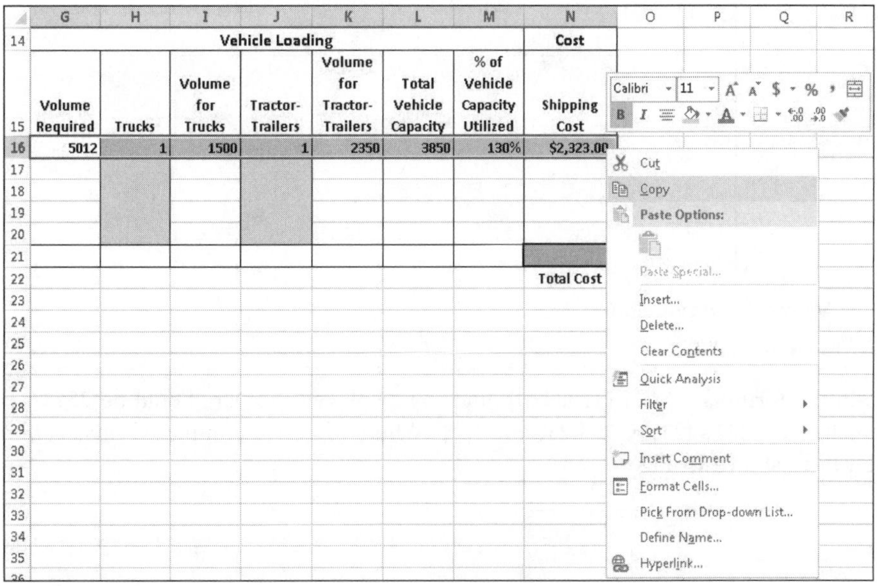

	G	H	I	J	K	L	M	N	O	P	Q	R
14			Vehicle Loading					Cost				
15	Volume Required	Trucks	Volume for Trucks	Tractor-Trailers	Volume for Tractor-Trailers	Total Vehicle Capacity	% of Vehicle Capacity Utilized	Shipping Cost				
16	5012	1	1500	1	2350	3850	130%	$2,323.00				
17												
18										Cut		
19										Copy		
20										Paste Options:		
21												
22								Total Cost				
23										Paste Special...		
24										Insert...		
25										Delete...		
26										Clear Contents		
27										Quick Analysis		
28										Filter		▶
29										Sort		▶
30										Insert Comment		
31										Format Cells...		
32										Pick From Drop-down List...		
33										Define Name...		
34										Hyperlink...		
35												

Source: Used with permission from Microsoft Corporation

FIGURE D-11   Copying formulas

Next, select cells G17 through N20, which are in the four rows beneath row 16. Either press Enter or click Paste in the Clipboard group. The formulas from row 16 should be copied to the rest of the destination cities (see Figure D-12).

	G	H	I	J	K	L	M	N
14	Vehicle Loading							Cost
15	Volume Required	Trucks	Volume for Trucks	Tractor-Trailers	Volume for Tractor-Trailers	Total Vehicle Capacity	% of Vehicle Capacity Utilized	Shipping Cost
16	5012	1	1500	1	2350	3850	130%	$2,323.00
17	3513	1	1500	1	2350	3850	91%	$874.00
18	2776	1	1500	1	2350	3850	72%	$2,300.00
19	6630	1	1500	1	2350	3850	172%	$1,242.00
20	7785	1	1500	1	2350	3850	202%	$4,163.00
21								
22								Total Cost

Source: Used with permission from Microsoft Corporation

FIGURE D-12   Formulas from row 16 successfully copied to rows 17 through 20

You have one row of formulas to complete: the Totals row. You will use the AutoSum function to sum up one column, and then copy the formula to the rest of the columns *except* cell M21. This cell is not actually a total, but an overall capacity utilization rate.

To enter the sum of cells G16 through G20 in cell G21, select cells G16 through G21, then click AutoSum in the Editing group on the Home tab of the Ribbon (see Figure D-13).

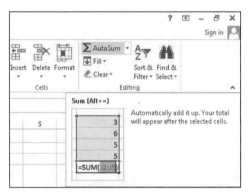

Source: Used with permission from Microsoft Corporation

FIGURE D-13   AutoSum button in the Editing group

Cell G21 should now contain the formula =SUM(G16:G20), and the displayed answer should be 25716. Now you can copy cell G21 to cells H21, I21, J21, K21, L21, and N21. When you have completed this section of the table, it should have the values shown in Figure D-14.

	G	H	I	J	K	L	M	N
14	Vehicle Loading							Cost
15	Volume Required	Trucks	Volume for Trucks	Tractor-Trailers	Volume for Tractor-Trailers	Total Vehicle Capacity	% of Vehicle Capacity Utilized	Shipping Cost
16	5012	1	1500	1	2350	3850	130%	$2,323.00
17	3513	1	1500	1	2350	3850	91%	$874.00
18	2776	1	1500	1	2350	3850	72%	$2,300.00
19	6630	1	1500	1	2350	3850	172%	$1,242.00
20	7785	1	1500	1	2350	3850	202%	$4,163.00
21	25716	5	7500	5	11750	19250		$10,902.00
22								Total Cost

Source: Used with permission from Microsoft Corporation

FIGURE D-14   Totals cells completed

The last formula to enter is for cell M21. This is not a total, but an overall percentage of Vehicle Capacity Utilized for all the vehicles used. This calculation uses the same formula as the cell above it, so you can simply copy cell M20 to cell M21. The formula for this cell is =G21/L21, which is Volume Required divided by Total Vehicle Capacity, expressed as a percentage. Your completed spreadsheet should look like Figure D-15.

	A	B	C	D	E	F	G	H	I	J	K	L	M	N
14		Distance/Demand Table			Store Demand				Vehicle Loading					Cost
15		Distance Table (from Memphis Plant)	Miles	EB	CT	TM	Volume Required	Trucks	Volume for Trucks	Tractor-Trailers	Volume for Tractor-Trailers	Total Vehicle Capacity	% of Vehicle Capacity Utilized	Shipping Cost
16		Philadelphia Store	1010	140	96	86	5012	1	1500	1	2350	3850	130%	$2,323.00
17		Atlanta Store	380	76	81	63	3513	1	1500	1	2350	3850	91%	$874.00
18		Miami Store	1000	56	64	52	2776	1	1500	1	2350	3850	72%	$2,300.00
19		Chicago Store	540	115	130	150	6630	1	1500	1	2350	3850	172%	$1,242.00
20		Los Angeles Store	1810	150	135	180	7785	1	1500	1	2350	3850	202%	$4,163.00
21						Totals:	25716	5	7500	5	11750	19250	134%	$10,902.00
22		Fill Legend:			Changing Cells									Total Cost
23					Optimization Cell									

Source: Used with permission from Microsoft Corporation

FIGURE D-15   Completed Calculations and Results section

Working the Model Manually

Now that you have a working model, you could manipulate the number of trucks and tractor-trailers manually to obtain a solution to the shipping problem. You would need to observe the following rules (or constraints):

1. Assign enough Total Vehicle Capacity to meet the Volume Required for each destination. (In other words, you cannot exceed 100% of Vehicle Capacity Utilized.)

2. The total number of trucks and tractor-trailers you assign cannot exceed the number available in your fleet.

Try to assign your trucks and tractor-trailers to meet your shipping requirements, and note the total shipping costs—you may get lucky and come up with an optimal solution. The tractor-trailers are more cost efficient than the trucks, but the problem is complicated by the fact that you want to achieve the best capacity utilization as well. In some instances, the trucks may be a better fit. Figure D-16 shows a sample solution determined from working the problem manually.

Source: Used with permission from Microsoft Corporation

FIGURE D-16   Manual attempt to solve the vehicle loading problem optimally

This probably looks like a good solution—after all, you have not violated any of your constraints, and you have a 94% average vehicle capacity utilization. But is it the most cost-effective solution for your company? This is where Solver comes in.

Setting Up Solver Using the Solver Parameters Window

To access the Solver pane, click the Data tab on the Ribbon, then click Solver in the Analysis group on the far right side of the Ribbon. The Solver Parameters window appears (see Figure D-17).

> **N O T E**
>
> Solver in Excel 2010 and 2013 has changed significantly from earlier versions of Excel. It allows three different calculation methods, and it allows you to specify an amount of time and number of iterations to perform before Excel ends the calculation. Refer to Microsoft Help for more information.

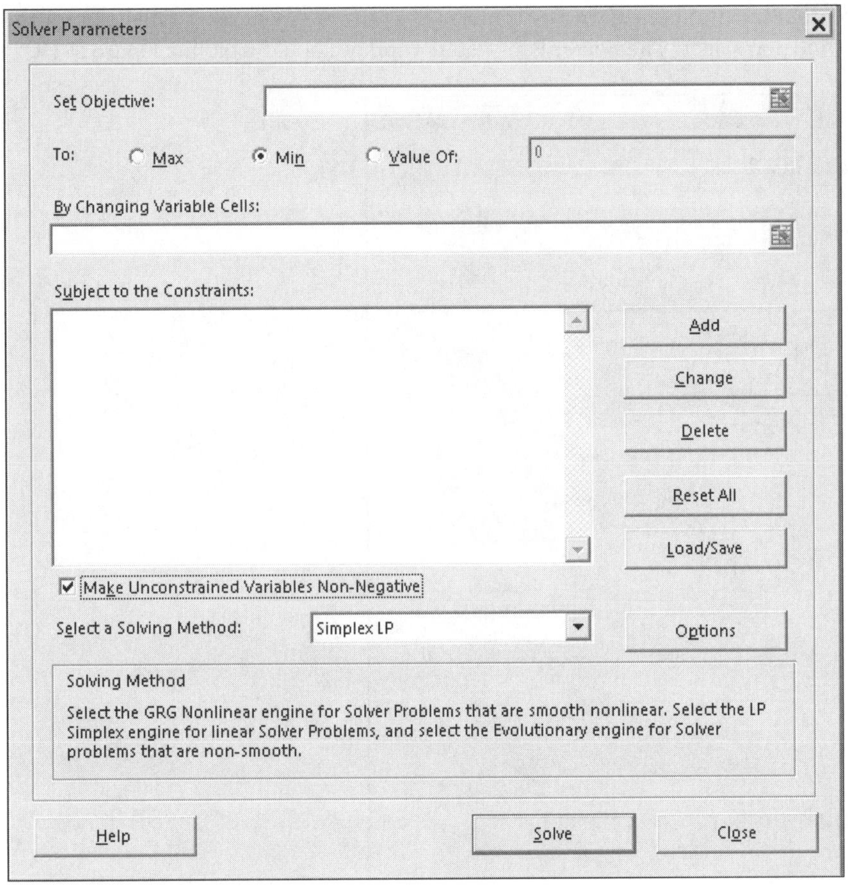

Source: Used with permission from Microsoft Corporation

FIGURE D-17   Solver Parameters window

The Solver Parameters window in Excel 2013 looks intimidating at first. However, to solve linear optimization problems, you have to satisfy only three sets of conditions by filling in the following fields:

- Set Objective—Specify the optimization cell.
- By Changing Variable Cells—Specify the changing cells in your worksheet.
- Subject to the Constraints—Define all of the conditions and limitations that must be met when seeking the optimal solution.

The following sections explain these fields in detail. You may also need to click the Options button and select one or more options for solving the problem. Most of the cases in this book are linear problems, so you can set the solving method to Simplex LP, as shown in Figure D-17. If this method does not work in later cases, you can select the GRG Nonlinear or Evolutionary method to try to solve the problem. Note that the GRG Nonlinear and Evolutionary solving methods are available only in Excel 2010 and 2013.

Optimization Cell and Changing Cells

To use Solver successfully, you must first specify the cell you want to optimize—in this case, the total shipping cost, or cell N21. To fill the Set Objective field, click the button at the right edge of the field, and then click cell N21 in the spreadsheet. You could also type the cell address in the window, but selecting the cell in the spreadsheet reduces your chance of entering the wrong cell address. Next, specify whether you want Solver to seek the maximum or minimum value for cell N21. Because you want to minimize the total shipping cost, click the radio button next to Min.

Next, tell Solver which cell values it will change to determine the optimal solution. Use the By Changing Variable Cells field to specify the range of cells that you want Solver to manipulate. Again, click the button at the right edge of the field, select the cells that contain the numbers of trucks (H16 to H20), and then hold down the

Ctrl key and select the cells that contain the numbers of tractor-trailers (J16 to J20). If you used a fill color for the changing cells, they will be easy to find and select. The Solver Parameters window should look like Figure D-18.

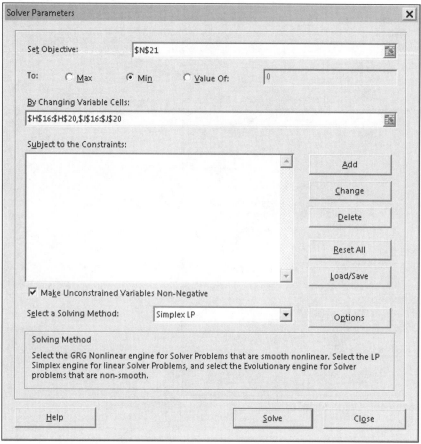

Source: Used with permission from Microsoft Corporation

FIGURE D-18   Solver Parameters window with the objective cell and changing cells entered

Note that Solver has added absolute cell references (the $ signs before the column and row designators) for the cells you have specified. Solver will also add these references to the constraints you define. Solver adds the references to preserve the links to the cells in case you revise the worksheet in the future. In fact, you will make changes to the worksheet later in the tutorial.

Defining and Entering Constraints

For Solver to successfully determine the optimum solution for the shipping problem, you need to specify what constraints or rules it must observe to calculate the solution. Without constraints, Solver theoretically might calculate that the best solution is not to ship anything, resulting in a cost of zero. Furthermore, if you failed to define variables as positive numbers, Solver would select "negative trucks" to maximize "negative costs." Finally, the vehicles are indivisible units—you cannot assign a fraction of a vehicle for a fraction of the cost, so you must define your changing cells as integers to satisfy this constraint.

Aside from the preceding logical constraints, you have operational constraints as well. You cannot assign more vehicles than you have in your fleet, and the vehicles you assign must have at least as much total capacity as your shipping volume.

Before entering the constraints in the Solver Parameters window, it is a good idea to write them down in regular language. You must enter the following constraints for this model:

- All trucks and tractor-trailers in the changing cells must be integers greater than or equal to zero.
- The sums of trucks and tractor-trailers assigned (cells H21 and J21) must be less than or equal to the available trucks and tractor-trailers (cells F5 and F6, respectively).

- The Total Vehicle Capacity for the vehicles assigned to each store (cells L16 to L20) must be greater than or equal to the Volume Required to be shipped to each store (cells G16 to G20, respectively).

You are ready to enter the constraints as equations or inequalities in the Add Constraint window. To begin, click the Add button in the Solver Parameters window. In the window that appears (see Figure D-19), click the button at the right edge of the Cell Reference box, select cells H16 to H20, and then click the button again. Next, click the drop-down menu in the middle field and select > =. Then go to the Constraint field and type 0. Finally, click Add; otherwise, the constraint you defined will not be added to the list defined in the Solver Parameters window.

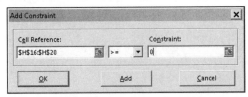

Source: Used with permission from Microsoft Corporation
FIGURE D-19   Add Constraint window

You can continue to add constraints in the Add Constraint window. For this example, enter the constraints shown in the completed Solver Parameters window in Figure D-20. When you finish, click Add to save the last constraint, then click Cancel in the Add Constraint window to return to the Solver Parameters window.

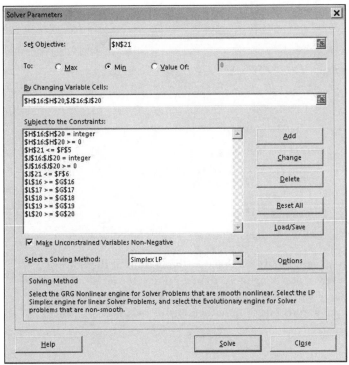

Source: Used with permission from Microsoft Corporation
FIGURE D-20   Completed Solver Parameters window

If you have difficulty reading the constraints listed in Figure D-20, use the following list instead:

- $H$16:$H$20 = integer
- $H$16:$H$20 >= 0
- $H$21 <= $F$5
- $J$16:$J$20 = integer
- $J$16:$J$20 >= 0

- $J$21 <= $F$6
- $L$16 >= $G$16
- $L$17 >= $G$17
- $L$18 >= $G$18
- $L$19 >= $G$19
- $L$20 >= $G$20

You should also click the Options button in the Solver Parameters window and check the Options window shown in Figure D-21. You can use this window to set the maximum amount of time and iterations you want Solver to run before stopping. Make sure that both options are set at 100 for now, but remember that Solver may need more time and iterations for more complex problems. To get the best solution, you should set the Integer Optimality (%) to zero. Click OK to close the window.

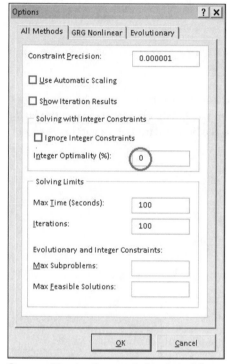

Source: Used with permission from Microsoft Corporation

FIGURE D-21   Solver Options window with Integer Optimality set to zero

You are ready to run Solver to find the optimal solution. Click Solve at the bottom of the Solver Parameters window. Solver might require only a few seconds or more than a minute to run all the possible iterations—the status bar at the bottom of the Excel window displays iterations and possible solutions continuously until Solver finds an optimal solution or runs out of time (see Figure D-22).

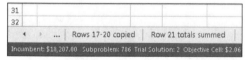

Source: Used with permission from Microsoft Corporation

FIGURE D-22   Excel status bar showing Solver running through possible solutions

A new window will appear eventually, indicating that Solver has found an optimal solution to the problem (see Figure D-23). The portion of the spreadsheet that displays the assigned vehicles and shipping cost should be visible below the Solver Results window. Solver has assigned nine of the 12 trucks and all six tractor-trailers, for a total shipping cost of $17,398. The earlier manual attempt to solve the problem (see Figure D-16) assigned all 12 trucks and four tractor-trailers, for a total shipping cost of $18,122. Using Solver in this situation saved your company $724.

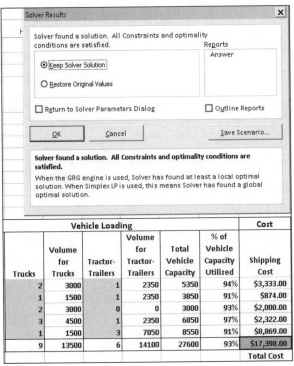

Source: Used with permission from Microsoft Corporation

FIGURE D-23   Solver Results window

If the Solver Results window does not report an optimal solution to the problem, it will report that the problem could not be solved given the changing cells and constraints you specified. For instance, if you had not had enough vehicles in your fleet to carry the required shipping volume to all the destinations, the Solver Results window might have looked like Figure D-24. In the figure, your vehicle fleet was reduced to 10 trucks and five tractor-trailers, so Solver could not find a solution that satisfied the shipping volume constraints.

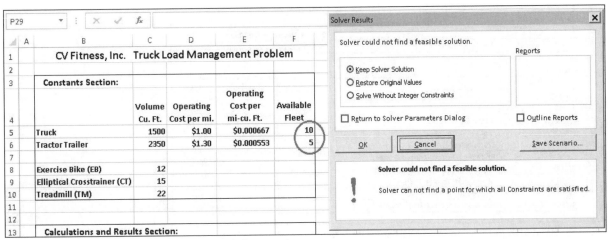

Source: Used with permission from Microsoft Corporation

FIGURE D-24   Solver could not find a feasible solution with a reduced vehicle fleet

Fortunately, Solver did find an optimal solution. To update the spreadsheet with the new optimal values for the changing cells and optimization cell, click OK in the Solver Results window. You can also create an Answer Report by clicking the Answer option in the Solver Results window (see Figure D-25) and then clicking OK.

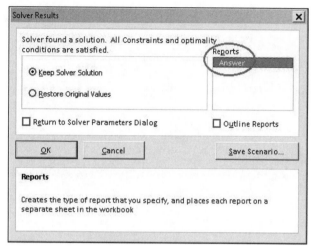

Source: Used with permission from Microsoft Corporation

FIGURE D-25   Creating an Answer Report

Excel will create a report in a separate sheet called Answer Report 1. The Answer Report is shown in Figures D-26 and D-27.

	A	B	C	D	E	F
1	Microsoft Excel 15.0 Answer Report					
2	Worksheet: [CV Fitness Trucking Problem--Working Copy-12th Edition.xlsx]Solver Solution					
3	Report Created: 7/3/2013 12:26:25 PM					
4	Result: Solver found a solution. All Constraints and optimality conditions are satisfied.					
5	Solver Engine					
6	Engine: Simplex LP					
7	Solution Time: 1.092 Seconds.					
8	Iterations: 3 Subproblems: 1416					
9	Solver Options					
10	Max Time 100 sec, Iterations 100, Precision 0.000001					
11	Max Subproblems Unlimited, Max Integer Sols Unlimited, Integer Tolerance 0%, Assume NonNegative					
12						
13	Objective Cell (Min)					
14		Cell	Name	Original Value	Final Value	
15		$N$21	Totals: Shipping Cost	$18,122.00	$17,398.00	
16						
17	Variable Cells					
18		Cell	Name	Original Value	Final Value	Integer
19		$H$16	Philadelphia Store Trucks	2	2	Integer
20		$H$17	Atlanta Store Trucks	1	1	Integer
21		$H$18	Miami Store Trucks	2	2	Integer
22		$H$19	Chicago Store Trucks	3	3	Integer
23		$H$20	Los Angeles Store Trucks	4	1	Integer
24		$J$16	Philadelphia Store Tractor-Trailers	1	1	Integer
25		$J$17	Atlanta Store Tractor-Trailers	1	1	Integer
26		$J$18	Miami Store Tractor-Trailers	0	0	Integer
27		$J$19	Chicago Store Tractor-Trailers	1	1	Integer
28		$J$20	Los Angeles Store Tractor-Trailers	1	3	Integer
29						

Source: Used with permission from Microsoft Corporation

FIGURE D-26   Top portion of the Answer Report

	Cell	Name	Cell Value	Formula	Status	Slack
30	Constraints					
31	Cell	Name	Cell Value	Formula	Status	Slack
32	$H$21	Totals: Trucks	9	$H$21<=$F$5	Not Binding	3
33	$J$21	Totals: Tractor-Trailers	6	$J$21<=$F$6	Binding	0
34	$L$16	Philadelphia Store Total Vehicle Capacity	5350	$L$16>=$G$16	Not Binding	338
35	$L$17	Atlanta Store Total Vehicle Capacity	3850	$L$17>=$G$17	Not Binding	337
36	$L$18	Miami Store Total Vehicle Capacity	3000	$L$18>=$G$18	Not Binding	224
37	$L$19	Chicago Store Total Vehicle Capacity	6850	$L$19>=$G$19	Not Binding	220
38	$L$20	Los Angeles Store Total Vehicle Capacity	8550	$L$20>=$G$20	Not Binding	765
39	$H$16	Philadelphia Store Trucks	2	$H$16>=0	Binding	0
40	$H$17	Atlanta Store Trucks	1	$H$17>=0	Binding	0
41	$H$18	Miami Store Trucks	2	$H$18>=0	Binding	0
42	$H$19	Chicago Store Trucks	3	$H$19>=0	Binding	0
43	$H$20	Los Angeles Store Trucks	1	$H$20>=0	Binding	0
44	$J$16	Philadelphia Store Tractor-Trailers	1	$J$16>=0	Binding	0
45	$J$17	Atlanta Store Tractor-Trailers	1	$J$17>=0	Binding	0
46	$J$18	Miami Store Tractor-Trailers	0	$J$18>=0	Binding	0
47	$J$19	Chicago Store Tractor-Trailers	1	$J$19>=0	Binding	0
48	$J$20	Los Angeles Store Tractor-Trailers	3	$J$20>=0	Binding	0
49	$H$16:$H$20=Integer					
50	$J$16:$J$20=Integer					
51						

Source: Used with permission from Microsoft Corporation

FIGURE D-27   Bottom portion of the Answer Report

The Answer Report gives you a wealth of information about the solution. The top portion displays the original and final values of the Objective cell. The second part of the report displays the original and final values of the changing cells. The last part of the report lists the constraints. Binding constraints are those that reached their maximum or minimum value; nonbinding constraints did not.

Perhaps a savings of $724 does not seem significant—however, this problem does not have a specified time frame. The example probably represents one week of shipments for CV Fitness. The store demands will change from week to week, but you could use Solver each time to optimize the truck assignments. In a 50-week business year, the savings that Solver helps you find in shipping costs could be well over $30,000!

Go to the File tab to print the worksheets you created. Save the Excel file as **CV Fitness Trucking Problem.xlsx**, then select the Save As command in the File tab to create a new file called **CV Fitness Trucking Problem 2.xlsx**. You will use the new file in the next section.

EXTENDING THE EXAMPLE

Like all successful companies, CV Fitness looks for ways to grow its business and optimize its costs. Your management team is considering two changes:

- Opening two new stores and expanding the vehicle fleet if necessary
- Improving product design and packaging to reduce the shipping volume of the treadmill from 22 cubic feet to 17 cubic feet

You have been asked to modify your model to see the new requirements for each change separately. The two new stores would be in Denver and Phoenix, and they are 1,040 and 1,470 miles from the Memphis plant, respectively. If necessary, open the CV Fitness Trucking Problem 2.xlsx file, then right-click row 21 at the left worksheet border. Click Insert to enter a new row between rows 20 and 21. Repeat the steps to insert a second new row. Your spreadsheet should look like Figure D-28. Do not worry about the borders for now—you can fix them later.

	A	B	C	D	E	F	G	H	I	J	K	L	M	N
13		Calculations and Results Section:												
14		Distance/Demand Table			Store Demand				Vehicle Loading					Cost
15		Distance Table (from Memphis Plant)	Miles	EB	CT	TM	Volume Required	Trucks	Volume for Trucks	Tractor-Trailers	Volume for Tractor-Trailers	Total Vehicle Capacity	% of Vehicle Capacity Utilized	Shipping Cost
16		Philadelphia Store	1010	140	96	86	5012	2	3000	1	2350	5350	94%	$3,333.00
17		Atlanta Store	380	76	81	63	3513	1	1500	1	2350	3850	91%	$874.00
18		Miami Store	1000	56	64	52	2776	2	3000	0	0	3000	93%	$2,000.00
19		Chicago Store	540	115	130	150	6630	3	4500	1	2350	6850	97%	$2,322.00
20		Los Angeles Store	1810	150	135	180	7785	1	1500	3	7050	8550	91%	$8,869.00
21														
22														
23						Totals:	25716	9	13500	6	14100	27600	93%	$17,398.00
24		Fill Legend:			Changing Cells									Total Cost
25					Optimization Cell									

Source: Used with permission from Microsoft Corporation

FIGURE D-28 Distance/Demand table with two blank rows inserted for the new stores

Enter the two new stores in cells B21 and B22, enter their distances in cells C21 and C22, and enter the Store Demands in cells D21 through F22, as shown in Figure D-29. When you complete this part of the table, fix the borders to include the two new stores. Select the area in the table you want to fix, click the No Borders button to clear the old borders, highlight the area to which you want to add the border, and then click the Outside Borders button.

	A	B	C	D	E	F
13		Calculations and Results Section:				
14		Distance/Demand Table			Store Demand	
15		Distance Table (from Memphis Plant)	Miles	EB	CT	TM
16		Philadelphia Store	1010	140	96	86
17		Atlanta Store	380	76	81	63
18		Miami Store	1000	56	64	52
19		Chicago Store	540	115	130	150
20		Los Angeles Store	1810	150	135	180
21		Denver Store	1040	74	67	43
22		Phoenix Store	1470	41	28	37
23						Totals:
24		Fill Legend:			Changing Cells	
25					Optimization Cell	

Source: Used with permission from Microsoft Corporation

FIGURE D-29 Distance/Demand table with new store locations and demands entered

Next, copy the formulas from cells G20 to N20 to the two new rows in the Vehicle Loading and Cost sections of the table. Select cells G20 to N20, right-click, and click Copy on the menu. Then select cells G21 to N22 and click Paste in the Clipboard group. Your table should look like Figure D-30.

	G	H	I	J	K	L	M	N
13								
14			Vehicle Loading					Cost
15	Volume Required	Trucks	Volume for Trucks	Tractor-Trailers	Volume for Tractor-Trailers	Total Vehicle Capacity	% of Vehicle Capacity Utilized	Shipping Cost
16	5012	2	3000	1	2350	5350	94%	$3,333.00
17	3513	1	1500	1	2350	3850	91%	$874.00
18	2776	2	3000	0	0	3000	93%	$2,000.00
19	6630	3	4500	1	2350	6850	97%	$2,322.00
20	7785	1	1500	3	7050	8550	91%	$8,869.00
21	2839	1	1500	3	7050	8550	33%	$5,096.00
22	1726	1	1500	3	7050	8550	20%	$7,203.00
23	25716	9	13500	6	14100	27600	93%	$17,398.00
24								Total Cost

Source: Used with permission from Microsoft Corporation

FIGURE D-30 Formulas from row 20 copied into rows 21 and 22

Note that most cells in the Totals row have not changed—their formulas need to be updated to include the values in rows 21 and 22. To quickly check which cells you need to update, display the formulas in the Totals row. Hold down the Ctrl key and press the ~ key (on most keyboards, this key is next to the "1" key). The Vehicle Loading and Cost sections now display formulas in the cells (see Figure D-31).

	G	H	I	J	K	L	M	N
13								
14			Vehicle Loading					Cost
15	Volume Required	Trucks	Volume for Trucks	Tractor-Trailers	Volume for Tractor-Trailers	Total Vehicle Capacity	% of Vehicle Capacity Utilized	Shipping Cost
16	=D16*$C$8+E16*$C$9+F16*$C$10	2	=H16*$C$5	1	=I16*$C$6	=I16+K16	=G16/L16	=H16*C16*$D$5+J16*C16*$D$6
17	=D17*$C$8+E17*$C$9+F17*$C$10	1	=H17*$C$5	1	=I17*$C$6	=I17+K17	=G17/L17	=H17*C17*$D$5+J17*C17*$D$6
18	=D18*$C$8+E18*$C$9+F18*$C$10	2	=H18*$C$5	0	=I18*$C$6	=I18+K18	=G18/L18	=H18*C18*$D$5+J18*C18*$D$6
19	=D19*$C$8+E19*$C$9+F19*$C$10	3	=H19*$C$5	1	=I19*$C$6	=I19+K19	=G19/L19	=H19*C19*$D$5+J19*C19*$D$6
20	=D20*$C$8+E20*$C$9+F20*$C$10	1	=H20*$C$5	3	=I20*$C$6	=I20+K20	=G20/L20	=H20*C20*$D$5+J20*C20*$D$6
21	=D21*$C$8+E21*$C$9+F21*$C$10	1	=H21*$C$5	3	=I21*$C$6	=I21+K21	=G21/L21	=H21*C21*$D$5+J21*C21*$D$6
22	=D22*$C$8+E22*$C$9+F22*$C$10	1	=H22*$C$5	3	=I22*$C$6	=I22+K22	=G22/L22	=H22*C22*$D$5+J22*C22*$D$6
23	=SUM(G16:G20)	=SUM(H16:H20)	=SUM(I16:I20)	=SUM(J16:J20)	=SUM(K16:K20)	=SUM(L16:L20)	=G23/L23	=SUM(N16:N20)
24								Total Cost

Source: Used with permission from Microsoft Corporation

FIGURE D-31 Vehicle Loading and Cost sections with formulas displayed in the cells

You must update any Totals cells that do not include the contents of rows 21 and 22. For example, you need to update the Totals cells G23 through L23 and cell N23. Cell M23 is not really a total; it is a cumulative ratio formula, so you do not need to update the cell. Use the following formulas to revise the Totals cells:

- Cell G23: =SUM(G16:G22)
- Cell H23: =SUM(H16:H22)
- Cell I23: =SUM(I16:I22)
- Cell J23: =SUM(J16:J22)
- Cell K23: =SUM(K16:K22)
- Cell L23: =SUM(L16:L22)
- Cell N23: =SUM(N16:N22)

The updated sections should look like Figure D-32.

	G	H	I	J	K	L	M	N
14			Vehicle Loading					Cost
15	Volume Required	Trucks	Volume for Trucks	Tractor-Trailers	Volume for Tractor-Trailers	Total Vehicle Capacity	% of Vehicle Capacity Utilized	Shipping Cost
16	5012	2	3000	1	2350	5350	94%	$3,333.00
17	3513	1	1500	1	2350	3850	91%	$874.00
18	2776	2	3000	0	0	3000	93%	$2,000.00
19	6630	3	4500	1	2350	6850	97%	$2,322.00
20	7785	1	1500	3	7050	8550	91%	$8,869.00
21	2839	1	1500	3	7050	8550	33%	$5,096.00
22	1726	1	1500	3	7050	8550	20%	$7,203.00
23	30281	11	16500	12	28200	44700	68%	$29,697.00
24								Total Cost

Source: Used with permission from Microsoft Corporation

FIGURE D-32 Vehicle Loading and Cost sections with the formulas updated

You are ready to use Solver to determine the optimal vehicle assignment. Click Solver in the Analysis group of the Data tab. You should notice immediately that you must revise the changing cells to include the two new stores; you must also change some of the constraints and add others. Solver has already updated the Objective cell from $N$21 to $N$23 and has updated the $H$23<=$F$5 and $J$23<=$F$6 constraints for vehicle fleet size. To update the changing cells, click the button to the right of the By Changing Variable Cells field and select the cells again, or edit the formula in the window by changing cell address $H$20 to $H$22 and cell address $J$20 to $J$22.

To change a constraint, select the one you want to change, and then click Change (see Figure D-33).

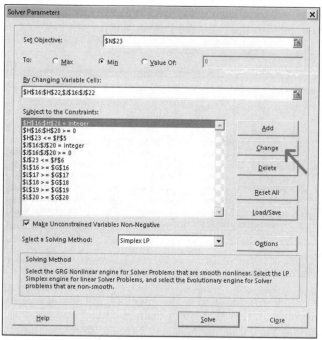

Source: Used with permission from Microsoft Corporation
FIGURE D-33   Selecting a constraint to change

When you click Change, the Change Constraint window appears. Click the Cell Reference button; the selected cells will appear on the spreadsheet with a moving marquee around them (see Figure D-34). Highlight the new group of cells; when the new range appears in the Cell Reference field, click OK. The Solver Parameters window appears with the constraint changed.

			Vehicle Loading				Cost
Volume Required	Trucks	Volume for Trucks	Tractor-Trailers	Volume for Tractor-Trailers	Total Vehicle Capacity	% of Vehicle Capacity Utilized	Shipping Cost
5012	2	3000	1	2350	5350	94%	$3,333.00
3513	1	1500	1	2350	3850	91%	$874.00
2776	2	3000	0	0	3000	93%	$2,000.00
6630	3	4500	1	2350	6850	97%	$2,322.00
7785	1	1500	3	7050	8550	91%	$8,869.00
2839	1	1500	3	7050	8550	33%	$5,096.00
1726	1	1500	3	7050	8550	20%	$7,203.00
30281	11	16500	12	28200	44700	68%	$29,697.00
							Total Cost

Source: Used with permission from Microsoft Corporation
FIGURE D-34   Adding cells H21 and H22 to the Trucks constraint cell range

You also need to update or add the following constraints:

- Update $J$16:$J$20 >=0 to $J$16:$J$22 >=0.
- Update $H$16:$H$20 = integer to $H$16:$H$22 = integer. When changing integer constraints, you must click "int" in the middle field of the Change Constraint window; otherwise, you will receive an error message.
- Update $J$16:$J$20 = integer to $J$16:$J$22 = integer.
- Add constraint $L$21 >= $G$21 (see Figure D-35).
- Add constraint $L$22 >= $G$22.

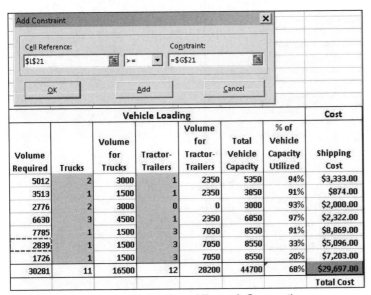

Vehicle Loading							Cost
Volume Required	Trucks	Volume for Trucks	Tractor-Trailers	Volume for Tractor-Trailers	Total Vehicle Capacity	% of Vehicle Capacity Utilized	Shipping Cost
5012	2	3000	1	2350	5350	94%	$3,333.00
3513	1	1500	1	2350	3850	91%	$874.00
2776	2	3000	0	0	3000	93%	$2,000.00
6630	3	4500	1	2350	6850	97%	$2,322.00
7785	1	1500	3	7050	8550	91%	$8,869.00
2839	1	1500	3	7050	8550	33%	$5,096.00
1726	1	1500	3	7050	8550	20%	$7,203.00
30281	11	16500	12	28200	44700	68%	$29,697.00
							Total Cost

Source: Used with permission from Microsoft Corporation

FIGURE D-35   Adding a constraint using the Add Constraint window

You are ready to solve the shipping problem to include the new stores in Denver and Phoenix. Figure D-36 shows the updated Solver Parameters window.

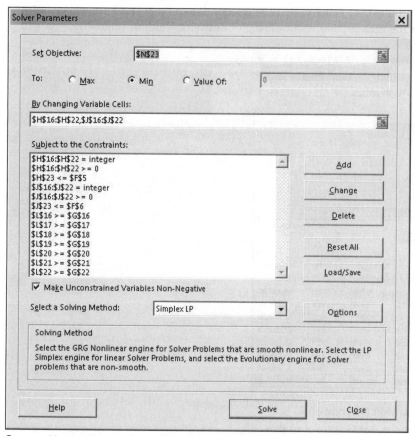

Source: Used with permission from Microsoft Corporation

FIGURE D-36   Solver parameters updated for shipping to seven stores

Before you run Solver again, you might want to attempt to assign the vehicles manually, because your fleet may not be large enough to handle two more stores. In this case, you will quickly realize that the vehicle fleet is at least one truck or tractor-trailer short of the minimum required to ship the needed volume. You can confirm this by running Solver (see Figure D-37).

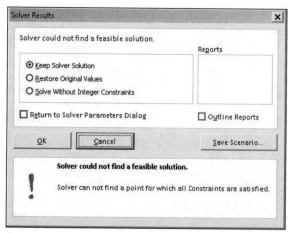

Source: Used with permission from Microsoft Corporation

FIGURE D-37   Vehicle fleet does not meet minimum requirements

The Solver Results window confirms that your truck fleet is too small, so change the value in cell F5 from 12 to 13 to add another truck to your fleet, and then run Solver again. As you add more stores and vehicles to make the problem more complex, Solver will take longer to run, especially on older computers. You may have to wait a minute or more for Solver to finish its iterations and find an answer (see Figure D-38). In this example, Solver recommends that you use 13 trucks and six tractor-trailers.

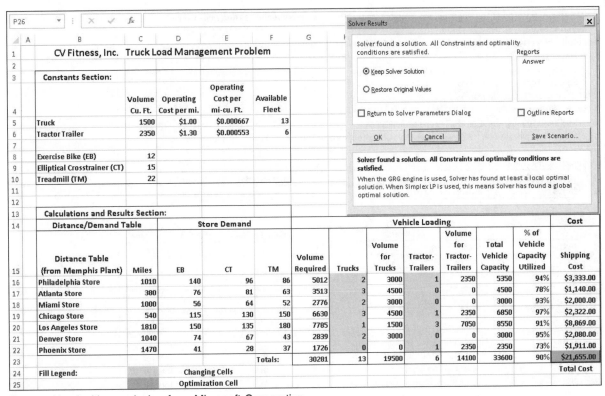

Source: Used with permission from Microsoft Corporation

FIGURE D-38   Solver's solution

Select Answer in the Reports list to add an Answer Report to the workbook, and then click OK. You can keep or delete the old Answer Report 1 tab from the earlier workbook. The new Answer Report is in a new worksheet named Answer Report 2.

You can meet the shipping requirements by adding one more truck, but is it really the most cost-effective solution? What if you add a tractor-trailer instead? Set the number of trucks back to 12, and add a tractor-trailer by entering 7 instead of 6 in cell F6. Run Solver again.

This time Solver finds a less expensive solution, as shown in Figures D-39 and D-40. At first it does not make sense—how can adding a more expensive vehicle (a tractor-trailer) reduce the overall expense? In fact, the additional tractor-trailer has replaced two trucks. With seven tractor-trailers, you only need 11 trucks instead of the original 13.

	A	B	C	D	E	F
2	Worksheet: [CV Fitness Trucking Problem--Working Copy-12th Edition.xlsx]Problem 2 Truck Mix Changed					
3	Report Created: 7/3/2013 2:12:45 PM					
4	Result: Solver found a solution. All Constraints and optimality conditions are satisfied.					
5	Solver Engine					
6	Engine: Simplex LP					
7	Solution Time: 15.646 Seconds.					
8	Iterations: 6 Subproblems: 15498					
9	Solver Options					
10	Max Time 100 sec, Iterations 100, Precision 0.000001					
11	Max Subproblems Unlimited, Max Integer Sols Unlimited, Integer Tolerance 0%, Assume NonNegative					
12						
13	Objective Cell (Min)					
14	Cell	Name		Original Value	Final Value	
15	$N$23	Totals: Shipping Cost		$21,655.00	$21,389.00	
16						
17	Variable Cells					
18	Cell	Name		Original Value	Final Value	Integer
19	$H$16	Philadelphia Store Trucks		2	2	Integer
20	$H$17	Atlanta Store Trucks		3	1	Integer
21	$H$18	Miami Store Trucks		2	2	Integer
22	$H$19	Chicago Store Trucks		3	3	Integer
23	$H$20	Los Angeles Store Trucks		1	1	Integer
24	$H$21	Denver Store Trucks		2	2	Integer
25	$H$22	Phoenix Store Trucks		0	0	Integer
26	$J$16	Philadelphia Store Tractor-Trailers		1	1	Integer
27	$J$17	Atlanta Store Tractor-Trailers		0	1	Integer
28	$J$18	Miami Store Tractor-Trailers		0	0	Integer
29	$J$19	Chicago Store Tractor-Trailers		1	1	Integer
30	$J$20	Los Angeles Store Tractor-Trailers		3	3	Integer
31	$J$21	Denver Store Tractor-Trailers		0	0	Integer
32	$J$22	Phoenix Store Tractor-Trailers		1	1	Integer

Source: Used with permission from Microsoft Corporation

FIGURE D-39   Answer Report 3 displays a more cost-effective solution

	G	H	I	J	K	L	M	N
13								
14			Vehicle Loading					Cost
15	Volume Required	Trucks	Volume for Trucks	Tractor-Trailers	Volume for Tractor-Trailers	Total Vehicle Capacity	% of Vehicle Capacity Utilized	Shipping Cost
16	5012	2	3000	1	2350	5350	94%	$3,333.00
17	3513	1	1500	1	2350	3850	91%	$874.00
18	2776	2	3000	0	0	3000	93%	$2,000.00
19	6630	3	4500	1	2350	6850	97%	$2,322.00
20	7785	1	1500	3	7050	8550	91%	$8,869.00
21	2839	2	3000	0	0	3000	95%	$2,080.00
22	1726	0	0	1	2350	2350	73%	$1,911.00
23	30281	11	16500	7	16450	32950	92%	$21,389.00
24								Total Cost

Source: Used with permission from Microsoft Corporation

FIGURE D-40   Seven tractor-trailers and 11 trucks are the optimal mix

You have a solution for the expansion to seven stores. Save your workbook, and then create a new workbook using the Save As command. Name the new workbook **CV Fitness Trucking Problem 3.xlsx**.

Next, evaluate the potential cost savings if the company redesigns its treadmill product and packaging to reduce the shipping volume from 22 cubic feet to 17 cubic feet. Your engineers report that the redesign will cost approximately $10,000. If you can save at least $500 per shipment, the project will pay for itself in less than six months (20 weekly shipments).

Go to cell C10 on the worksheet, replace 22 with 17, and run Solver again. When Solver finds the solution, select Answer to create another Answer Report, and then click OK. See Figure D-41.

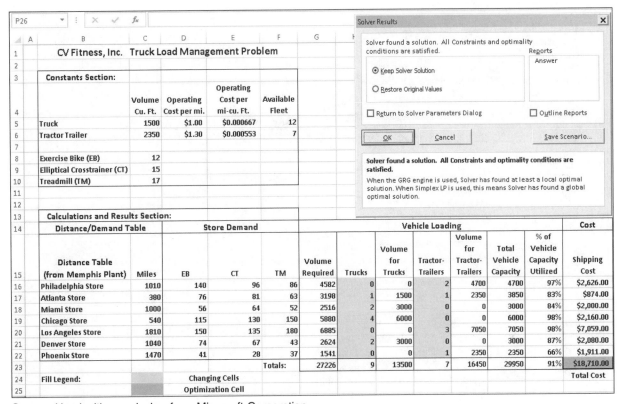

Source: Used with permission from Microsoft Corporation

FIGURE D-41 Solver solution with redesigned treadmill and packaging

Check the Answer Report to see the cost difference between shipping the old treadmills and the redesigned models (see Figure D-42). The cost savings for one shipment is $2,679, which is more than five times the minimum savings you needed. You should go ahead with the project.

	A	B	C	D	E	F
1	Microsoft Excel 15.0 Answer Report					
2	Worksheet: [CV Fitness Trucking Problem--Working Copy-12th Edition.xlsx]Problem 3 Treadmill Repackage					
3	Report Created: 7/3/2013 2:29:19 PM					
4	Result: Solver found a solution. All Constraints and optimality conditions are satisfied.					
5	Solver Engine					
6	Engine: Simplex LP					
7	Solution Time: 9.969 Seconds.					
8	Iterations: 8 Subproblems: 10238					
9	Solver Options					
10	Max Time 100 sec, Iterations 100, Precision 0.000001					
11	Max Subproblems Unlimited, Max Integer Sols Unlimited, Integer Tolerance 0%, Assume NonNegative					
12						
13	Objective Cell (Min)					
14		Cell	Name	Original Value	Final Value	
15		$N$23	Totals: Shipping Cost	$21,389.00	$18,710.00	
16						
17	Variable Cells					
18		Cell	Name	Original Value	Final Value	Integer
19		$H$16	Philadelphia Store Trucks	2	0	Integer
20		$H$17	Atlanta Store Trucks	1	1	Integer
21		$H$18	Miami Store Trucks	2	2	Integer
22		$H$19	Chicago Store Trucks	3	4	Integer
23		$H$20	Los Angeles Store Trucks	1	0	Integer
24		$H$21	Denver Store Trucks	2	2	Integer
25		$H$22	Phoenix Store Trucks	0	0	Integer
26		$J$16	Philadelphia Store Tractor-Trailers	1	2	Integer
27		$J$17	Atlanta Store Tractor-Trailers	1	1	Integer
28		$J$18	Miami Store Tractor-Trailers	0	0	Integer
29		$J$19	Chicago Store Tractor-Trailers	1	0	Integer
30		$J$20	Los Angeles Store Tractor-Trailers	3	3	Integer
31		$J$21	Denver Store Tractor-Trailers	0	0	Integer
32		$J$22	Phoenix Store Tractor-Trailers	1	1	Integer

Source: Used with permission from Microsoft Corporation

FIGURE D-42 Answer Report for the treadmill redesign

When you finish examining the Answer Report, save your file and then close it. To close the workbook, click the File tab and then click Close (see Figure D-43).

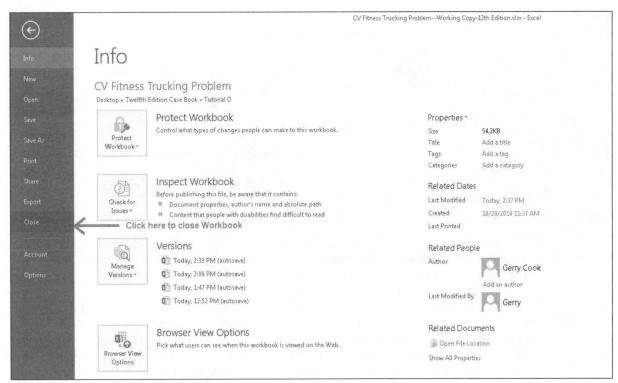

Source: Used with permission from Microsoft Corporation

FIGURE D-43   Closing the Excel workbook

USING SOLVER ON A NEW PROBLEM

A common problem in manufacturing businesses is deciding on a product mix for different items in the same product family. Sensuous Scents Inc. makes a premium collection of perfume, cologne, and body spray for sale in large department stores and boutiques. The primary ingredient is ambergris, a valuable digestive excretion from whales that is harvested without harming the animals. Ambergris costs more than $9,000 per pound and is very difficult to obtain in large quantities; Sensuous Scents can obtain only about 20 pounds of ambergris each year. The other ingredients—deionized water, ethanol, and various additives—are available in unlimited quantities for a reasonable cost.

You have been asked to create a spreadsheet model for Solver to determine the optimal product mix that maximizes Sensuous Scents' net income after taxes.

Setting up the Spreadsheet

The sections in this spreadsheet are different from those in the preceding trucking problem. You will create a Constants section, a Bill of Materials section for the three products, a Quantity Manufactured section that contains the changing cells, a Calculations section (to calculate ambergris usage, manufacturing costs, and sales revenue per product line), and an Income Statement section to determine the net income after taxes, which will be the optimization cell.

AT THE KEYBOARD

Start a new file called **Sensuous Scents Inc.xlsx** and set up the spreadsheet. *You can also download the spreadsheet skeleton if you prefer; it will save you time.* To access this skeleton file, select Tutorial D from your data files, and then select **Sensuous Scents Inc.xlsx**.

Spreadsheet Title and Constants Section

Your spreadsheet title and Constants section should look like Figure D-44. A discussion of the section entries follows the figure.

◢ A	B	C	D	E	F
1	Sensuous Scents Inc. Product Mix				
2					
3	**Constants:**		**Body Spray**	**Cologne**	**Perfume**
4	Sales Price per bottle		$11.95	$21.00	$53.00
5	Conversion Cost per Unit (Direct Labor plus Manufacturing Overhead)		$2.60	$6.50	$13.00
6	Minimum Sales Demand		60000	25000	12000
7	Income Tax Rate	0.32			
8	Sales, General and Administrative Expenses per Dollar Revenue	0.30			
9	Available Ambergris (lbs)	20			
10	Cost per lb, Deionized Water	$0.50			
11	Cost per lb, Ethanol	$1.00			
12	Cost per lb, other Additives	$182.00			
13	Cost per lb, Ambergris	$9,072.00			

Source: Used with permission from Microsoft Corporation

FIGURE D-44   Spreadsheet title and Constants section for Sensuous Scents Inc.

- Sales Price per bottle—These values are the sales prices for each of the three products.
- Conversion Cost per Unit—These values are the direct labor costs plus the manufacturing overhead costs budgeted per unit manufactured. A conversion cost is often used in industries that manufacture liquid products.
- Minimum Sales Demand—These values reflect the forecast minimum sales demand that you must supply to your customers. These values will be used later as constraints.
- Income Tax Rate—The rate is 32% of your pretax income. No taxes are paid on losses.
- Sales, General and Administrative Expenses per Dollar Revenue—This value is an estimate of the non-manufacturing costs that Sensuous Scents will incur per dollar of sales revenue. These expenses are subtracted from the Gross Profit value in the Income Statement section to obtain Net Income before taxes.
- Available Ambergris (lbs.)—This value is the amount of ambergris that Sensuous Scents obtained this year for production.
- Cost per lb., Deionized Water—This value is the current cost per pound of deionized water.
- Cost per lb., Ethanol—This value is the current cost per pound of ethanol.
- Cost per lb., other Additives—Scent products contain other additives and fixatives to enhance or preserve the fragrance. This value is the cost per pound of the other additives.
- Cost per lb., Ambergris—This value is the current market price per pound of naturally harvested ambergris. Again, no whales are harmed to obtain the ambergris.

The rest of the cells are filled with a gray background to indicate that you will not use their values or formulas. The section is arranged this way to maintain one column per product all the way down the spreadsheet, which will simplify writing the formulas later.

Bill of Materials Section

Your spreadsheet should contain a Bill of Materials section, as shown in Figure D-45. The section entries are explained after the figure. A bill of materials is a list of raw materials and ingredients required to make one unit of a product.

◢ A	B	C	D	E	F
14					
15	**Bill of Materials:**		**Body Spray**	**Cologne**	**Perfume**
16	Deionized Water (lb)		0.4	0.1	0.05
17	Ethanol (lb)		0.1	0.02	0.01
18	Other Additives (lb)		0.01	0.001	0.0001
19	Ambergris (lb)		0.0001	0.00018	0.00055

Source: Used with permission from Microsoft Corporation

FIGURE D-45   Bill of Materials section

- Deionized Water (lb.)—The amount of deionized water required to make one unit of each product
- Ethanol (lb.)—The amount of ethanol required to make one unit of each product
- Other Additives (lb.)—The amount of other additives required to make one unit of each product
- Ambergris (lb.)—The amount of ambergris required to make one unit of each product

Extremely small quantities of ambergris and other additives are required to make one bottle of each product. Also, each product requires a different amount of ambergris. Check the values to make sure you entered the correct number of decimal places.

Quantity Manufactured (Changing Cells) Section

This model contains a separate Changing Cells section called Quantity Manufactured, as shown in Figure D-46. This section contains the cells that you want Solver to manipulate to achieve the highest net income after taxes.

	A	B	C	D	E	F
20						
21		Quantity Manufactured (Changing Cells)		Body Spray	Cologne	Perfume
22		Units Produced		60000	25000	12000

Source: Used with permission from Microsoft Corporation

FIGURE D-46   Quantity Manufactured (changing cells) section

Cells D22, E22, and F22 are yellow to indicate that Solver will change them to reach an optimal solution. To begin, enter the minimum sales demand in these cells, which will remind you to specify the minimum demand constraints from the Constants section in the Solver Parameters window.

Calculations Section

Your model should contain the Calculations section shown in Figure D-47.

	A	B	C	D	E	F	G
23							
24		Calculations:		Body Spray	Cologne	Perfume	Totals
25		Lbs of Ambergris Used					
26		Manufacturing Cost per Unit (Materials Costs plus Conversion Cost)					
27		Total Manufacturing Costs per Product Line					
28		Sales Revenues per Product Line					

Source: Used with permission from Microsoft Corporation

FIGURE D-47   Calculations section

The section contains the following calculations:

- Lbs. of Ambergris Used—This value is the pounds of ambergris per unit from the Bill of Materials section, multiplied by Units Produced from the Quantity Manufactured section for each of the three products. The Totals cell (G25) is the sum of cells D25, E25, and F25. Use the value in this cell to specify the constraint that you have only 20 pounds of ambergris available to use for raw materials (Constants section, cell C9).
- Manufacturing Cost per Unit (Materials Costs plus Conversion Cost)—To get this value, write a formula that multiplies the unit cost for each of the four product ingredients by the amount per unit specified in the bill of materials, multiplied by Units Produced. The total materials costs for the four ingredients are added together, and then the Conversion Cost per Unit is added from the Constants section to obtain the Manufacturing Cost per Unit. Enter the following formula for the Body Spray Manufacturing Cost per Unit in cell D26:

    =$C$10*D16+$C$11*D17+$C$12*D18+$C$13*D19+D5

    Use absolute cell references for the cells that hold values for costs per pound ($C$10, $C$11, $C$12, and $C$13). By doing so, you can copy the body spray formula to the Manufacturing Cost per Unit cells for the cologne and perfume values (cells E26 and F26). The Totals cell (G26) is not used in this row—you can fill the cell in gray to indicate that it is not used.
- Total Manufacturing Costs per Product Line—This value is the Manufacturing Cost per Unit multiplied by Units Produced from the Quantity Manufactured section. The Totals cell (G27) is the sum of cells D27, E27, and F27. You will use the value in the Totals cell in the Income Statement section.

- Sales Revenues per Product Line—This value is the Sales Price per bottle from the Constants section multiplied by Units Produced from the Quantity Manufactured section. The Totals cell (G28) is the sum of cells D28, E28, and F28. You will use the value in this cell in the Income Statement section.

Income Statement Section

The last section you need to construct is the Income Statement, as shown in Figure D-48. An explanation of the needed formulas follows the figure.

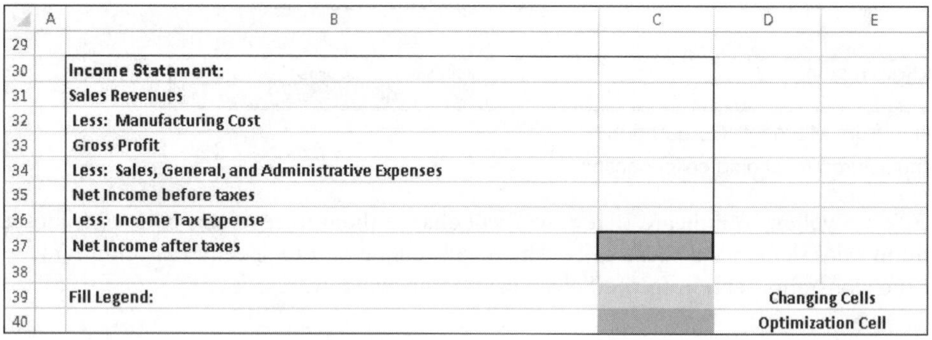

Source: Used with permission from Microsoft Corporation

FIGURE D-48   Income Statement section with fill legend

- Sales Revenues—This value is the total sales revenues from the Calculations section (cell G28).
- Less: Manufacturing Cost—This value is the total manufacturing costs from the Calculations section (cell G27).
- Gross Profit—This value is the Sales Revenues minus the Manufacturing Cost.
- Less: Sales, General, and Administrative Expenses—This value is the Sales Revenues multiplied by the Sales, General, and Administrative Expenses per Dollar Revenue from the Constants section (cell C8).
- Net Income before taxes—This value is the Gross Profit minus the Sales, General, and Administrative Expenses.
- Less: Income Tax Expense—If the Net Income before taxes is greater than zero, this value is the Net Income before taxes multiplied by the Income Tax Rate in the Constants section. If Net Income before taxes is zero or less, the Income Tax Expense is zero.
- Net Income after taxes—This value is the Net Income before taxes minus the Income Tax Expense. You will use this value as your optimization cell because you want to maximize Net Income after taxes.

Setting up Solver

You need to satisfy the following conditions when running Solver:

- Your objective is to maximize Net Income after taxes (cell C37).
- Your changing cells are the Units Produced (cells D22, E22, and F22).
- Observe the following constraints:

    - You must produce at least the Minimum Sales Demand for each product (cells D6, E6, and F6).
    - Your total Lbs. of Ambergris Used (cell G25) cannot exceed the Available Ambergris (cell C9).
    - You cannot produce negative units of any product (enter constraints for the changing cells to be greater than or equal to zero).
    - You can produce only whole units of any product (enter constraints for the changing cells to be integers).

Run Solver and create an Answer Report when Solver finds the solution. When you complete the program, print your spreadsheet with the Solver solution, and print the Answer Report. Save your work and close Excel.

TROUBLESHOOTING SOLVER

Solver is a fairly complex software program. This section helps you address common problems you may encounter when attempting to run Solver.

Using Whole Numbers in Changing Cells

Before you run your first Solver model or rerun a previous model, always enter a positive whole number in each of the changing cells. If you have not already defined maximum and minimum constraints for the values in the changing cells, enter 1 in each cell before running Solver.

Getting Negative or Fractional Answers

If you receive negative or fractional answers when running Solver, you may have neglected to specify one or more of the changing cells as non-negative integers. Alternatively, if you are working on a cost minimization problem and you fail to specify the optimization cell as non-negative, you may receive a negative answer for the cost. Sometimes Solver will also warn you that you have one or more unbounded constraints (see Figure D-49).

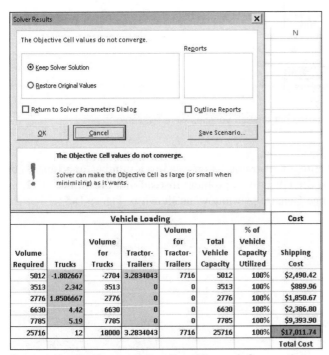

Source: Used with permission from Microsoft Corporation

FIGURE D-49   Solver has an "unbounded" objective function because you did not specify non-negative integer constraints

Creating Overconstrained Models

If Solver cannot find a solution because it cannot meet the constraints you defined, you will receive an error message. When this happens, Solver may even violate the integer constraints you defined in an attempt to find an answer, as shown in Figure D-50. To create this error, the number of trucks available was reduced from 12 to 10.

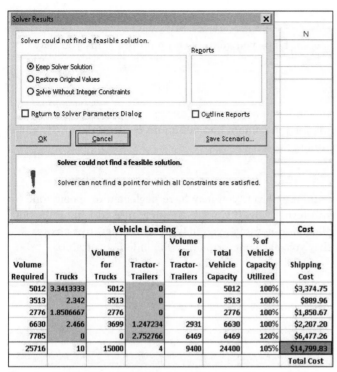

Source: Used with permission from Microsoft Corporation

FIGURE D-50   Solver could not find a feasible solution because not enough vehicles were available

Setting a Constraint to a Single Amount

Sometimes you may want to enter an exact amount into a constraint, as opposed to a number in a range. For example, if you wanted to assign exactly 11 trucks in the CV Fitness problem instead of a maximum of 12, you would select the equals (=) operator in the Change Constraint window, as shown in Figure D-51.

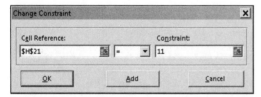

Source: Used with permission from Microsoft Corporation

FIGURE D-51   Constraining a value to a specific amount

Setting Changing Cells to Integers

Throughout the tutorial, you were directed to set the changing cells to integers in the Solver constraints. In many business situations, there is a logical reason for demanding integer solutions, but this approach does have disadvantages. Forcing integers can sometimes increase the amount of time Solver needs to find a feasible solution. In addition, Solver sometimes can find a solution using real numbers in the changing cells instead of integers. If Solver cannot find a feasible solution or reports that it has reached its calculation time limit, consider removing the integer constraints from the changing cells and rerunning Solver to see if it finds an optimal solution that makes sense.

Restarting Solver with New Constraints

Suppose you want to start over with a completely new set of constraints. In the Solver Parameters window, click Reset All. You will be asked to confirm that you want to reset all the Solver options and cell selections (see Figure D-52).

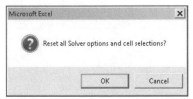

Source: Used with permission from Microsoft Corporation

FIGURE D-52   Reset options warning

If you want to clear all the Solver settings, click OK. An empty Solver Parameters window appears with all the former entries deleted, as shown in Figure D-53. You can then set up a new model.

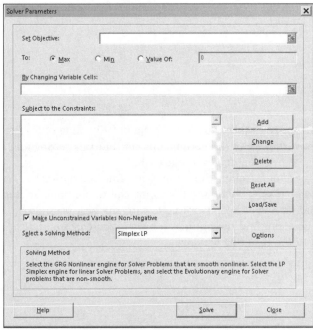

Source: Used with permission from Microsoft Corporation

FIGURE D-53   Solver Parameters window after selecting Reset All

> **NOTE**
>
> Be certain that you want to select the Reset All option before you use it. If you only want to edit, delete, or add a constraint, use the Add, Delete, or Change button for that constraint.

Using the Solver Options Window

Solver has several internal settings that govern its search for an optimal answer. Click the Options button in the Solver Parameters window to see the default selections for these settings, as shown in Figure D-54.

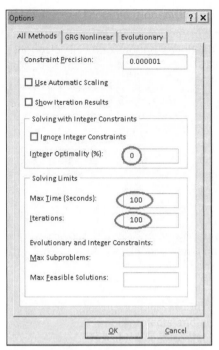

Source: Used with permission from Microsoft Corporation
FIGURE D-54   Solver Options window

You should not need to change the settings in the Options window except for the default value of 5% for Integer Optimality. When it is set at 5%, Solver will get within 5% of the optimal answer, but this setting might not give you the lowest cost or highest income. Change the setting to 0 and click OK.

In more complex problems that have a dozen or more constraints, Solver may not find the optimal solution within the default 100 seconds or 100 iterations. If so, a window will prompt you to continue or stop (see Figure D-55). If you have time, click Continue and let Solver keep working toward the best possible solution. If Solver works for several minutes and still does not find the optimal solution, you can stop by pressing the Ctrl and Break keys together. Click Stop in the resulting window.

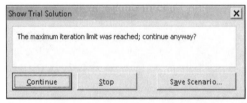

Source: Used with permission from Microsoft Corporation
FIGURE D-55   Prompt that appears when Solver reaches its maximum iteration limit

If you think that Solver needs more time and iterations to reach an optimal solution, you can increase the Max Time and Iterations, but you should probably keep both values under 32,000.

Printing Cell Formulas in Excel

Earlier in the tutorial, you learned how to display cell formulas in your spreadsheet cells. Hold down the Ctrl key and then press the ~ key (on most keyboards, this key is next to the "1" key). You can change the cell widths to see the entire formula by clicking and dragging the column by the dividing lines between the column letters. See Figure D-56.

	G	H	I	J	K	L	M	N
13								
14	Vehicle Loading							Cost
15	Volume Required	Trucks	Volume for Trucks	Tractor-Trailers	Volume for Tractor-Trailers	Total Vehicle Capacity	% of Vehicle Capacity Utilized	Shipping Cost
16	=D16*$C$8+E16*$C$9+F16*$C$10	1	=H16*$C$5	1	=J16*$C$6	=I16+K16	=G16/L16	=H16*C16*$D$5+J16*C16*$D$6
17	=D17*$C$8+E17*$C$9+F17*$C$10	1	=H17*$C$5	1	=J17*$C$6	=I17+K17	=G17/L17	=H17*C17*$D$5+J17*C17*$D$6
18	=D18*$C$8+E18*$C$9+F18*$C$10	1	=H18*$C$5	1	=J18*$C$6	=I18+K18	=G18/L18	=H18*C18*$D$5+J18*C18*$D$6
19	=D19*$C$8+E19*$C$9+F19*$C$10	1	=H19*$C$5	1	=J19*$C$6	=I19+K19	=G19/L19	=H19*C19*$D$5+J19*C19*$D$6
20	=D20*$C$8+E20*$C$9+F20*$C$10	1	=H20*$C$5	1	=J20*$C$6	=I20+K20	=G20/L20	=H20*C20*$D$5+J20*C20*$D$6
21	=SUM(G16:G20)	=SUM(H16:H20)	=SUM(I16:I20)	=SUM(J16:J20)	=SUM(K16:K20)	=SUM(L16:L20)	=G21/L21	=SUM(N16:N20)

Source: Used with permission from Microsoft Corporation

FIGURE D-56   Spreadsheet with formulas displayed in the cells

To print the formulas, click the File tab and select Print. To restore the screen to its normal appearance and display values instead of formulas, press Ctrl+~ again; the key combination is actually a toggle switch. If you changed the column widths in the formula view, you might have to resize the columns after you change back.

"Fatal" Errors in Solver

When you run Solver, you might sometimes receive a message like the one shown in Figure D-57.

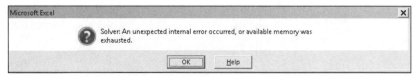

Source: Used with permission from Microsoft Corporation

FIGURE D-57   Fatal error in Solver

Solver usually attempts to find a solution or reports why it cannot. When Solver reports a fatal error, the root cause is difficult to troubleshoot. Possible causes include merged cells on the spreadsheet or printing multiple Answer Reports after running Solver multiple times. A common solution to this error has been to remove the Solver add-in, close Excel, reopen it, and then reinstall Solver. If you encounter a fatal error when using this book, check with your instructor.

Sometimes Solver will generate strange results. Even when your cell formulas and constraints match the ones your instructor has created, Solver's answers might not match the "book" answers. You might have entered your constraints into Solver in a different order, you may have changed some of the options in Solver, or you may have specified real numbers instead of integers for the constraints (or vice versa). Also, the solving method you selected and the amount of time you gave Solver to work can affect the final answer. If your solution is close to the one posted by your instructor, but not exactly the same, show the instructor your setup in the Solver Parameters window. Solver is a powerful tool, but it is not infallible—ask your instructor for guidance if necessary.

DUCHESS CRUISES SHIP ASSIGNMENT PROBLEM

Decision Support Using Microsoft Excel Solver

PREVIEW

Duchess Cruises is an international cruise line with headquarters in Long Beach, California. The company started in 1969 with one ship that offered a Caribbean cruise, and the business has grown steadily since. Duchess currently operates a fleet of 30 cruise ships and offers excursions to the Caribbean, Alaska, the Mediterranean Sea, the Baltic Sea, and Hawaii. The 30 ships are spread across five classes of different sizes and efficiency. Until recently, Duchess Cruises used manual methods to assign its ships to cruise destinations. Jane McDowall, the new vice president of operations, believes that Duchess Cruises is not optimizing its operating costs using the old method of assignment.

Jane is also investigating two new business opportunities for the company. The builder of the original Island Duchess models (the smallest and least efficient class) has approached Duchess Cruises with a proposal to retrofit the ships with new high-efficiency engines that can burn a variety of fuels. The new engines would significantly reduce operating costs for the Island Duchess class. In addition, the Marketing department has studied potential new markets and proposed that Duchess Cruises begin to offer Australian cruises.

You have been hired as an MIS consultant to develop a DSS model for Duchess Cruises. Your completed model will be used to assign the fleet to its cruise venues at a minimal cost. You will also modify the model to examine how upgrading the Island Duchess ships will affect operating costs and determine whether Duchess Cruises can expand into Australia with its current fleet.

PREPARATION

- Review the spreadsheet concepts discussed in class and in your textbook.
- Your instructor may assign Excel exercises to help prepare you for this case.
- Tutorial D explains how to set up and use Solver for maximization and minimization problems. The CV Fitness exercise in the tutorial should be particularly helpful.
- Review the file-saving instructions in Tutorial C—saving an extra copy of your work on a USB thumb drive or online is always a good idea.
- Review Tutorial F to brush up on your presentation skills.

BACKGROUND

You will use your Excel skills to build a decision model and determine how many ships from each class should be assigned to the Duchess Cruises venues. The model requires the following data, which the management team has compiled:

- Data for the five classes of ships in the Duchess Cruises fleet:
  - Gross tonnage
  - Passenger capacity
  - Number of balcony berths, which are staterooms with an outside balcony

- Sailing cost per nautical mile and per passenger-mile
- Number of ships in each class
- Ticket pricing for each cruise venue
- Balcony upgrade pricing for each cruise venue
- Weekly docking fees per gross ton for each cruise venue
- Length of each cruise venue in nautical miles

The Marketing department has also given you the following information about passenger and balcony upgrade demand for each cruise venue:

- The expected weekly passenger bookings
- The expected weekly balcony upgrades booked

To satisfy the booking demand, your Solver model will assign the needed number of ships by class to each cruise venue while minimizing total weekly operating costs. The model will also calculate total weekly revenues from passenger tickets and balcony upgrades as well as the total weekly operating costs.

Duchess Cruises Fleet

Duchess Cruises started operations with a single Island Duchess ship and added eight more ships in the same class as the business grew. Over the next 40 years, Duchess added four more classes, each larger and more luxurious, and with more efficient sailing costs:

- Coral Duchess, five ships
- Emerald Duchess, nine ships
- Grand Duchess, four ships
- Millennium Duchess, three ships

The Island Duchess ships have been well maintained and updated, but they have the least efficient engines of the five classes. Duchess Cruises is considering the shipbuilder's offer to update the power plants of the Island Duchess class.

ASSIGNMENT 1: CREATING SPREADSHEET MODELS FOR DECISION SUPPORT

In this assignment, you create spreadsheets that model the business decision Duchess Cruises is seeking. In Assignment 1A, you create a spreadsheet and attempt to assign the fleet manually to minimize the total weekly operating cost. In Assignment 1B, you copy the spreadsheet to a new worksheet and then set up and run Solver to minimize the total weekly operating cost. In Assignment 1C, you copy the Solver solution to a new worksheet and modify it to evaluate whether the company should upgrade the Island Duchess class of ships. You then rerun Solver to determine if the upgrades further improve the total weekly operating cost and the upgrade investment meets the company's goals for payback and internal rate of return. In Assignment 1D, you copy the modified Solver spreadsheet, further modify it to add Australia as a cruise venue, and then determine whether additional ships will be needed to service the new market and which ship class will provide the most cost-efficient solution.

This section helps you set up each of the following spreadsheet components before entering the cell formulas:

- Constants
- Calculations and Results
- Income Statement

The Calculations and Results section is the heart of the decision model. You will set up sections for length of cruise, weekly demand, ship assignments by class, capacity and utilization, and costs. The spreadsheet rows represent the cruise venues. The Ship Assignments section will hold the range of changing cells for Solver to manipulate. The Total Cruise Costs cell serves as the optimization cell for all the assignments. The Income Statement section calculates gross profit, which is required to calculate the positive cash flows in the Investment Analysis section. You add this section in Assignment 1C to calculate simple payback and internal rate of return for the Island Duchess engine upgrades.

Assignment 1A: Creating the Spreadsheet—Base Case

A discussion of each spreadsheet section follows. This information helps you set up each section of the model and learn the logic of the formulas in the spreadsheet. If you choose to enter the data directly, follow the cell structure shown in the figures. *You can also download the spreadsheet skeleton if you prefer.* To access the base spreadsheet skeleton, select Case 8 from your data files, and then select **Duchess Cruises Skeleton.xlsx.**

Constants Section

First, build the skeleton of your spreadsheet. Set up the spreadsheet title and Constants section, as shown in Figure 8-1. An explanation of the column items follows the figure.

	A	B	C	D	E	F	G	H	I	J	K	L
1		**Duchess Cruises Ship Assignment Problem**										
2												
3		Constants Section:										
4		Ship Data:										
5		Ship Class	Gross Tonnage	Passenger Capacity	Number of Balcony Berths	Sailing Cost per Nautical Mile	Sailing Cost per Passenger-Mile	Number of ships in class		Fill Legend		
6		Island Duchess	75,000	1500	700	$1,700.00	$1.133	9			Changing Cells	
7		Coral Duchess	90,000	1900	1200	$2,000.00	$1.053	5			Optimization Cell	
8		Emerald Duchess	94,000	2220	1300	$2,200.00	$0.991	9				
9		Grand Duchess	110,000	3000	1600	$2,800.00	$0.933	4				
10		Millennium Duchess	120,000	3450	2000	$3,000.00	$0.870	3				
11		Ticket Prices/Docking Fees by Destination:										
12		Cruise Location	Average Ticket Price	Balcony Upgrade Price	Weekly Docking Fee per Gross Ton							
13		Eastern Caribbean	$1,000	$400.00	$1.00							
14		Western Caribbean	$900	$300.00	$0.80							
15		Alaska	$1,800	$600.00	$2.00							
16		Mediterranean	$2,500	$800.00	$4.00							
17		Baltic	$3,600	$1,000.00	$6.00							
18		Hawaii	$1,800	$600.00	$3.00							

Source: Used with permission from Microsoft Corporation

FIGURE 8-1   Spreadsheet title and Constants section

- Spreadsheet title—Enter the spreadsheet title in cell B1, and then merge and center the title across cells B1 through H1.
- Constants Section and Ship Data table—Enter the column headings shown in cells B5 through H5.
- Ship Class—Enter each of the five ship classes listed in cells B6 through B10.
- Gross Tonnage—Enter each of the gross tonnages listed in cells C6 through C10.
- Passenger Capacity—Enter each of the five passenger capacities listed in cells D6 through D10.
- Number of Balcony Berths—This value is the total number of passenger berths for cabins that have an outside balcony. Balcony cabins are in high demand for cruise venues with scenic routes, such as the Inside Passage of Alaska. Enter each of the five values in cells E6 through E10.
- Sailing Cost per Nautical Mile—This value is the estimated cost of fuel, supplies, food, labor, and maintenance per nautical mile sailed for each of the five ship classes. Enter each of the five values listed in cells F6 through F10.
- Sailing Cost per Passenger-Mile—This value is the Sailing Cost per Nautical Mile divided by the passenger capacity of the ship class. Normally, you do not include formulas in the Constants section, but this column lets you quickly see the cost efficiency of each ship class. Each ship class has been more cost efficient than its predecessor. Also, you will examine an engine upgrade to the Island Duchess class later; the formula will automatically update its value when you enter the new Sailing Cost per Nautical Mile given in the upgrade. Enter the formula for each of the five classes in cells G6 through G10. To save time, write the formula for cell G6, and then copy the formula to cells G7 through G10.

- Number of ships in class—Duchess Cruises has five classes of ships. The ships in a class have a common size and design, and operate at a common degree of efficiency. Enter the number of ships in each class in cells H6 through H10.
- Ticket Prices/Docking Fees by Destination—Enter the column headings shown in cell B11 and cells B12 through E12.
- Cruise Location—Enter the six cruise venues in cells B13 through B18.
- Average Ticket Price—Enter the passenger ticket prices for the six cruise venues in cells C13 through C18.
- Balcony Upgrade Price—This value is the amount each passenger pays for a balcony cabin upgrade. Enter the price for the six cruise venues in cells D13 through D18.
- Weekly Docking Fee per Gross Ton—Each port that a cruise ship visits charges a daily fee for docking. Some docking fees are based on the length of the ship, but some ports charge according to the ship's gross tonnage. This value is the total docking fees per ton for all the ports visited on the cruise. Enter these values in cells E13 through E18.
- Fill Legend—This section helps you identify the changing cells and the optimization cell in the Calculations and Results section. Enter "Fill Legend" in cell J5, fill cell J6 in yellow, fill cell J7 in orange, and then enter "Changing Cells" in cell K6 and "Optimization Cell" in cell K7.

Calculations and Results Section

The Calculations and Results section (see Figure 8-2) contains the cruise lengths, weekly passenger bookings, and weekly balcony upgrade bookings obtained from the Marketing department. Although these values are constants, keeping them in the Calculations and Results section facilitates writing and copying the formulas in the Ship Assignments, Passenger Capacity, Balcony Capacity, and Costs columns. This section includes the Ship Assignments table, which contains the changing cells; calculations for passenger and balcony berth utilization; and cost calculations. An explanation of these sections and columns follows the figure.

	A	B	C	D	E	F	G	H	I	J	K	L	M	N	O	P	Q
19																	
20		Calculations and Results Section:		Weekly Demand				Ship Assignments			Passenger Capacity		Balcony Capacity		Costs		
21		Cruise Location	Length of Cruise Nautical Miles	Weekly Passenger Bookings	Weekly Balcony Upgrades Booked	Island Duchesses Assigned	Coral Duchesses Assigned	Emerald Duchesses Assigned	Grand Duchesses Assigned	Millennium Duchesses Assigned	Total Passenger Capacity	% of Passenger Capacity Utilized	Total Balcony Capacity	% of Balcony Capacity Utilized	Weekly Docking Fees	Weekly Sailing Costs	
22		Eastern Caribbean	900	10150	5200	1	1	1	1	1							
23		Western Caribbean	1000	9862	4750	1	1	1	1	1							
24		Alaska	1200	11162	7600	1	1	1	1	1							
25		Mediterranean	1100	8430	5120	1	1	1	1	1							
26		Baltic	1400	6300	3150	1	1	1	1	1							Total Cruise
27		Hawaii	850	8560	5130	1	1	1	1	1							Costs
28			Total/Avg														

Source: Used with permission from Microsoft Corporation

FIGURE 8-2   Calculations and Results section

- Table Headings—If you did not use the skeleton spreadsheet, enter the column headings shown in cells B20 through P21 in Figure 8-2. Note that the section headings in row 20 have merged and centered cells.
- Cruise Location—Cells B22 through B27 hold the names of the six cruise venues offered by Duchess Cruises.
- Length of Cruise Nautical Miles—Cells C22 through C27 hold the length of each cruise in nautical miles.
- Weekly Passenger Bookings—Cells D22 through D27 hold the average number of passenger tickets booked each week for the six cruise venues.
- Weekly Balcony Upgrades Booked—Cells E22 through E27 hold the average number of balcony upgrades booked each week for the six cruise venues.
- Ship Assignments section—Cells F20 through J27 are the heart of the Solver model: the changing cells. The cells hold the number of ships in each class that Solver will assign to the six cruise venues. Enter "1" in each of these cells for now, and fill the cells with a background color to indicate that they are the changing cells for Solver. To fill the cells, select them and then click the Fill Color button in the Font group on the Home tab. (The button's icon is a can of paint.) In the spreadsheet skeleton, the fill color is yellow.
- Total Passenger Capacity—Cells K22 through K27 hold the total passenger capacity for each cruise location. This value is calculated by multiplying the number of ships in each class by its

passenger capacity listed in the Constants section (cells D6 through D10), then summing the total capacities for the five classes of ships. Be sure to use *absolute* cell references for the passenger capacity values from the Constants section so that you have to write the formula only for the first cell (K22); you can then copy the formula to cells K23 through K27.

- % of Passenger Capacity Utilized—Cells L22 through L27 hold the percentage of passenger capacity used for each cruise venue. The value is calculated by dividing Weekly Passenger Bookings by Total Passenger Capacity. This column provides a quick visual check to ensure that cruises are not overbooked.

- Total Balcony Capacity—Cells M22 through M27 hold the total balcony capacity for each cruise location. This value is calculated by multiplying the number of ships in each class by its balcony berth capacity listed in the Constants section (cells E6 through E10), then summing the total balcony capacities for the five classes of ships. Be sure to use *absolute* cell references for the balcony berth capacity values from the Constants section so that you have to write the formula only for the first cell (M22); you can then copy the formula to cells M23 through M27.

- % of Balcony Capacity Utilized— Cells N22 through N27 hold the percentage of balcony capacity used for each cruise venue. The value is calculated by dividing Weekly Balcony Upgrades Booked by Total Balcony Capacity. This column provides a quick visual check to ensure that cruise balcony upgrades are not overbooked.

- Weekly Docking Fees—Cells O22 through O27 hold the weekly docking fees charged to all ships assigned to each cruise location. The docking fees are calculated by summing the multiplied products of the number of ships assigned in each class and their respective gross tonnage, then multiplying that sum by the Weekly Docking Fee per Gross Ton in the Constants section. Remember to use *absolute* cell references for the gross tonnages listed in the Constants section (cells C6 through C10) so that you have to write the formula only for the first cell (O22). You can then copy the formula to cells O23 through O27.

- Weekly Sailing Costs—Cells P22 through P27 hold the weekly sailing costs charged to all ships assigned to each cruise location. These costs are calculated by summing the multiplied products of the number of ships assigned in each class and their respective Sailing Cost per Nautical Mile, then multiplying that sum by the Length of Cruise in Nautical Miles. Remember to use *absolute* cell references for the Sailing Cost per Nautical Mile listed in the Constants section (cells F6 through F10) so that you have to write the formula only for the first cell (P22). You can then copy the formula to cells P23 through P27.

- Total/Avg—In row 28, total the cell entries for rows 22 through 27 in every column *except* column L (% of Passenger Capacity Utilized) and column N (% of Balcony Capacity Utilized), and place the totals in cells D28 through K28, M28, O28, and P28. Cells L28 and N28 are averages, not totals. For cell L28, divide the total Weekly Passenger Bookings (cell D28) by the Total Passenger Capacity (cell K28) and format the result as a percentage. Similarly, for cell N28, divide the total Weekly Balcony Upgrades Booked (cell E28) by the Total Balcony Capacity (cell M28), and format the result as a percentage. These two averages are the overall utilization rates of the cruise fleet. Ideally, these rates should be as close as possible to 100 percent for lowest total cost.

- Total Cruise Costs—Cell Q28 holds the total costs, which are the sum of the total Weekly Docking Fees (cell O28) and the Total Weekly Sailing Costs (cell P28). Cell Q28 is the optimization cell. Fill the cell background with the same color you used previously for cell J7.

If you wrote your formulas correctly, the Calculations and Results section should look like Figure 8-3.

A	B	C	D	E	F	G	H	I	J	K	L	M	N	O	P	Q
19																
20	Calculations and Results Section:		Weekly Demand				Ship Assignments				Passenger Capacity		Balcony Capacity		Costs	
21	Cruise Location		Length of Cruise Nautical Miles	Weekly Passenger Bookings	Weekly Balcony Upgrades Booked	Island Duchesses Assigned	Coral Duchesses Assigned	Emerald Duchesses Assigned	Grand Duchesses Assigned	Millennium Duchesses Assigned	Total Passenger Capacity	% of Passenger Capacity Utilized	Total Balcony Capacity	% of Balcony Capacity Utilized	Weekly Docking Fees	Weekly Sailing Costs	
22	Eastern Caribbean		900	10150	5200	1	1	1	1	1	12070	84%	6800	76%	$489,000	$10,530,000	
23	Western Caribbean		1000	9862	4750	1	1	1	1	1	12070	82%	6800	70%	$391,200	$11,700,000	
24	Alaska		1200	11162	7600	1	1	1	1	1	12070	92%	6800	112%	$978,000	$14,040,000	
25	Mediterranean		1100	8430	5120	1	1	1	1	1	12070	70%	6800	75%	$1,956,000	$12,870,000	
26	Baltic		1400	6300	3150	1	1	1	1	1	12070	52%	6800	46%	$2,934,000	$16,380,000	Total Cruise
27	Hawaii		850	8560	5130	1	1	1	1	1	12070	71%	6800	75%	$1,467,000	$9,945,000	Costs
28		Total/Avg		54464	30950	6	6	6	6	6	72420	75%	40800	76%	$8,215,200	$75,465,000	$83,680,200

Source: Used with permission from Microsoft Corporation

FIGURE 8-3   Completed Calculations and Results section

Income Statement Section

The Income Statement section (see Figure 8-4) is actually a projection of weekly gross profits, and is based on a selection of ships that will be assigned either manually or by Solver. An explanation of the line items follows the figure.

	A	B	C
29			
30		Income Statement Section:	
31		Passenger Revenues	
32		Balcony Revenues	
33		Total Revenues	
34		less: Total Cruise Costs	
35		Weekly Gross Profit	

Source: Used with permission from Microsoft Corporation

FIGURE 8-4   Income Statement section

- Passenger Revenues—This value is calculated by multiplying the weekly passenger bookings for each cruise location (cells D22 through D27) by their respective average ticket prices (cells C13 through C18), then totaling the ticket revenues for the six cruise locations.
- Balcony Revenues—This value is calculated by multiplying the weekly balcony upgrades booked (cells E22 through E27) by their respective balcony upgrade prices (cells D13 through D18), then totaling the balcony upgrade revenues for the six cruise locations.
- Total Revenues—This value is the sum of Passenger Revenues and Balcony Revenues.
- Less: Total Cruise Costs—This value is the Total Cruise Costs from cell Q28.
- Weekly Gross Profit—This value is the Total Revenues minus the Total Cruise Costs. This cell could also be used as an optimization cell; if so, you would maximize the value rather than minimizing Total Cruise Costs. In this case, both cells yield the same optimal answer, so it is not necessary to use Solver for both maximization and minimization. However, you will use the Weekly Gross Profit value when examining the Australian cruise opportunity later in this case.

If your formulas are correct, the initial Income Statement section should look like Figure 8-5.

	A	B	C
29			
30		Income Statement Section:	
31		Passenger Revenues	$98,280,400
32		Balcony Revenues	$18,389,000
33		Total Revenues	$116,669,400
34		less: Total Cruise Costs	$83,680,200
35		Weekly Gross Profit	$32,989,200

Source: Used with permission from Microsoft Corporation

FIGURE 8-5   Initial Income Statement section

The initial Income Statement section correctly reflects the revenues expected from the passenger and balcony upgrade bookings, but the total cruise costs are not correct because the ships have not been assigned yet.

Attempting a Manual Solution

First, attempt to assign the fleet manually. You have several good reasons for doing this. For example, you can make sure that your model is working correctly before you set up Solver to run. Also, assigning the fleet manually demonstrates which constraints you must meet to solve the problem. For instance, if any of your passenger or balcony utilization rates are more than 100 percent, you have not assigned enough ships to a cruise location to accommodate demand. Therefore, one constraint is that the total passenger capacity for the ships assigned to a cruise location must be greater than the passenger bookings. Another constraint is that the total balcony capacity for the ships assigned must be greater than the balcony upgrades booked. Given the fleet size, you can probably assign the ships manually while meeting all of your constraints. However, are the resulting total cruise costs the least expensive solution? Running the problem manually provides an initial total cruise cost to which you can compare your Solver solution later. The Solver optimization tool should provide a less expensive solution than assigning the fleet manually.

When attempting to assign the ships manually in the Ship Assignments section (cells F22 through J27), you should assign the most cost-efficient ship classes first, which is why the Sailing Cost per Passenger-Mile data is included in the Constants section. Assign the three Millennium Duchesses first, then the four Grand Duchesses, and so on. You must satisfy *both* the passenger and balcony upgrade demands for each cruise location—in other words, the values in cells K22 through K27 and M22 through M27 must be equal to or greater than the values in cells D22 through D27 and E22 through E27, respectively. If you satisfy the passenger and balcony demands correctly, none of the utilization rates in cells L22 through L27 and N22 through N27 will exceed 100 percent. In addition, the total ships assigned for each class (cells F28 through J28) cannot exceed the number of ships in each class (cells H6 to H10). You might need some extra time to arrive at a solution that satisfies all the constraints, but don't get discouraged; the problem has many feasible solutions. In the next assignment, Solver will require more than 10 minutes of calculations to find an optimal solution.

When you reach a manual solution that satisfies the preceding constraints, save your workbook. Name the worksheet **Duchess Cruises Manual Solution**, and then right-click the worksheet name tab (see Figure 8-6). Click Move or Copy, and then copy the worksheet, as shown in Figure 8-7. Rename the new copy **Duchess Cruises Solver**. You will use the new worksheet to complete the next part of the assignment.

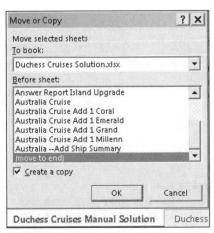

Source: Used with permission from Microsoft Corporation

FIGURE 8-6   Right-clicking the worksheet name tab

Source: Used with permission from Microsoft Corporation

FIGURE 8-7   Move or Copy menu

Assignment 1B: Setting Up and Running Solver

Before using the Solver Parameters window, you should jot down the parameters you must define and their cell addresses. Here is a suggested list:

- The cell you want to minimize (Total Cruise Costs, cell Q28)
- The cells you want Solver to manipulate to obtain the optimal solution (Ship Assignments, cells F22 through J27)

- The constraints you must define:

  - All the ship assignment cells must contain non-negative integers.
  - The total number of ships assigned from each class (cells F28 through J28) cannot exceed the number of ships in each class (cells H6 through H10).
  - The total passenger capacity for each cruise location (cells K22 through K27) must be equal to or greater than the weekly passenger bookings for each cruise location (cells D22 through D27).
  - The total balcony capacity for each cruise location (cells M22 through M27) must be equal to or greater than the weekly balcony upgrades booked (cells E22 through E27).

Next, set up your problem. In the Analysis group on the Data tab, click Solver; the Solver Parameters window appears, as shown in Figure 8-8. Enter Q28 in the Set Objective text box, or click the button to the right of the box and then click cell Q28. Click the Min button on the next line, designate your Changing Cells (cells F22 through J27), and add the constraints from the preceding list. Make sure that the "Make Unconstrained Variables Non-Negative" check box is selected, and leave Simplex LP selected as the default solving method. If you need help defining your constraints, refer to Tutorial D.

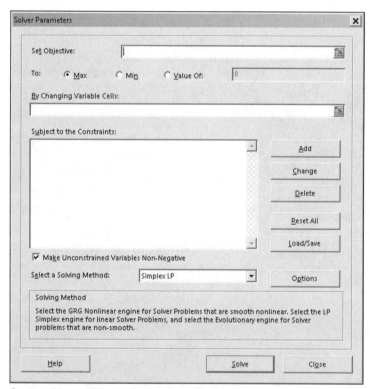

Source: Used with permission from Microsoft Corporation

FIGURE 8-8   The Solver Parameters window

Next, you should click the Options button and check the Options window that appears (see Figure 8-9). The default Integer Optimality is 5%; change it to 0% to get a better answer. Make sure that the Constraint Precision is set to the default value of .000001. With 30 changing cells, Solver will probably require 10 minutes of calculations to reach an optimal solution. To make sure that Solver does not exceed a time limit, enter 3600 for the Max Time (in seconds) so Solver can run for up to one hour. When you finish setting the options, click OK to return to the Solver Parameters window.

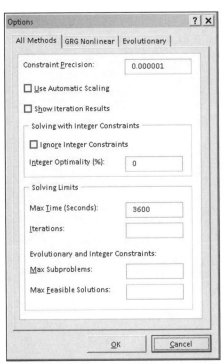

Source: Used with permission from Microsoft Corporation
FIGURE 8-9   The Solver Options window

Run Solver by clicking the Solve button. You can see the progress of the calculations by observing the green status bar at the bottom of the screen. Click Answer Report when Solver finds a solution that satisfies the constraints. When you finish, print the entire workbook, including the Solver Answer Report Sheet. To save the workbook, click the File tab and then click Save. For the rest of the case, you either can use the Save As command to create new Excel workbooks or continue copying and renaming the worksheets in one workbook. Both options offer distinct advantages, but having all of your worksheets and Answer Reports in one Excel workbook allows you to compare different solutions easily, as well as prepare summary reports.

Before continuing, examine the ship assignments that Solver chose for minimizing the total cost. If you set up Solver correctly, you should see a significant reduction in total cost from your manual assignment. You should also see that Solver assigned fewer ships, especially from the older Island Duchess class. This information means that Duchess Cruises can take some ships out of service and upgrade them to be more cost efficient, or it can add another cruise venue. You will explore those possibilities in the following assignments.

Assignment 1C: Engine Upgrades to the Island Duchess Class

Using the model you created in Assignment 1A, Duchess Cruises can also determine the cost benefit of upgrading the older ships in the Island Duchess class. The original shipbuilder of the class has approached the management team with a proposal to retrofit the nine older ships with new eco-efficient engines. The cost of the upgrade is $20 million per ship; the new engines will add 5,000 tons to each ship's gross tonnage, but they are designed to make the Sailing Cost per Passenger-Mile lower than that of any other ship class in service. Spending $180 million represents a major investment for Duchess Cruises, whose management team requires a payback of less than two years with an internal rate of return of more than 30 percent. You have been asked to modify the worksheet to determine whether Duchess Cruises should agree to the proposed upgrade. Recalculate the total cruise costs with Solver and determine the payback and internal rate of return.

Copy your Duchess Cruises Solver worksheet, and rename the copy **Island Duchess Upgrade Solver**. Before running Solver again, you must perform the following tasks:

- Insert the new gross tonnage for the Island Duchess class (80,000 tons) into cell C6, and the new Sailing Cost per Nautical Mile ($1,200) into cell F6 of the Constants section, as shown in Figure 8-10.
- Build a new section titled Investment Analysis Section at the bottom of the spreadsheet, as shown in Figure 8-11.

Source: Used with permission from Microsoft Corporation

FIGURE 8-10   The changes in the Constants section for the Island Duchess upgrade

Source: Used with permission from Microsoft Corporation

FIGURE 8-11   The Investment Analysis section added to the worksheet

The Investment Analysis section enables your model to calculate the simple payback of the investment in years and the investment's internal rate of return. Some simplifications have been made to the concepts of cash flow. For instance, this model assumes that the weekly gross profit actually represents cash inflow—in reality, non-cash operating expenses, such as depreciation of the ships' asset value, would be added back to get actual cash inflows. Also, any administrative expenses directly attributed to the investment would be subtracted from the cash inflows. This case has not included administrative expenses in calculations for the Income Statement section.

The Investment Analysis section contains the following cell entries:

- Cost of Upgrades—This value is the amount of money required to upgrade all nine ships of the Island Duchess class. The amount should be entered as a negative number to represent a cash outflow.
- Old Weekly Gross Profit—This value is the Weekly Gross Profit from cell C35 of the Duchess Cruises Solver worksheet in the previous section (Assignment 1B).
- New Weekly Gross Profit—This value is transferred from cell C35 in the Income Statement section of the current worksheet.
- Gain in Weekly Profit—This value is the New Weekly Gross Profit minus the Old Weekly Gross Profit.
- Annual Profit Gain (50 wk)—Because Duchess Cruises operates 50 weeks a year, this value is the Gain in Weekly Profit multiplied by 50.
- Payback Period—For equal annual cash inflows, the payback period is the investment (cell C38, Cost of Upgrades) divided by the annual cash inflows (cell C42, Annual Profit Gain). The payback period is a simple screening tool used in capital budgeting. Duchess Cruises will not seriously consider an investment that has a payback period of more than two years.
- Cash Outflow—This value is taken from cell C38 (Cost of Upgrades).
- Cash Inflow Year 1—This value is taken from cell C42, which is Annual Profit Gain (50 wk).

- Cash Inflow Year 2—This value is also taken from cell C42. Duchess Cruises is not considering cash inflows beyond Year 2 for this investment.
- IRR—The internal rate of return is the result of applying the IRR formulas to cells C44 through C46. To obtain the IRR, enter the following into cell C47: =IRR(C44:C46).

If you built the Investment Analysis section correctly, it should look like Figure 8-12. Your values for Old Weekly Gross Profit could be slightly different from those shown here, depending on how long you let Solver run to obtain an optimal solution in Assignment 1B.

	A	B	C	D	E	F	G	H	I
29									
30		Income Statement Section:							
31		Passenger Revenues	$98,280,400						
32		Balcony Revenues	$18,389,000						
33		Total Revenues	$116,669,400						
34		less: Total Cruise Costs	$63,203,200						
35		Weekly Gross Profit	$53,466,200						
36									
37		Investment Analysis Section:							
38		Cost of Upgrades	($180,000,000)	Energy Efficient Engine upgrade cost is $20,000,000 per Island Duchess					
39		*Old Weekly Gross Profit	$51,139,200	*This value taken from the original Solver solution					
40		New Weekly Gross Profit	$53,466,200						
41		Gain in Weekly Profit	$2,327,000		NOTE: Solver has *not* been run yet				
42		Annual Profit Gain (50 wk)	$116,350,000		on this worksheet. These numbers				
43		Payback Period (years)	1.55		are only the result of entering the				
44		Cash Outflow	($180,000,000)		formulas for the cells correctly.				
45		Cash Inflow Year 1	$116,350,000						
46		Cash Inflow Year 2	$116,350,000						
47		IRR	18.97%						
48									
49									
50									
51									

Source: Used with permission from Microsoft Corporation

FIGURE 8-12   The Investment Analysis section with formulas added (before running Solver)

Because the Sailing Cost per Nautical Mile has been changed in the Constants section for the upgraded Island Duchess ships, you must rerun Solver to obtain a new optimal ship assignment. Allow Solver at least 10 minutes to arrive at an optimal solution.

Examine the new Solver solution, and compare the total operating costs and gross profit to the same values from the earlier Solver solution. Does the solution meet the goals of less than two years payback and an IRR of greater than 30 percent? Also, note that Solver has redistributed the fleet to different cruise venues. Do you see a difference between how many Island Duchess ships are assigned now and how many were assigned with the original Solver worksheet?

Assignment 1D: Adding Australia to the Cruise Venues

Assume that Duchess Cruises has completed the engine upgrades to the Island Duchess class. Copy your Island Duchess Upgrade Solver worksheet, and rename the new worksheet **Australia Cruise**. You will use this worksheet to add Australia to the company's list of cruise locations and determine whether Duchess Cruises has enough ship capacity to handle the new demand.

You should remove the Investment Analysis section from the bottom of the new worksheet. Click and drag the mouse down the row numbers on the left margin of the worksheet, then right-click to see the context menu shown in Figure 8-13. Select Delete to delete the appropriate rows.

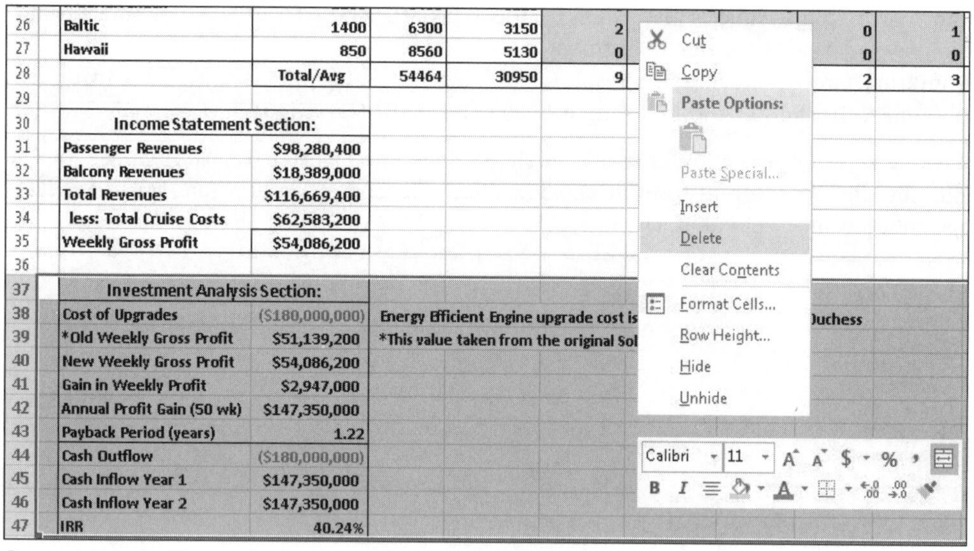

26	Baltic		1400	6300	3150	2		0		1
27	Hawaii		850	8560	5130	0		0		0
28		Total/Avg		54464	30950	9		2		3
29										
30		Income Statement Section:								
31	Passenger Revenues	$98,280,400								
32	Balcony Revenues	$18,389,000								
33	Total Revenues	$116,669,400								
34	less: Total Cruise Costs	$62,583,200								
35	Weekly Gross Profit	$54,086,200								
36										
37		Investment Analysis Section:								
38	Cost of Upgrades	($180,000,000)	Energy Efficient Engine upgrade cost is					uchess		
39	*Old Weekly Gross Profit	$51,139,200	*This value taken from the original Sol							
40	New Weekly Gross Profit	$54,086,200								
41	Gain in Weekly Profit	$2,947,000								
42	Annual Profit Gain (50 wk)	$147,350,000								
43	Payback Period (years)	1.22								
44	Cash Outflow	($180,000,000)								
45	Cash Inflow Year 1	$147,350,000								
46	Cash Inflow Year 2	$147,350,000								
47	IRR	40.24%								

Context menu items: Cut, Copy, Paste Options:, Paste Special..., Insert, Delete, Clear Contents, Format Cells..., Row Height..., Hide, Unhide

Source: Used with permission from Microsoft Corporation

FIGURE 8-13 Deleting the Investment Analysis section from the worksheet

Next, you must modify the worksheet to add the Australian cruise venue and costs. Add another row to the Ticket Prices/Docking Fees section and the Calculations and Results section, and then enter the data shown in Figure 8-14. To display the entire sheet on your screen, you can delete the empty row between the spreadsheet title and the Constants section.

	A	B	C	D	E	F	G	H	I
1		Duchess Cruises Ship Assignment Problem--Australia Cruise Added							
2		Constants Section:							
3		Ship Data:							
4		Ship Class	Gross Tonnage	Passenger Capacity	Number of Balcony Berths	Sailing Cost per Nautical Mile	Sailing Cost per Passenger-Mile	Number of ships in class	Fill Legend
5		Island Duchess	80,000	1500	700	$1,200.00	$0.800	9	Changing Cells
6		Coral Duchess	90,000	1900	1200	$2,000.00	$1.053	5	Optimization Cell
7		Emerald Duchess	94,000	2220	1300	$2,200.00	$0.991	9	
8		Grand Duchess	110,000	3000	1600	$2,800.00	$0.933	4	
9		Millennium Duchess	120,000	3450	2000	$3,000.00	$0.870	3	
10		Ticket Prices/Docking Fees by Destination:							
11		Cruise Location	Average Ticket Price	Balcony Upgrade Price	Weekly Docking Fee per Gross Ton				
12		Eastern Caribbean	$1,000	$400	$1.00				
13		Western Caribbean	$900	$300	$0.80				
14		Alaska	$1,800	$600	$2.00				
15		Mediterranean	$2,500	$800	$4.00				
16		Baltic	$3,600	$1,000	$6.00				
17		Hawaii	$1,800	$600	$3.00				
18		Australia	$2,700	$900	$4.00				
19									

Calculations and Results Section:

Row	Cruise Location	Length of Cruise Nautical Miles	Weekly Passenger Bookings	Weekly Balcony Upgrades Booked	Island Duchesses Assigned	Coral Duchesses Assigned	Emerald Duchesses Assigned	Grand Duchesses Assigned	Millennium Duchesses Assigned	Total Passenger Capacity	% of Passenger Capacity Utilized	Total Balcony Capacity	% of Balcony Capacity Utilized	Weekly Docking Fees	Weekly Sailing Costs	
22	Eastern Caribbean	900	10150	5200	3	0	0	2	0	10500	97%	5300	98%	$460,000	$8,280,000	
23	Western Caribbean	1000	9862	4750	4	1	1	0	0	10120	97%	5300	90%	$403,200	$9,000,000	
24	Alaska	1200	11162	7600	0	2	4	0	0	12680	88%	7600	100%	$1,112,000	$15,360,000	
25	Mediterranean	1100	8430	5120	0	1	0	0	2	8800	96%	5200	98%	$1,320,000	$8,800,000	
26	Baltic	1400	6300	3150	2	0	0	0	1	6450	98%	3400	93%	$1,680,000	$7,560,000	
27	Hawaii	850	8560	5130	0	0	4	0	0	8880	96%	5200	99%	$1,128,000	$7,480,000	Total Cruise
28	Australia	1350	9000	5200	0	0	0	0	0	0	#DIV/0!	0	#DIV/0!	$0	$0	Costs
29	Total/Avg		63464	36150	9	4	9	2	3	57430	111%	32000	113%	$6,103,200	$56,480,000	$62,583,200

Source: Used with permission from Microsoft Corporation

FIGURE 8-14 Worksheet modified to add Australian cruise venue

Copy the formulas from row 27 down to row 28. If you enter all zeros for the cruise ships in row 28, a #DIV/0! (division by zero) error appears for the utilization rates in cells L28 and N28. You can fix the error later. Next, you must edit the Total/Avg formulas in cells D29 through K29, M29, O29, and P29 to include the values in the cells of row 28.

You then must assign ships to the Australian cruise venue. Your Solver solution from the Island Duchess upgrade should have left three unassigned ships: one Coral Duchess and two Grand Duchesses. Assign those ships in the Ship Assignments section, as shown in Figure 8-15.

		Weekly Demand		Ship Assignments					Passenger Capacity		Balcony Capacity		Costs		
Calculations and Results Section:		Length of Cruise Nautical Miles	Weekly Passenger Bookings	Weekly Balcony Upgrades Booked	Island Duchesses Assigned	Coral Duchesses Assigned	Emerald Duchesses Assigned	Grand Duchesses Assigned	Millennium Duchesses Assigned	Total Passenger Capacity	% of Passenger Capacity Utilized	Total Balcony Capacity	% of Balcony Capacity Utilized	Weekly Docking Fees	Weekly Sailing Costs	
Cruise Location																
Eastern Caribbean	900	10150	5200	3	0	0	2	0	10500	97%	5300	98%	$460,000	$8,280,000	
Western Caribbean	1000	9862	4750	4	1	1	0	0	10120	97%	5300	90%	$403,200	$9,000,000	
Alaska	1200	11162	7600	0	2	4	0	0	12680	88%	7600	100%	$1,112,000	$15,360,000	
Mediterranean	1100	8430	5120	0	1	0	0	2	8800	96%	5200	98%	$1,320,000	$8,800,000	
Baltic	1400	6300	3150	2	0	0	0	1	6450	98%	3400	93%	$1,680,000	$7,560,000	
Hawaii	850	8560	5130	0	0	4	0	0	8880	96%	5200	99%	$1,128,000	$7,480,000	Total Cruise
Australia	1350	9000	5200	0	1	0	2	0	7900	114%	4400	118%	$1,240,000	$10,260,000	Costs
	Total/Avg	63464	36150	9	5	9	4	3	65330	97%	36400	99%	$7,343,200	$66,740,000	$74,083,200

Income Statement Section:	
Passenger Revenues	$122,580,400
Balcony Revenues	$23,069,000
Total Revenues	$145,649,400
less: Total Cruise Costs	$74,083,200
Weekly Gross Profit	$71,566,200

One Coral Duchess ship and two Grand Duchess ships are available to serve the Australian cruise location.

Source: Used with permission from Microsoft Corporation

FIGURE 8-15　Ships added to the Australian cruise venue

When you open Solver, the parameters window should update any row changes to the existing constraints. However, because you have added a new row to the changing cells, you must edit the line shown under By Changing Variable Cells and update the constraints for integer and non-negative numbers, as shown in Figure 8-16. You must also add the passenger and balcony capacity constraints for row 28, as shown in Figure 8-17.

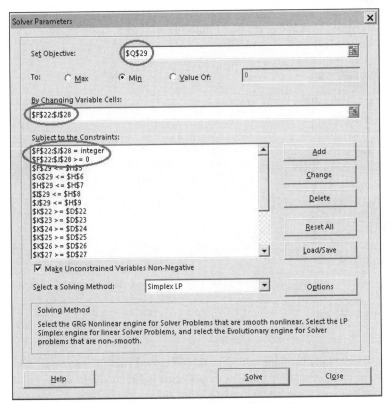

Source: Used with permission from Microsoft Corporation

FIGURE 8-16　Parameters and constraints

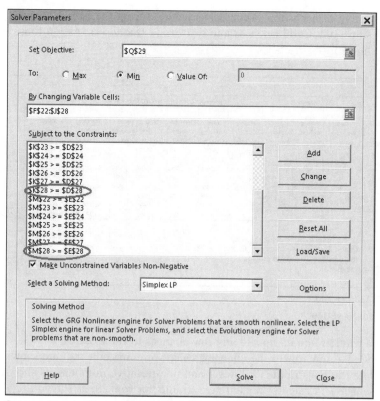

Source: Used with permission from Microsoft Corporation

FIGURE 8-17   Constraints for Australia passenger and balcony capacity

When you run Solver, it will quickly determine that it cannot find a feasible solution, as shown in Figure 8-18.

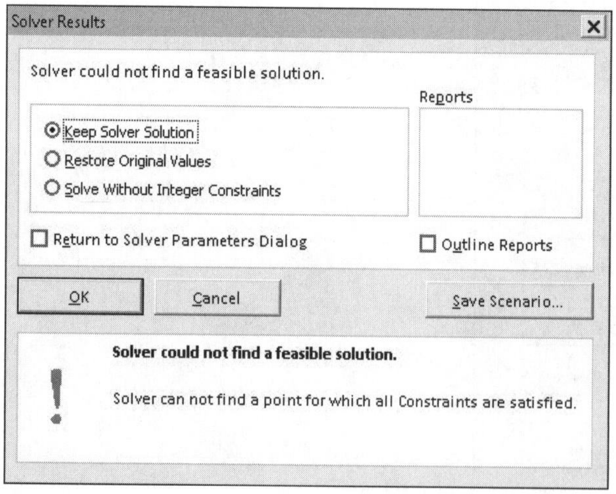

Source: Used with permission from Microsoft Corporation

FIGURE 8-18   Solver cannot find a feasible solution

Examine the worksheet. Even with the three available ships added to the proposed Australian venue, both the passenger and balcony utilization rates are well above 100 percent (see Figure 8-15). It might seem pointless to have run Solver for this scenario, but a slim possibility existed that Solver could have rearranged the ships to find a feasible solution. Given these results, you know that Duchess Cruises would have to add at least one more ship to its fleet to service the Australian cruise venue. But, which ship would give Duchess Cruises the lowest costs and highest gross profit? Solver can help determine the best ship to add to the fleet.

Copy the Australia Cruise worksheet and rename the copy **Australia Cruise—Add 1 Island**. Add one more Island Duchess ship to the number of ships in the class, as shown in Figure 8-19.

	Ship Class	Gross Tonnage	Passenger Capacity	Number of Balcony Berths	Sailing Cost per Nautical Mile	Sailing Cost per Passenger-Mile	Number of ships in class		Fill Legend	
1	Duchess Cruises Ship Assignment Problem--Australia Cruise--add one Island Duchess									
2	Constants Section:									
3	Ship Data:									
4										
5	Island Duchess	80,000	1500	700	$1,200.00	$0.800	10			Changing Cells
6	Coral Duchess	90,000	1900	1200	$2,000.00	$1.053	5			Optimization Cell
7	Emerald Duchess	94,000	2220	1300	$2,200.00	$0.991	9			
8	Grand Duchess	110,000	3000	1600	$2,800.00	$0.933	4			
9	Millennium Duchess	120,000	3450	2000	$3,000.00	$0.870	3			

Source: Used with permission from Microsoft Corporation

FIGURE 8-19   Adding one Island Duchess ship to the fleet in the Constants section

Rerun Solver to find a good solution. Copy the Australia Cruise worksheet four more times to add one Coral Duchess ship, one Emerald Duchess, one Grand Duchess, and one Millennium Duchess, respectively. Run Solver to evaluate each of these four alternatives, and then create a new worksheet named **Australia—Add Ship Summary**. Format the sheet as shown in Figure 8-20.

	Ship to be added:	Cruise Line Weekly Total Cruise Cost	Cruise Line Weekly Gross Profit	Extra Ships Made Available (not assigned by Solver solution)			
1	Summary of Financial Results--Additional Ship Selected for Australia Cruise Expansion						
2							
3							
4	Island Duchess						
5	Coral Duchess						
6	Emerald Duchess						
7	Grand Duchess						
8	Millennium Duchess						

Source: Used with permission from Microsoft Corporation

FIGURE 8-20   Australia—Add Ship Summary worksheet

Enter the weekly total cruise cost and weekly gross profit for each of the five alternatives into the Add Ship Summary worksheet. The last column, "Extra Ships Made Available (not assigned by Solver solution)," is important as well. Check the ship totals for each class at the bottom of the five alternative worksheets. For some of the alternatives, Solver did not assign one ship to a class. This information could be important when you consider whether to add a smaller or larger ship to the fleet, as it provides future capacity for increased demand. Enter any ships made available by the Solver solution into the last column of the Add Ship Summary worksheet.

ASSIGNMENT 2: USING THE WORKBOOK FOR DECISION SUPPORT

You have built a series of worksheets to determine the best ship assignment solutions with and without the Australian cruise venue. You now complete the case by using your solutions and Answer Reports to make your recommendations in a memorandum.

Use Microsoft Word to write a memo to the management team at Duchess Cruises. State the results of your analysis, your recommendations about adding the Australian cruise venue, and your recommendations for which ship should be added to the Duchess fleet to service the venue.

- Set up your memo as described in Tutorial E.
- In the first paragraph, briefly describe the situation and state the purpose of your analysis.
- Next, summarize the results of your analysis and state your recommendations.
- Support your recommendation with appropriate screen shots or Excel objects from the Excel workbook. (Tutorial C explains how to copy and paste Excel objects.)

ASSIGNMENT 3: GIVING AN ORAL PRESENTATION

Your instructor may request that you summarize your analysis and recommendations in an oral presentation. If so, prepare a presentation for Jane McDowall and other managers that lasts 10 minutes or less. When preparing your presentation, use PowerPoint slides or handouts that you think are appropriate. Tutorial F explains how to prepare and give an effective oral presentation.

DELIVERABLES

Prepare the following deliverables for your instructor:

- A printout of the memo
- Printouts of your worksheets and Answer Reports
- Your Word document, Excel workbook, and PowerPoint presentation on electronic media or sent to your course site as directed by your instructor

Staple the printouts together with the memo on top. If you have more than one Excel workbook file for your case, write your instructor a note that describes the different files.

ED AND CARLA'S RETIREMENT PLAN

Decision Support Using Microsoft Excel Solver

PREVIEW

Ed and Carla are a self-employed couple who have been saving for retirement for several years. Ed has a private law practice and Carla is a family care physician. They have accumulated $200,000 in a professionally managed individual retirement account, but are dissatisfied with the growth rate of their savings. Ed thinks that he and Carla should set up their own retirement portfolio of mutual funds and manage it themselves. While Carla agrees with Ed, she is concerned about the risk associated with the funds they choose and how much to invest in each one.

Ed has researched numerous mutual funds and identified 10 that he thinks meet their investment needs. The funds differ in composition, sources, return, and risk. Ed and Carla have discussed their investment strategy and the constraints they want to impose on their portfolio. They want to optimize their allocation of funds and obtain the best overall return for the selected risk level. Because you are an MIS consultant and a friend of the couple, Ed and Carla have asked you to build a spreadsheet model that determines the best allocation of money for the funds in their retirement savings portfolio.

PREPARATION

- Review the spreadsheet concepts discussed in class and in your textbook.
- Your instructor may assign Excel exercises to help prepare you for this case.
- Tutorial D explains how to set up and use Solver for maximization and minimization problems.
- Review the file-saving instructions in Tutorial C—saving an extra copy of your work on a USB thumb drive or online is always a good idea.
- Review Tutorial F to brush up on your presentation skills.

BACKGROUND

You will use your Excel skills to build a decision model and determine how much money Ed and Carla should invest in 10 mutual funds to maximize their return on investment. To build the model, you need the following data, which Ed has provided:

- Ed and Carla's goals and constraints for their portfolio:

  - The amount of money to be invested initially
  - Maximum portfolio standard deviation for the past 5 and 10 years
  - Maximum percentage of portfolio in stock, bond, or sector mutual funds
  - Maximum percentage of portfolio in foreign mutual funds
  - Minimum percentage of portfolio in any particular fund
  - Maximum percentage of portfolio in any particular fund

- Performance data for the 10 funds Ed has selected:

  - Name of each fund
  - Type of fund (stock, bond, or sector)
  - Source of fund (domestic or foreign)
  - Annual return of each fund for the past 5 and 10 years
  - Standard deviation of the return of each fund for the past 5 and 10 years

Your Solver model must calculate the expected annual return for the portfolio based on fund returns from the past 5 and 10 years, and the expected standard deviation of return (or risk) for the portfolio based on 5- and 10-year fund standard deviations. These calculations will be the cells that the Solver model optimizes. The model must also calculate the portfolio's balance to meet the constraints of the model. This balance comprises the percentages of the portfolio invested in equities, bonds, sector funds, and foreign funds, as well as the smallest and largest percentages of the portfolio invested in a particular fund.

Mutual Funds

Stocks and bonds are financial instruments that serve two purposes: They allow companies to acquire capital for operations and expansion, and they give private citizens and institutions an opportunity to invest their money. Investing comes with risk; the value of a stock or bond can change during daily trading on the securities markets. One way to manage investment risk is through diversification—spreading your investment money across many different stocks and bonds. An investor's collection of stocks and bonds is called a portfolio. If an investor does not have a lot of money to start his or her portfolio, it is difficult to diversify effectively and minimize investment risk.

To address the needs of private and institutional investors, mutual funds appeared in the United States in the late 1890s. A mutual fund is a professionally managed vehicle that pools money from many investors to purchase stocks, bonds, and other investments for its portfolio. Investors can buy shares of a mutual fund with smaller investments, yet enjoy the diversification offered by the fund's portfolio. Shares of mutual funds are bought and sold on the securities markets, and the value of a share is computed at the end of the trading day. (An exception is exchange-traded funds, which are bought, sold, and revalued throughout the trading day.) Ed and Carla could have chosen to build a portfolio of individual stocks and bonds, but they think that a portfolio of mutual funds is a less risky investment.

Ed and Carla have selected the following types of mutual funds for investment:

- Stock fund—A stock mutual fund invests primarily in common stocks of foreign and domestic companies. Common stocks are the riskiest form of investment, but over time they provide the highest return. Stocks are also called "equities," which refers to the ownership of company assets reflected in the owner's equity portion of each company's balance sheet. Stock funds typically focus on growth of the share value from the companies' success, income from stock dividends declared, or both. Other types of stock funds include "value" funds, which buy undervalued stocks relative to the companies' performance, and capitalization or "cap" funds, which define each company's total outstanding share value in the market.

- Bond fund—A bond mutual fund invests primarily in corporate or public-sector bonds issued by companies or governments. Unlike a stock, a bond is a long-term loan that a company issues with a specified interest rate. The rate can be fixed or variable, and the bonds can be bought or sold like stocks. The primary difference between a bond and a stock is that the latter represents ownership in a company, while a bond is simply a loan that must be repaid. The return from bond funds is typically lower than that from stock funds over time, but the market risk is less. Keep in mind, however, that bond funds are not risk-free investments.

- Sector fund—A sector fund targets a specific industry and purchases stocks of competing companies in that sector. For example, an energy sector fund has most of its stock holdings in oil and natural gas companies as well as public utilities. A leisure sector fund would invest in travel and entertainment companies. Depending on the targeted industry, sector funds generally outperform stock funds, but they carry greater risk because of their deliberate lack of diversification. Although some sector funds specialize in bonds, most are equity (stock) based.

- Foreign funds—Foreign or international funds consist of stocks, bonds, or a mixture of both. If foreign markets are moving in a different direction from the U.S. market, investing money in foreign funds can help provide balance in a portfolio.

Most mutual funds do not consist exclusively of stocks or bonds, but are a mixture of stocks, bonds, money market shares, and short-term commercial notes. For the purposes of this case, however, a stock fund and a sector fund have a preponderance of common stocks (equities), and a bond fund consists chiefly of bonds.

ASSIGNMENT 1: CREATING A SPREADSHEET FOR DECISION SUPPORT

In this assignment, you create spreadsheets that model the investment decision Ed and Carla are seeking. In Assignment 1A, you create a spreadsheet to model the portfolio and attempt to distribute the funds manually while meeting the investment constraints. In Assignment 1B, you set up Solver to maximize portfolio returns both for 5-year and 10-year risk constraints. In Assignment 1C, you set up and run Solver to minimize 10-year risk given a specified minimum return on the portfolio. In Assignment 2, you use the spreadsheet models to develop a table of summary results for Ed and Carla.

This section helps you set up each of the following spreadsheet components before entering the cell formulas:

- Constants and Requirements
- Calculations
- Portfolio Balance
- Results

The Calculations section is the heart of the decision model. You will set up each mutual fund in its own column; one row will be the range of changing cells for Solver to manipulate. Selected cells in the Results section will serve as your optimization cell, depending on the requirements of the assignment.

Assignment 1A: Creating the Spreadsheet—Base Case

A discussion of each spreadsheet section follows. This information helps you set up each section of the model and learn the logic of the formulas in the Calculations, Portfolio Balance, and Results sections. If you choose to enter the data directly, follow the cell structure shown in the figures. *You can also download the spreadsheet skeleton if you prefer; it will save you time.* To access the base spreadsheet skeleton, select Case 9 in your data files, and then select **Ed and Carla Retirement Plan Skeleton.xlsx**.

Constants and Requirements Section

First, build the skeleton of your spreadsheet. Set up the spreadsheet title and Constants and Requirements section, as shown in Figure 9-1. An explanation of the line items follows the figure.

▲	A	B	C	D	E	F	G	H
1		Ed and Carla's Retirement Plan						
2								
3		**Constants and Requirements**						
4		Money available to invest	$200,000					
5		Maximum Portfolio Standard Deviation--5 yr	12%					
6		Maximum Portfolio Standard Deviation--10 yr	10%					
7		Maximum percentage of portfolio in equities	70%		Note: Equities include Stocks and Sector Funds			
8		Maximum percentage of portfolio in bonds	35%					
9		Maximum percentage of portfolio in sector funds	25%		Fill Legend			
10		Maximum percentage of portfolio in foreign funds	25%			Changing Cells		
11		Minimum percentage of portfolio in any one fund	5%			Optimization Cell		
12		Maximum percentage of portfolio in any one fund	15%					

Source: Used with permission from Microsoft Corporation

FIGURE 9-1   Spreadsheet title and Constants and Requirements section

- Spreadsheet title—Enter the spreadsheet title into cell B1, and then merge and center the title across cells B1 through D1.
- Money available to invest—This value is the amount of money that Ed and Carla have available to invest in the mutual funds ($200,000).
- Maximum Portfolio Standard Deviation, 5 Yr—This value is the maximum weighted standard deviation of return for the past 5 years that Ed and Carla require for the portfolio. The actual value depends on the allocation of money to the various mutual funds, and will be calculated in the Portfolio Balance portion of the Calculations section.
- Maximum Portfolio Standard Deviation, 10 Yr—This value is the maximum weighted standard deviation of return for the past 10 years that Ed and Carla require for the portfolio. The actual value depends on the allocation of money to the various mutual funds, and will be calculated in the Portfolio Balance portion of the Calculations section.

- Maximum percentage of portfolio in equities—This value is the maximum percentage of the portfolio value that Ed and Carla have specified for investments in stock or sector funds. This percentage was originally set at 70 percent of the total portfolio value, but it will be changed in later assignments of the case in which Ed and Carla are willing to tolerate more risk.

- Maximum percentage of portfolio in bonds—This value is the maximum percentage of the portfolio value that Ed and Carla have specified for investment in bond funds. They want to limit their risk in equities, but do not want the bond portion to be more than 35 percent of the total portfolio value. This value will be changed in later assignments of this case in which Ed and Carla are willing to tolerate more risk.

- Maximum percentage of portfolio in sector funds—This value is the maximum percentage of the portfolio value that Ed and Carla have specified for investment in sector funds. Sector funds are the riskier option of the equity funds because of their concentration in one industry, so Ed and Carla have set the desired maximum percentage at 25 percent. They may reset this maximum later to examine the return associated with greater risk.

- Maximum percentage of portfolio in foreign funds—This value is the maximum percentage of the portfolio value that Ed and Carla have specified for investment in equity or bond funds with foreign sources of capital. Foreign funds are not necessarily riskier than domestic funds, but because Ed and Carla are not as well informed about global markets, they have set the desired maximum percentage at 25 percent.

- Minimum percentage of portfolio in any one fund—This value is the minimum percentage of portfolio value that Ed and Carla want to invest in any particular fund. They want to spread their investment money across all 10 funds, with a 5 percent minimum in each fund to start. This amount may change as Ed and Carla explore different alternatives for risk and return.

- Maximum percentage of portfolio in any one fund—This value is the maximum percentage of portfolio value that Ed and Carla are willing to invest in a particular fund. They have set this maximum at 15 percent to start, but it will change as Ed and Carla explore alternative strategies.

You may have noticed by now that Ed and Carla's requirements for their portfolio strategy will become constraints when you set up Solver to find an optimal portfolio. These constraints can probably be met, but be aware that having too many constraints in an optimization problem can create a situation in which no solution is feasible.

Calculations and Portfolio Balance Section

This section is the heart of the worksheet. It contains numerical data for fund returns and standard deviations, the changing cells (money invested in each fund) and percentages of total investment, and valuable data that categorizes the funds by type and source. This data will be necessary when you build the formulas in the Portfolio Balance section (see Figure 9-2). An explanation of the line items follows the figure.

	A	B	C	D	E	F	G	H	I	J	K	L	M
13													
14	**Calculations**												
15	Name of Mutual Fund		Pinnacle Growth	Redrock Technology	Hamilton Income	Columbus Flagship	Singleton Global	Sewell Securities	New Markets Income	Enterprise Energy	Perkins Small Cap Value	Crown Real Estate	
16	Type of Fund		Stock	Sector	Bond	Stock	Stock	Bond	Income	Sector	Stock	Sector	
17	Source of funds		Domestic	Domestic	Domestic	Domestic	Foreign	Domestic	Foreign	Domestic	Domestic	Domestic	Totals
18	Money Invested in Each Fund		$0	$0	$0	$0	$0	$0	$0	$0	$0	$0	
19	Percent of Total Investment												
20	Annual Return–5 Year		7.93%	10.12%	5.64%	8.20%	9.83%	6.12%	7.66%	11.83%	10.78%	12.92%	
21	Standard Deviation of Return–5 Year		11%	16%	5%	12%	15%	6%	8%	14%	13%	17%	
22	Annual Return–10 Year		10.14%	15.44%	5.76%	12.28%	14.62%	6.02%	9.63%	16.89%	14.35%	19.10%	
23	Standard Deviation of Return–10 Year		9%	14%	4%	10%	13%	5%	7%	11%	11%	15%	
24	**Portfolio Balance**												
25	Percentage of Portfolio in Equities												
26	Percentage of Portfolio in Bonds												
27	Percentage of Portfolio in Sector Funds												
28	Percentage of Portfolio in Foreign Funds												
29	Smallest percentage of portfolio in any one fund												
30	Largest percentage of portfolio in any one fund												

Source: Used with permission from Microsoft Corporation

FIGURE 9-2   Calculations and Portfolio Balance section

- Name of Mutual Fund—Row 15 contains the names of the 10 mutual funds selected for investment.

- Type of Fund—The three types of mutual funds examined are stocks, bonds, and sector funds, as entered in row 16. Stock and sector funds are considered equity funds.

- Source of funds—The entries in row 17 report whether the source of the mutual funds is domestic or foreign.

- Money Invested in Each Fund—Row 18 is the heart of the Solver model: the changing cells. These values are the amounts of money that you invest manually or that Solver assigns to each fund. Cells C18 through L18 are backfilled in yellow to distinguish them as changing cells. Enter $20,000 in each of these cells for now. Cell M18 is a Totals cell. Use the AUTOSUM function to sum the values from the changing cells into cell M18.
- Percent of Total Investment—Cells C19 through L19 are the percentages of total investment that each fund represents. To calculate these percentages, divide the Money Invested in each Fund (cells C18 through L18) by the money available to invest (cell C4). If you use the absolute cell reference for C4 ($C$4), you need to write the formula only for cell C19 and then copy it to the other nine cells in the row.
- Annual Return, 5 Year—Cells C20 through L20 contain the percentages of annual return for each fund averaged over the past five years.
- Standard Deviation of Return, 5 Year—Cells C21 through L21 contain the standard deviation of the annual return on investment for each fund over the past five years. Higher standard deviations of return indicate greater variations in the annual return over the 5-year period.
- Annual Return, 10 Year—Cells C22 through L22 contain the percentages of annual return for each fund averaged over the past 10 years.
- Standard Deviation of Return, 10 Year—Cells C23 through L23 contain the standard deviation of the annual return on investment for each fund over the past 10 years. Higher standard deviations of return indicate greater variations in the annual return over the 10-year period.
- Percentage of Portfolio in Equities—The value in cell C25 is the sum of the percentage values from row 19 for the identified stock or sector funds in row 16.
- Percentage of Portfolio in Bonds—The value in cell C26 is the sum of the percentage values from row 19 for the funds identified as bonds in row 16. The values in cells C25 and C26 should total 100 percent.
- Percentage of Portfolio in Sector Funds—The value in cell C27 is the sum of the percentage values from row 19 for the identified sector funds in row 16.
- Percentage of Portfolio in Foreign Funds—The value in cell C28 is the sum of the percentage values from row 19 for the identified foreign funds in row 17.
- Smallest percentage of portfolio in any one fund—The value in cell C29 is the smallest percentage of money invested in any fund from row 19. Use the MIN function for the range of cells from C19 through L19.
- Largest percentage of portfolio in any one fund—The value in cell C30 is the largest percentage of money invested in any fund from row 19. Use the MAX function for the range of cells from C19 through L19.

If you set up the Calculations and Portfolio Balance section correctly, it should look like Figure 9-3.

	A	B	C	D	E	F	G	H	I	J	K	L	M
13													
14		Calculations											
15		Name of Mutual Fund	Pinnacle Growth	Redrock Technology	Hamilton Income	Columbus Flagship	Singleton Global	Sewell Securities	New Markets Income	Enterprise Energy	Perkins Small Cap Value	Crown Real Estate	
16		Type of Fund	Stock	Sector	Bond	Stock	Stock	Bond	Bond	Sector	Stock	Sector	Totals
17		Source of funds	Domestic	Domestic	Domestic	Domestic	Foreign	Domestic	Foreign	Domestic	Domestic	Domestic	
18		Money Invested in Each Fund	$20,000	$20,000	$20,000	$20,000	$20,000	$20,000	$20,000	$20,000	$20,000	$20,000	$200,000
19		Percent of Total Investment	10.00%	10.00%	10.00%	10.00%	10.00%	10.00%	10.00%	10.00%	10.00%	10.00%	
20		Annual Return–5 Year	7.93%	10.12%	5.64%	8.20%	9.83%	6.12%	7.66%	11.83%	10.78%	12.92%	
21		Standard Deviation of Return–5 Year	11%	16%	5%	12%	15%	6%	8%	14%	13%	17%	
22		Annual Return–10 Year	10.14%	15.44%	5.76%	12.28%	14.62%	6.02%	9.63%	16.89%	14.35%	19.10%	
23		Standard Deviation of Return–10 Year	9%	14%	4%	10%	13%	5%	7%	11%	11%	15%	
24		Portfolio Balance											
25		Percentage of Portfolio in Equities	70%										
26		Percentage of Portfolio in Bonds	30%										
27		Percentage of Portfolio in Sector Funds	30%										
28		Percentage of Portfolio in Foreign Funds	20%										
29		Smallest percentage of portfolio in any one fund	10%										
30		Largest percentage of portfolio in any one fund	10%										

Source: Used with permission from Microsoft Corporation

FIGURE 9-3   Completed Calculations and Portfolio Balance section

Results Section

The Results section contains the calculated and weighted portfolio returns and standard deviations based on the past 5 and 10 years of data from the Calculations section. Using weighted averages for portfolio returns is statistically sound. Using a weighted average of standard deviations to evaluate mutual fund portfolio risk is subject to debate, but for the purposes of this case, it is a reasonable approximation.

N O T E—CALCULATING A PORTFOLIO'S WEIGHTED STANDARD DEVIATION

This case uses a weighted average calculation of portfolio standard deviation based on the percentage (or weight) of the portfolio invested in each fund. This calculation is a simplification. Modern financial risk theory seeks to determine the risk relationships between investments through complex calculations. Because mutual funds are collections of many stocks, bonds, and other monetary instruments, the mathematical determination of risk between mutual funds is even more difficult and theoretical. The object of this exercise is to build an optimization model rather than explore financial theory, so using a weighted average method for portfolio standard deviation is reasonable, and works as well as any other alternative.

An explanation of the line items in the Results section follows Figure 9-4.

	A	B	C
31			
32		**Results**	
33		**Weighted Expected Annual Return based on 5 yr**	
34		**Weighted Standard Deviation based on 5 yr**	
35		**Weighted Expected Annual Return based on 10 yr**	
36		**Weighted Standard Deviation based on 10 yr**	

Source: Used with permission from Microsoft Corporation

FIGURE 9-4   Results section

- Weighted Expected Annual Return based on 5 yr—The value in cell C33 is the weighted average of all 5-year annual returns from the funds in row 20. To obtain this value, write a formula to multiply the percentage of total investment from each cell in row 19 by the 5-year annual return from each cell in row 20, and then add all 10 products together. In other words, write a formula of =C19*C20 + D19*D20 + E19*E20 +...+ L19*L20. Although this is a long formula, the values are vertically aligned in rows, and an Excel function called SUMPRODUCT is available to help you execute this task without much work. The syntax for this function is =SUMPRODUCT(C19: L19, C20:L20). The function multiplies each value in the first array (C19:L19) by its corresponding value in the second array (C20:L20), and then computes the sum of all the products. This formula is particularly useful for completing the other three calculations in the Results section. Because row 19 is used in every calculation in the Results section, you may want to use an absolute cell reference for the row in your SUMPRODUCT formula so you can copy it easily into the other Results cells. Note that the cell is filled with an orange background to indicate that it is the optimization cell for the first part of the Solver assignment.
- Weighted Standard Deviation based on 5 yr—The value in cell C34 is the weighted average of all funds' standard deviations of return for the past five years. To obtain this value, write a formula to multiply the percentage of total investment from each cell in row 19 by the 5-year standard deviation of return from each cell in row 21, and then add all 10 products together. In other words, write a formula of =C19*C21 + D19*D21 + E19*E21 +...+ L19*L21. You can greatly simplify this formula by using the SUMPRODUCT function, as described in the previous paragraph. The weighted standard deviation is a simplified estimate of portfolio risk, as described in the preceding note.
- Weighted Expected Annual Return based on 10 yr—The value in cell C35 is the weighted average of all funds' 10-year annual returns from row 22. To obtain this value, write a formula to multiply the percentage of total investment from each cell in row 19 by the 10-year annual return from each cell in row 22, and then add all 10 products together. In other words, write a formula of =C19*C22 + D19*D22 + E19*E22 +...+ L19*L22. You can greatly simplify this formula by using the SUMPRODUCT function.
- Weighted Standard Deviation based on 10 yr—The value in cell C36 is the weighted average of all funds' standard deviations of return for the past 10 years. To obtain this value, write a formula to multiply the percentage of total investment from each cell in row 19 by the 10-year standard deviation of return from each cell in row 23, and then add all 10 products together. In other words, write a formula of =C19*C23 + D19*D23 + E19*E23 +...+ L19*L23. You can greatly simplify this formula by using the SUMPRODUCT function. The weighted standard deviation is a simplified estimate of portfolio risk, as described in the preceding note.

If you wrote the formulas correctly, your Results section should look like Figure 9-5.

	A	B	C
31			
32		**Results**	
33		**Weighted Expected Annual Return based on 5 yr**	9.10%
34		**Weighted Standard Deviation based on 5 yr**	11.70%
35		**Weighted Expected Annual Return based on 10 yr**	12.42%
36		**Weighted Standard Deviation based on 10 yr**	9.90%

Source: Used with permission from Microsoft Corporation

FIGURE 9-5   Completed Results section

Finding a Manual Solution

Now that you have completed the worksheet, attempt to find a manual solution that meets *all* the requirements listed in the Constants and Requirements sections. For example, entering $20,000 in each of the changing cells yields a good 5-year weighted expected annual return, but it violates the maximum amount of 25 percent that can be invested in sector funds. Try to adjust the amounts in the funds while maintaining the total investment at $200,000. Do not spend an excessive amount of time trying to optimize your portfolio manually; just find a solution that meets the requirements. In the next section, Solver will be able to optimize the portfolio after you set up the Solver parameters.

After you have found a manual solution, rename the worksheet **Ed-Carla Manual Guess**, copy the worksheet, and rename the copy **Ed-Carla 5 Yr Risk-Solver**.

Assignment 1B: Setting Up and Running Solver

Ed and Carla first want you to optimize the portfolio for the largest 5-year annual return that meets a maximum 5-year portfolio standard deviation of 12 percent. Before you set up Solver, you should jot down the parameters you must define and their cell addresses:

- The cell you want to optimize (Weighted Expected Annual Return based on 5 yr) and whether you want to minimize or maximize it
- The changing cells you want Solver to manipulate to obtain the optimal solution
- The following constraints:

    - Each changing cell must contain a positive number because you cannot make a negative investment.
    - All the requirements listed in the Constants and Requirements section *except* the 10-year maximum portfolio standard deviation. You must meet the percentage maximums for each type of fund, the percentage minimums for investment in each fund, and the 5-year maximum portfolio standard deviation *only*.
    - You cannot allocate a total that exceeds the initial investment ($200,000).

You should define your constraints using the values in the Portfolio Balance and Results sections, and compare the calculated values to those in the Requirements section.

Next, go to the Analysis group in the Data tab and click Solver to set up your problem (see Figure 9-6). Designate your changing cells and add the constraints described in the preceding list and the Constants and Requirements sections. Retain the default solving method, Simplex LP.

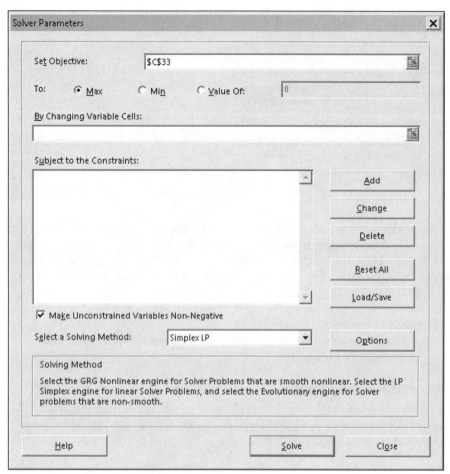

Source: Used with permission from Microsoft Corporation

FIGURE 9-6   The Solver Parameters window

Click the Options button to open the Solver Options window. Click the All Methods tab if necessary, and confirm that the Constraint Precision is set to .000001, the Integer Optimality % is set to 0, and the Use Automatic Scaling box is checked (see Figure 9-7). Close the Options window, and attempt to run Solver.

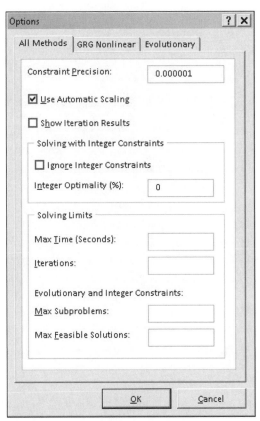

Source: Used with permission from Microsoft Corporation
FIGURE 9-7   The Solver Options window

Solver immediately displays an error message stating that the linearity conditions have not been met (see Figure 9-8).

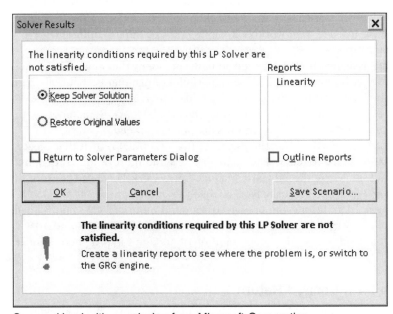

Source: Used with permission from Microsoft Corporation
FIGURE 9-8   Nonlinearity error message

Because the problem contains nonlinear products and conditional statements, you must select a nonlinear solving method. Click Cancel, and then open the Solver Parameters window again. Choose GRG Nonlinear in the Select a Solving Method field, and click the Options button. Click the GRG Nonlinear tab and set the Convergence value to .0000000001 (see Figure 9-9).

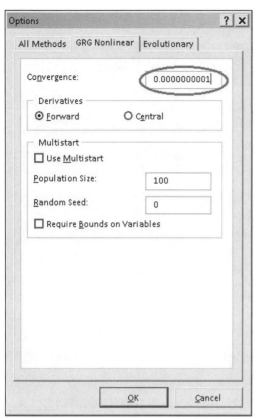

Source: Used with permission from Microsoft Corporation
FIGURE 9-9   GRG Nonlinear tab in the Options window

Run Solver and ask for the Answer Report when Solver finds a solution that satisfies the constraints. Print the entire workbook, including the Solver Answer Report Sheet. Save the workbook by clicking the File tab and then clicking Save. For the rest of the case, you either can use the Save As command to create new Excel workbooks or continue copying and renaming the worksheets in one workbook. Both options offer distinct advantages, but having all of your worksheets in one Excel workbook allows you to compare different solutions easily, as well as prepare summary reports.

Before continuing to the next part of the assignment, examine the amounts of money that Solver assigned to each fund in the portfolio. If you set up Solver correctly, you should see a portfolio return that is about one-half percent greater than your manual guess. The increased return may not seem like much money, but over the course of 20 years or more, the difference could amount to several hundred thousand dollars. Remember, however, that any investment carries risk, and that the estimated returns are not guaranteed.

Next, Ed and Carla want you to optimize the portfolio for the greatest 10-year annual return that meets a maximum 10-year portfolio standard deviation of 10 percent. The changing cells and most of the constraints will remain the same for these new calculations. Copy the Ed-Carla Manual Guess worksheet, and rename the new worksheet **Ed-Carla 10 Yr Risk-Solver**. When setting up Solver, change the optimization cell to C35 (Weighted Expected Annual Return based on 10 yr). Undo the orange background fill in cell C33 and fill cell C35 with the orange background instead. In Solver, you can ignore the constraint for Weighted Standard Deviation based on 5 yr in cell C34, but you need to include a constraint of 10 percent for Weighted Standard Deviation based on 10 yr (cell C36).

Assignment 1C: Risk Minimization for a Desired Return

Ed and Carla have decided that they want to target a 10-year annual return of 12 percent. Therefore, you need to modify the preceding worksheet to minimize the 10-year portfolio standard deviation, given a target 10-year annual return of 12 percent. Ed and Carla have also decided to make their portfolio more aggressive. They want to allow up to 75 percent of the portfolio to be invested in equities, up to 30 percent to be invested in sector funds, and up to 30 percent to be invested in foreign funds. Finally, they want to stipulate that no single fund has a required minimum investment.

Copy the "Ed-Carla 10 Yr Risk-Solver" worksheet and rename the new one **Ed-Carla 10 Yr Min Return**. Change the worksheet title and the Constants and Requirements section, as shown in Figure 9-10. An explanation of the changes follows the figure.

	A	B	C	D
1		Ed and Carla's Retirement Plan--10 Year Minimum Return 12%		
2				
3		**Constants and Requirements**		
4		Money available to invest	$200,000	
5		Maximum Portfolio Standard Deviation--5 yr	12%	
6		Minimum Portfolio Return--10 yr	12%	
7		Maximum percentage of portfolio in equities	75%	
8		Maximum percentage of portfolio in bonds	35%	
9		Maximum percentage of portfolio in sector funds	30%	
10		Maximum percentage of portfolio in foreign funds	30%	
11		Minimum percentage of portfolio in any one fund	0%	
12		Maximum percentage of portfolio in any one fund	20%	

Source: Used with permission from Microsoft Corporation

FIGURE 9-10   New worksheet title and entries in the Constants and Requirements section

- Worksheet title—Enter the title shown in Figure 9-10 in cells B1 through D1, then center and merge the title.
- Minimum Portfolio Return—10 yr—Change the title in cell B6 and enter 12% in cell C6.
- Maximum percentage of portfolio in equities—Change the value in cell C7 from 70% to 75%.
- Maximum percentage of portfolio in sector funds—Change the value in cell C9 from 25% to 30%.
- Maximum percentage of portfolio in foreign funds—Change the value in cell C10 from 25% to 30%.
- Minimum percentage of portfolio in any one fund—Change the value in cell C11 from 5% to zero.

In the Results section of the worksheet, move the fill background and border from cell C35 to cell C36, as shown in Figure 9-11.

	A	B	C
31			
32		**Results**	
33		Weighted Expected Annual Return based on 5 yr	9.26%
34		Weighted Standard Deviation based on 5 yr	11.62%
35		Weighted Expected Annual Return based on 10 yr	12.58%
36		Weighted Standard Deviation based on 10 yr	9.81%

Source: Used with permission from Microsoft Corporation

FIGURE 9-11   New optimization cell in the Results section

The 10-year weighted standard deviation becomes the new optimization cell, so you must modify the Solver formula to minimize the value in cell C36. You must also add a new constraint so the 10-year weighted expected annual return (cell C35) is greater than or equal to the 10-year minimum portfolio return (cell C6). When you finish setting your Solver parameters, but *before* running Solver, make a copy of the worksheet and rename the new one **Ed-Carla 10 Yr Min Return 200K**. You will use this worksheet in case Solver assigns less than the total $200,000 available to the portfolio. Return to the "Ed-Carla 10 Yr Min Return" worksheet and run Solver to obtain the solution and Answer Report.

If you defined the constraint for the total investment to be less than or equal to the money available to invest ($200,000), Solver will probably arrive at its optimal solution while allocating less than Ed and Carla's total amount to invest. If so, you can "force" Solver to invest all the money. Return to the "Ed-Carla 10 Yr Min Return 200K" worksheet, open the Solver Parameters window, and change the constraint for the total portfolio value (cell M18) to equal the money available to invest (cell C4). If you use the "=" constraint, it forces Solver to find a solution that allocates all $200,000 to the portfolio. Run Solver again and create a new Answer Report.

ASSIGNMENT 2: USING THE WORKBOOK FOR DECISION SUPPORT

You have built a series of worksheets to help Ed and Carla allocate their retirement money wisely. You will now complete the case by using your solutions and Answer Reports to summarize your worksheet results and printouts in a table.

If you choose to build the summary table yourself, follow the cell structure shown in Figure 9-12. *You can also use the table in the spreadsheet skeleton if you prefer.* To access the base spreadsheet skeleton, select Case 9 from your data files, and then select **Ed and Carla Retirement Plan Skeleton.xlsx.** The worksheet tab is named "Summary of Results—Blank Table." You can copy this format directly into your Solver workbook.

Ed and Carla' s Retirement Plan--Summary Results

Situation Examined	Expected Annual Return	Portfolio Risk (Weighted Standard Deviation)
Manual Guess--5 year Return/Risk		
Solver--5 Year Return/Risk		
Solver--10 Year Return/Risk		
Solver--10 Year, Minimum 12% Return		
*Solver--10 Year, Minimum 12% Return, Invest all $200K		

*Note: Constraint entered to force Solver to invest all $200,000

Recommended Fund Allocations:

Name of Mutual Fund	Pinnacle Growth	Redrock Technology	Hamilton Income	Columbus Flagship	Singleton Global	Sewell Securities	New Markets Income	Enterprise Energy	Perkins Small Cap Value	Crown Real Estate	
Type of Fund	Stock	Sector	Bond	Stock	Stock	Bond	Bond	Sector	Stock	Sector	Total
Source of funds	Domestic	Domestic	Domestic	Domestic	Foreign	Domestic	Foreign	Domestic	Domestic	Domestic	Invested
Manual Guess--5 year Return/Risk											
Solver--5 Year Return/Risk											
Solver--10 Year Return/Risk											
Solver--10 Year, Minimum 12% Return											
Solver--10 Year, Minimum 12% Return, Invest all $200K											

Source: Used with permission from Microsoft Corporation

FIGURE 9-12 Summary table format

There are several ways to populate the summary table. For example, you can transcribe the data manually from printouts, but that option is time-consuming and you can easily make entry errors. You can also copy and paste values into the table using the Copy and Paste Values commands, as shown in Figures 9-13 and 9-14. Remember to use the Paste Values command instead of Paste; otherwise, the cell formatting in the summary table will be changed.

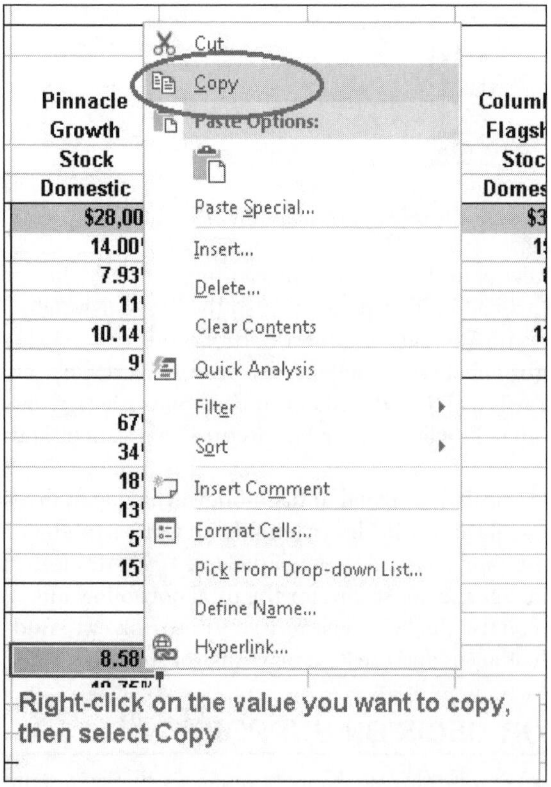

Source: Used with permission from Microsoft Corporation

FIGURE 9-13 Copy command

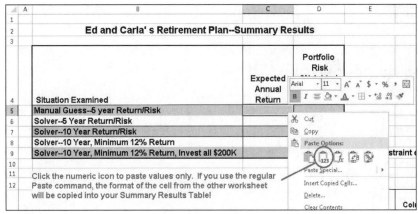

Source: Used with permission from Microsoft Corporation

FIGURE 9-14   Paste Values command

The third way to copy values into the summary table is slightly more difficult, but it creates active links to the values in the original worksheets. If the values in the original worksheet are changed, then the values in the summary table are updated automatically, which is useful when you work with accounting reports that are updated regularly. To create an active link for a cell value from another worksheet, take the following steps:

1. Click the cell in the summary table worksheet where you want to create the active link and enter "=", as shown in Figure 9-15.

2. Click the worksheet tab that contains the data value of interest, then click the cell with the value you want to link to your summary table cell, as shown in Figure 9-16. Notice that "='Ed-Carla Manual Guess'!C33" appears in the cell-editing window. This formula takes the value from cell C33 of the selected worksheet and inserts it in the cell you selected in the summary table worksheet.

3. To complete the link, press Enter. The linked value appears in the cell you selected in the summary table, as shown in Figure 9-17.

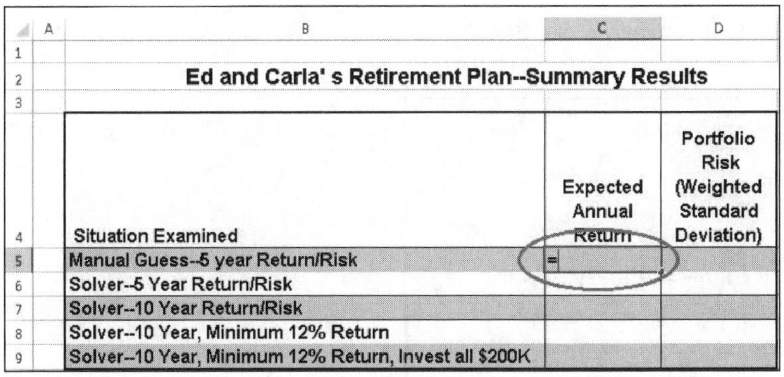

Source: Used with permission from Microsoft Corporation

FIGURE 9-15   Linking a value from a worksheet to a summary table: Step 1

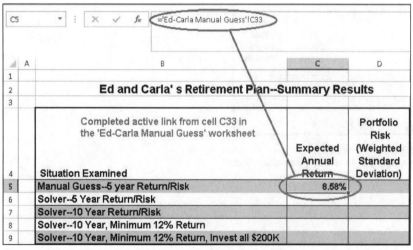

| C33 | ▼ | : | ✕ | ✓ | fx | ='Ed-Carla Manual Guess'!C33 |

▲	A	B	C	D
1		Ed and Carla's Retirement Plan--5 yr Risk--Manual Guess		
2				
3		**Constants and Requirements**		
4		Money available to invest	$200,000	
5		Maximum Portfolio Standard Deviation--5 yr	12%	
6		Maximum Portfolio Standard Deviation--10 yr	10%	
7		Maximum percentage of portfolio in equities	70%	
8		Maximum percentage of portfolio in bonds	35%	
9		Maximum percentage of portfolio in sector funds	25%	
10		Maximum percentage of portfolio in foreign funds	25%	
11		Minimum percentage of portfolio in any one fund	5%	
12		Maximum percentage of portfolio in any one fund	20%	
13				
14		**Calculations**		
15		**Name of Mutual Fund**	Pinnacle Growth	Redrock Technology
16		**Type of Fund**	Stock	Sector
17		**Source of funds**	Domestic	Domestic
18		Money Invested in Each Fund	$28,000	$10,000
19		Percent of Total Investment	14.00%	5.00%
20		Annual Return--5 Year	7.93%	10.12%
21		Standard Deviation of Return--5 Year	11%	16%
22		Annual Return--10 Year	10.14%	15.44%
23		Standard Deviation of Return--10 Year	9%	14%
24		**Portfolio Balance**		
25		Percentage of Portfolio in Equities	67%	
26		Percentage of Portfolio in Bonds	34%	
27		Percentage of Portfolio in Sector Funds	18%	
28		Percentage of Portfolio in Foreign Funds	13%	
29		Smallest percentage of portfolio in any one fund	5%	
30		Largest percentage of portfolio in any one fund	15%	
31				
32		**Results**		
33		Weighted Expected Annual Return based on 5 yr	8.58%	
34		Weighted Standard Deviation based on 5 yr	10.75%	
35		Weighted Expected Annual Return based on 10 yr	11.39%	
36		Weighted Standard Deviation based on 10 yr	9.01%	

Source: Used with permission from Microsoft Corporation

FIGURE 9-16   Linking a value from a worksheet to a summary table: Step 2

| C5 | ▼ | : | ✕ | ✓ | fx | ='Ed-Carla Manual Guess'!C33 |

▲	A	B	C	D
1				
2		**Ed and Carla' s Retirement Plan--Summary Results**		
3				
4		Completed active link from cell C33 in the 'Ed-Carla Manual Guess' worksheet	Expected Annual Return	Portfolio Risk (Weighted Standard Deviation)
		Situation Examined		
5		Manual Guess--5 year Return/Risk	8.58%	
6		Solver--5 Year Return/Risk		
7		Solver--10 Year Return/Risk		
8		Solver--10 Year, Minimum 12% Return		
9		Solver--10 Year, Minimum 12% Return, Invest all $200K		

Source: Used with permission from Microsoft Corporation

FIGURE 9-17   Linked value in the summary table

ASSIGNMENT 3: DOCUMENTING YOUR RECOMMENDATIONS IN A MEMO

When you have completed the summary table, you are ready to write a brief memorandum to Ed and Carla that documents the decision model and results. Use Microsoft Word to write the memo. Summarize the results of your analysis and include your recommendations for allocating investments in their portfolio. Your memo should conform to the following requirements:

- Set up your memo as described in Tutorial E.
- In the first paragraph, briefly describe the situation and state the purpose of your analysis.
- Next, summarize the results of your analysis.
- Support your recommendation by copying your summary table from the Excel workbook into the memo. (Tutorial C includes a brief description of how to copy and paste Excel objects.)
- In the future, you might suggest that Ed and Carla revisit the Solver analysis each year to compare the actual performance of the mutual fund portfolio with the model estimates.

DELIVERABLES

Prepare the following deliverables for your instructor:

1. A printout of the memo
2. Printouts of all your worksheets and Answer Reports
3. Your Word document and Excel workbook on electronic media or sent to your course site as directed by your instructor

Assemble your printouts and staple them together with the memo on top. If you used more than one Excel workbook file for your case, write your instructor a note that describes the different files.

PART 4

DECISION SUPPORT CASE
USING BASIC EXCEL FUNCTIONALITY

THE UNIVERSITY ARENA CONSTRUCTION DECISION

Decision Support Using Excel

PREVIEW

A university wants to build a new 5,000-seat arena for sporting events and other attractions, but it would need $60 million to fund the project. Can the cost be justified economically? In this case, you will use Microsoft Excel to help answer that question.

PREPARATION

- Review spreadsheet concepts discussed in class and in your textbook.
- Complete any exercises that your instructor assigns.
- Complete any parts of Tutorials C and D that your instructor assigns, or refer to them as necessary.
- Review the Microsoft Windows file-saving procedures in Tutorial C.
- Refer to Tutorials E and F as necessary.

BACKGROUND

A midsized state university has a 2,000-seat arena that was built in 1975. For many years, it was serviceable for sports events and other indoor events, but it is now considered inadequate by modern standards. The concession stands are small and antiquated, there are no hospitality areas, and the facilities for athletes are poor.

Top-notch high school athletes want to attend colleges and universities that have good locker rooms and exercise facilities. University officials want to upgrade the basketball, volleyball, and gymnastics programs, which means they need to improve amenities to attract better athletes.

A new arena could also be used for purposes other than sports. For example, a large city in the region has an arena that seats 15,000 people and can accommodate political conventions and concerts by superstar musical groups. However, most musical groups cannot fill such a large arena; a 5,000-seat venue would be sufficient for them.

University planners have designed the arena they want, and it could be ready to open by January 2015. The design would include 5,000 seats in the main seating area and 17 luxury boxes. The arena would also have a large hospitality room, a souvenir shop, three large concession stands, a "hall of fame" room, a multimedia production room, deluxe men's and women's locker rooms, modern exercise facilities, office space for coaches and other athletic department personnel, and a large parking lot.

The expected construction costs are $60 million. The university would need the state to issue the entire amount in bonds to finance the construction. If the bonds were approved, the arena would have to generate sufficient revenue to pay off the bonds.

The arena would serve as the home court for the men's and women's basketball teams, the women's volleyball team, and the men's gymnastics team. The basketball teams each play 30 games a season, which consists of 13 conference home games, 13 conference road games, and four games at neutral sites. The volleyball and gymnastics teams each have 13 home meets per season.

The men's and women's basketball teams have been mediocre in recent years, and attendance for both teams has declined. Attendance would improve if the teams won more games, but better athletes have to be recruited and then kept in school. Improved recruiting requires hiring more assistant coaches, hiring tutors for athletes with academic problems, more work on diet and conditioning, and hiring another trainer to help get athletes back on the court after recovering from injuries. The teams could then reasonably expect to make gradual improvements and win an additional game or two per year for several years in a row.

More wins would lead to better season ticket sales and better walkup ticket sales on the day of a game. Increased attendance would mean better concession income. Also, more wins would translate into increased enthusiasm for the teams within the community, which would make it easier to rent private luxury boxes to local businesses. People and businesses that purchase luxury boxes are essentially buying season tickets to men's and women's basketball games. The luxury boxes come with catered food and drinks, which would be an added expense of the arena's operations.

Besides hosting varsity sports, university officials have discussed other ideas for generating revenue at the arena:

- The National Basketball Association has a developmental league that gives players an opportunity to improve their games and make the big time. A regional team is joining the league and needs a place to play. The university's arena would be a good place for the team to rent. A five-year rental agreement could be negotiated to accommodate the team's schedule of 20 home games per season. The team would keep the gate receipts, but the university would receive the concession income.
- A series of electronic screens would encircle the concourse level and display advertisements. The arena's large scoreboards could also be used for electronic advertising.
- Naming rights could be sold to a corporation that wants to have the arena carry its name.
- A women's indoor football league has a team in the region that is looking for a place to play. A five-year rental agreement could be negotiated to accommodate the team's schedule of 10 home games per season. The team would keep the gate receipts but the university would receive the concession income. The university's president is not sure that he wants to host the football team. He thinks that the governor would want to have the team play in a state school's arena only if the economics were feasible.
- The university could increase its general fee for full-time students.
- The university could manage concerts and events on its own, but a more sensible solution would be to hire an entertainment marketing firm to organize and manage events at the arena. The university would pay a retainer to the firm each year and one-third of the gate receipts for concerts and other events. By hiring an outside firm, the university would not have to increase its own staff but would still get the lion's share of the gate receipts.

The university's chief financial officer (CFO) and president have asked you to create a model of the construction project in Excel. Your model must show the revenues and expenses that would result if the university builds the new arena. Ultimately, state officials need to be convinced that the arena makes economic sense before they agree to issue construction bonds as a state obligation.

The CFO gives you a short lesson in public finance: "Assume the principal of the bonds would be $60 million, with payments each month for 15 years. The interest rate will be determined later; for now, assume 3 percent per year. The state would issue the bonds. There will be extra revenue from the arena, but extra expenses, too. We expect the extra revenues to exceed the extra expenses. We will turn over the excess to the state. We hope that the excess is enough to make the construction bond payments, but if not, the state— that is, the voters—would have to cover the shortfall."

The university president provides a quick lesson in state politics: "The governor understands the need for a new arena and she wants us to have it, but she must justify having the voters pay for it. There are intangible benefits to having a shiny new arena. I can network better with businesspeople, alumni, government officials, and parents in the new place. Students of all sorts—not just student-athletes—want to attend a school whose sports teams win big games on television. We can call the effect of such factors the "value of intangible benefits," which everyone believes are real but cannot be quantified. I assume that the arena will make money, but probably not enough to fully cover the bond payment each month. The governor can sell a small shortfall to the voters as the value of intangibles. But we cannot sell a huge shortfall, especially if it happens early in the life of the arena. So, the bonds will not be issued if it looks like we'll have a giant shortfall, and then we won't have our arena. You need to tell me if I will have to sell the arena to the governor on the basis of its intangibles. The CFO and I will ask you to include several scenarios in your spreadsheet before I talk to the governor."

ASSIGNMENT 1: CREATING A SPREADSHEET FOR DECISION SUPPORT

In this assignment, you produce a spreadsheet that models the problem. Then, in Assignment 2, you will write a memorandum that explains your findings. In Assignment 3, you may be asked to prepare an oral presentation of your analysis.

A spreadsheet skeleton is available for you to use; it will save you time. If you want to use the spreadsheet skeleton, locate Case 10 in your data files and then select **Arena.xlsx**. Your worksheet should contain the following sections:

- Constants
- Inputs
- Summary of Key Results
- Calculations
- Cash Flow Statement

A discussion of each section follows.

Constants Section

Your spreadsheet should include the constants shown in Figure 10-1. An explanation of the line items follows the figure. (Your model runs for 15 years through 2029, but values through 2021 only are shown in Figure 10-1.)

	A	B	C	D	E	F	G	H	I
1	**UNIVERSITY $60 MILLION ARENA CONSTRUCTION DECISION**								
2									
3	**CONSTANTS**	**2014**	**2015**	**2016**	**2017**	**2018**	**2019**	**2020**	**2021**
4	Number of full time students	NA	15,000	15,100	15,200	15,300	15,400	15,500	15,600
5	Cost of assistant basketball coach	NA	$ 100,000	$ 103,000	$ 106,000	$ 109,000	$ 112,000	$ 115,000	$ 118,000
6	Luxury box rental fee	NA	$ 10,000	$ 10,200	$ 10,400	$ 10,600	$ 10,800	$ 11,000	$ 11,200
7	Concession income per fan	NA	$ 3.00	$ 3.25	$ 3.50	$ 3.75	$ 4.00	$ 4.25	$ 4.50
8	Expected volleyball and gymnastics attendance	NA	100	100	100	100	100	100	100
9	Pro basketball attendance per game	NA	500	500	500	500	500	0	0
10	Pro basketball rental	NA	$ 20,000	$ 20,000	$ 20,000	$ 20,000	$ 20,000	0	0
11	Women's football attendance per game	NA	2,000	2,000	2,000	2,000	2,000	0	0
12	Women's football rental	NA	$ 20,000	$ 20,000	$ 20,000	$ 20,000	$ 20,000	0	0
13	Naming rights revenue	NA	$ 400,000	$ 400,000	$ 400,000	$ 400,000	$ 400,000	$ 500,000	$ 500,000
14	Expected miscellaneous income	NA	$ 400,000	$ 410,000	$ 420,000	$ 430,000	$ 440,000	$ 450,000	$ 460,000
15	Expected utilities and maintenance cost	NA	$ 500,000	$ 510,000	$ 520,000	$ 530,000	$ 540,000	$ 550,000	$ 560,000

Source: Used with permission from Microsoft Corporation

FIGURE 10-1   Constants section

- Number of full time students—The expected number of full-time university students is shown through 2029. A modest increase is expected each year.
- Cost of assistant basketball coach—The expected yearly costs of a coach's salary and benefits have been estimated through 2029.
- Luxury box rental fee—A luxury box can be rented for the basketball season at the cost shown here.
- Concession income per fan—This amount is the average difference between what a fan pays for drinks and food and the cost to the university of purchasing the concessions.
- Expected volleyball and gymnastics attendance—Attendance for these sports is not very good, as shown.
- Pro basketball attendance per game—The expected attendance per game is shown for each of the five years being analyzed.
- Pro basketball rental—The basketball team would rent the arena at the cost shown for each of the five years being analyzed.
- Women's football attendance per game—The expected attendance per game is shown for each of the five years being analyzed.
- Women's football rental—The football team would rent the arena at the cost shown for each of the five years being analyzed.
- Naming rights revenue—The naming rights would be sold for a certain amount each year, as shown. This amount is expected to increase during the second half of the contract.

- Expected miscellaneous income—The university will make money from souvenir sales in the arena store and from selling multimedia recordings of some arena events. The excess of revenues over expenses for these items is expected to be substantial, as shown.
- Expected utilities and maintenance cost—This amount is the expected cost required to operate the new arena each year.

Inputs Section

Your spreadsheet should include the inputs shown in Figure 10-2. An explanation of the line items follows the figure.

	A	B	C	D	E
17	**INPUTS**				
18	Basketball Recruiting Effort (1=Same, 2=More, 3=Max)		Number of concerts in year		
19	Added Student Fee (L=Low, M=Medium, O=Outrageous)		Average concert attendance		
20	Do the Women's Football League? (Y/N)		Average concert ticket		
21	Expected bond interest rate (annual)				

Source: Used with permission from Microsoft Corporation

FIGURE 10-2   Inputs section

- Basketball Recruiting Effort—To win more games, the basketball teams need better recruiting efforts. If these efforts are expected to remain the same, enter 1. If recruiting efforts will increase by a moderate amount, enter 2. If the effort will increase by the maximum amount, enter 3. The entry applies to all years.
- Added Student Fee—To indicate the increase in the student fee, enter L for a low amount, M for a medium amount, or O for an outrageous amount. The entry applies to all years.
- Do the Women's Football League? (Y/N)—If the university allows the women's league to play in the arena, enter Y. If not, enter N. The entry applies to all years.
- Expected bond interest rate (annual)—Enter .03 if the annual rate is expected to be 3 percent, .04 if the rate is expected to be 4 percent, and so on. The rate applies to the full 15 years of the bond term. The cell should be formatted for percentage.
- Number of concerts in year—A firm would be hired to attract concerts and other events to the arena. Enter the number of event days for the year. The entry applies to each year.
- Average concert attendance—Enter the number of tickets expected to be sold for an average concert. The entry applies to each year. The maximum number would be 5,000, which is the seating capacity of the arena.
- Average concert ticket—Enter the expected average price of a concert ticket. The entry applies to each year. The cell should be formatted for currency.

The cells should be formatted appropriately for numbers, currency, or percentages. You should insert a comment in each of the cells and use conditional formatting in one or more cells. The existence of a comment is indicated by a diamond in the upper-right corner of a cell; place the mouse pointer over the diamond to see the comment. To enter a comment, right-click a cell and then choose Insert Comment from the menu. Conditional formatting is available from the Styles group on the Home tab; click the Conditional Formatting button and then click Highlight Cells Rules from the drop-down menu.

Summary of Key Results Section

Your worksheet should include the key results shown in Figure 10-3. The values are echoed from other parts of your spreadsheet. An explanation of the line items follows the figure. Your model runs for 15 years through 2029, but values through 2021 only are shown in Figure 10-3.

	A	B	C	D	E	F	G	H	I
23	**SUMMARY OF KEY RESULTS**	**2014**	**2015**	**2016**	**2017**	**2018**	**2019**	**2020**	**2021**
24	Intangibles: Arena net income less bond payment	NA							
25	Accumulated value of intangibles	NA		NA	NA	NA	NA	NA	NA

Source: Used with permission from Microsoft Corporation

FIGURE 10-3   Summary of Key Results section

- Intangibles: Arena net income less bond payment—The amount echoed here equals the arena's net income minus the year's bond payment. Each year, the arena will take in revenues and incur expenses. The excess of revenues over expenses may not be enough to cover the required bond payment. A shortfall during a year would be called the value of the arena's "intangibles."
- Accumulated value of intangibles—This value is the sum of the arena's net income minus bond payments for the 15 years of the analysis.

Calculations Section

The first few years of data in the Calculations section are shown in Figure 10-4. Your model runs for 15 years through 2029, but values through 2021 only are shown in the figure. Values are calculated by formula. You should not hard-code a value in a formula unless you are told to do so. Use cell addresses when referring to constants in formulas unless otherwise directed. Use absolute addressing properly. An explanation of the line items follows the figure.

	A	B	C	D	E	F	G	H	I
27	**CALCULATIONS**	**2014**	**2015**	**2016**	**2017**	**2018**	**2019**	**2020**	**2021**
28	Construction bond payments in year	NA							
29	Number of assistant basketball coaches	NA							
30	Cost of assistant basketball coaches	NA							
31	Other added basketball program cost	NA							
32	Potential change in basketball wins	NA							
33	Expected number of basketball wins	13							
34	Increase in season tickets sold	NA							
35	Number of basketball season tickets sold	500							
36	Cost of season basketball ticket	$ 200							
37	Season ticket revenue	NA							
38	Added student fee	NA							
39	Student fee revenue	NA							
40	Walkup basketball tickets sold	650							
41	Cost of walkup ticket	$ 15							
42	Walkup ticket revenue	NA							
43	Luxury boxes rented	NA							
44	Luxury box revenue	NA							
45	Luxury box expense	NA							
46	College basketball concession income	NA							
47	Volleyball and gymnastics concession income	NA							

Source: Used with permission from Microsoft Corporation

FIGURE 10-4 Calculations section

- Construction bond payments in year—This amount is computed by the PMT function, which requires the monthly interest rate, the number of months in the full term of the bond, and the total principal. The monthly rate is the annual interest rate divided by 12; the annual rate is taken from the Inputs section. The number of months is the number of years in the bond term multiplied by 12. The total principal is $60 million. The function returns the monthly payment, which should be multiplied by 12 to determine the yearly amount. The function returns a negative amount, so multiply the result by –1. The amount computed is the same for each year. (The PMT function is discussed in Tutorial E.)
- Number of assistant basketball coaches—This amount is a function of the basketball recruiting effort from the Inputs section. If the effort remains the same, the men's and women's teams will each have one assistant coach. If recruiting efforts increase by a moderate amount, each team will have two assistant coaches. If efforts increase by the maximum amount, each team will have three assistant coaches.
- Cost of assistant basketball coaches—This amount is a function of the number of coaches (from the preceding row in the Calculations section) and the cost of a coach (from the Constants section).
- Other added basketball program cost—This cost is for additional tutors, coaching, and related expenses. The amount is a function of the recruiting effort, which is taken from the Inputs section. If the recruiting effort remains the same, then no additional program costs will be incurred. If recruiting efforts increase by a moderate amount, the additional program costs will be $250,000 per year. If efforts increase by the maximum amount, the added program costs will be $400,000 per year.
- Potential change in basketball wins—This amount is the potential increase in wins for a team in one season. The basketball recruiting effort from the Inputs section is used to predict

improvement. If the recruiting effort remains the same, then no additional wins would be expected. If recruiting efforts increase by a moderate amount, one more win per season would be expected. If efforts increase by the maximum amount, two more wins per season would be expected.

- Expected number of basketball wins—The number of wins in a year is the number of wins in the prior year plus the potential change in wins for the year, which is taken from the preceding row of the Calculations section. The expected number of wins in a year should not exceed 25. (See Tutorial E for a discussion of the MIN function, which can be used to set this limit.)

- Increase in season tickets sold—The increase in season tickets sold for a team depends on the expected number of wins in the *prior* year. For each win over 12 in the prior year, 50 more season tickets will be sold for the current season.

- Number of basketball season tickets sold—The number of season tickets sold for a team in a year is a function of the number sold in the prior year plus the increase in the number sold in the current year. The latter number is taken from the preceding row of the Calculations section. The university wants to allow up to 2,500 walkup ticket sales each game, so the number of season tickets sold should not exceed 2,500. (See Tutorial E for a discussion of the MIN function.)

- Cost of season basketball ticket—This amount is a function of the prior year's season ticket cost and an expected 2 percent increase each year. In other words, the price in 2015 will be 2 percent more than the $200 charged in 2014, and the price in 2016 will be 2 percent more than the price in 2015. You can hard-code this 2 percent increase in your formulas. Use the ROUND function and no decimals to eliminate pennies. (This built-in function is discussed in Tutorial E.)

- Season ticket revenue—This amount is a function of the number of basketball season tickets sold (from the Calculations section), the cost of a season ticket (also from the Calculations section), and the number of teams (2), a factor you can hard-code in your formulas.

- Added student fee—This amount depends on the added student fee from the Inputs section. If the added amount is low, the increase is $10 each year. If the added amount qualifies as medium, the increase is $20 each year. If the added amount qualifies as outrageous, the increase is $50 each year. Recall that the added fee is allocated to the construction effort.

- Student fee revenue—This amount is a function of the number of full-time students enrolled for the year (from the Constants section) and the added student fee (from the preceding row of the Calculations section).

- Walkup basketball tickets sold—The number of walkup tickets sold for each game will increase by 50 for each win over 12 expected in the current year. The number of wins expected in the current year is taken from the Calculations section. The number of walkup tickets sold for a game cannot exceed 2,500.

- Cost of walkup ticket—Each year, this cost will increase by a dollar per ticket. The cost per ticket should not exceed $25.

- Walkup ticket revenue—This amount is a function of the number of home games (13), the number of teams (2), the cost of a ticket (from the preceding row of the Calculations section), and the number of tickets sold per game (also from the Calculations section). You can hard-code the number of teams and the number of games.

- Luxury boxes rented—This number is a function of the prior season's wins. For each win over 12 in the prior year, two boxes are sold for the year. No more than 17 boxes can be sold.

- Luxury box revenue—This amount is a function of the number of boxes rented for the year (from the preceding row of the Calculations section) and the fee for the year (from the Constants section).

- Luxury box expense—This amount includes all expenses for catering food and drink to the luxury boxes and for maintaining the boxes in good condition. Each year, this amount is a fixed cost of $5,000 plus $3,000 multiplied by the number of boxes rented. You can hard-code the $3,000 and $5,000 amounts.

- College basketball concession income—This yearly amount is a function of the number of tickets sold per game (season plus walkup, both from the Calculations section), concession income per fan (from the Constants section), the number of home games in a season (13), and the number of teams (2). You can hard-code the number of teams and games.

- Volleyball and gymnastics concession income—This yearly amount is a function of the average attendance per event (from the Constants section), concession income per fan (from the Constants section), the number of events in a season (13), and the number of teams (2). You can hard-code the number of teams and events.

The remaining calculations are discussed next. In Figure 10-5, column B contains amounts for 2014, column C contains amounts for 2015, and so on.

	A	B	C	D	E	F	G	H	I
48	Pro basketball concession income	NA							
49	Women's football concession income	NA							
50	Concert revenue	NA							
51	Concert organizer fee	NA							
52	Advertising revenue	NA							

Source: Used with permission from Microsoft Corporation

FIGURE 10-5   Calculations section (continued)

- Pro basketball concession income—This yearly amount is a function of the attendance per game (from the Constants section), concession income per fan (also from the Constants section), and the number of games in a season (20). You can hard-code the number of games.
- Women's football concession income—This amount is zero if the arena is not used for women's football. Otherwise, this yearly amount is a function of the attendance per game (from the Constants section), concession income per fan (also from the Constants section), and the number of games in a season (10). You can hard-code the number of games.
- Concert revenue—This amount is a function of the number of concerts per year, the average attendance per concert, and the average price of a concert ticket. These values are all taken from the Inputs section.
- Concert organizer fee—This amount is the sum of a yearly retainer of $120,000 plus one-third of concert revenue. Concert revenue is a value taken from the previous row of the Calculations section. You can hard-code the retainer value in your formulas.
- Advertising revenue—The university expects to earn $3,000 in advertising revenue per event. The arena would host 52 university sporting events a year, 20 professional basketball games a year, and 10 professional football games a year if the league were adopted. The number of concert events is a value taken from the Inputs section. You can hard-code the number of events and $3,000 as the per-game revenue.

Cash Flow Statement Section

Data for the first few years of the Cash Flow Statement section are shown in Figure 10-6. Your model runs through 2029, but values through 2021 only are shown in the figure. Values are calculated by formula, not hard-coded. Use cell addresses when referring to constants in formulas unless otherwise directed. Use absolute addressing properly. An explanation of the line items follows the figure.

	A	B	C	D	E	F	G	H	I
54	**CASH FLOW STATEMENT**	**2014**	**2015**	**2016**	**2017**	**2018**	**2019**	**2020**	**2021**
55	Revenue:								
56	Men's and women's basketball	NA							
57	Concessions	NA							
58	Student fee	NA							
59	Luxury boxes	NA							
60	Concerts	NA							
61	Advertising	NA							
62	Pro basketball rental	NA							
63	Women's football rental	NA							
64	Naming rights	NA							
65	Miscellaneous income	NA							
66	Total revenue:	NA							
67	Costs and expenses:								
68	Utilities and maintenance	NA							
69	Added basketball coaches	NA							
70	Other added basketball cost	NA							
71	Luxury box expense	NA							
72	Concert organizer fee	NA							
73	Total costs and expenses	NA							
74	Arena net income	NA							

Source: Used with permission from Microsoft Corporation

FIGURE 10-6   Cash Flow Statement section

- Men's and women's basketball—This amount is the sum of men's and women's basketball revenue both from season tickets and walkup tickets. The values are taken from the Calculations section.
- Concessions—This amount is the sum of all concession revenues, which are taken from the Calculations section.
- Student fee—This amount is from the Calculations section and can be echoed here.
- Luxury boxes—This revenue amount is from the Calculations section and can be echoed here.
- Concerts—The amount of concert revenue is taken from the Calculations section and can be echoed here.
- Advertising—The amount of advertising revenue is taken from the Calculations section and can be echoed here.
- Pro basketball rental—The rent that the team would pay the university is taken from the Constants section and can be echoed here.
- Women's football rental—The rent that the team would pay the university is taken from the Constants section and can be echoed here. If the league and university do not reach an agreement, this amount is zero.
- Naming rights—This revenue amount is from the Constants section and can be echoed here.
- Miscellaneous income—This amount is from the Constants section and can be echoed here.
- Total revenue—This amount is the sum of all preceding revenue amounts in this statement.
- Utilities and maintenance—This amount is from the Constants section and can be echoed here.
- Added basketball coaches—This amount is from the Calculations section and can be echoed here.
- Other added basketball cost—This amount is from the Calculations section and can be echoed here.
- Luxury box expense—This amount is from the Calculations section and can be echoed here.
- Concert organizer fee—This amount is from the Calculations section and can be echoed here.
- Total costs and expenses—This amount is the sum of all preceding costs and expenses in this statement.
- Arena net income—This amount is the difference between total revenue and total costs and expenses. The university hopes that the arena's net income will be adequate to cover the year's construction bond payment.

The rest of the cash flow statement is discussed next. In Figure 10-7, column B contains amounts for 2014, column C contains amounts for 2015, and so on.

	A	B	C	D	E	F	G	H	I
76	Intangibles: Arena net income less bond payment	NA							
77	Accumulated value of intangibles	NA		NA	NA	NA	NA	NA	NA

Source: Used with permission from Microsoft Corporation

FIGURE 10-7   Cash Flow Statement section (continued)

- Intangibles: Arena net income less bond payment—This amount is the difference between revenue minus costs and expenses (arena net income) and the year's bond payment. The bond payment is taken from the Calculations section. If the result is a positive number, the net income is sufficient to cover the bond payment, so the university president need not claim that the arena conferred valuable intangible benefits. If the amount is negative, the net income is not sufficient, and the state must help make the bond payment. In this case, the university president will need to claim that the arena conferred intangible benefits and that the state subsidy is justified.
- Accumulated value of intangibles—This amount is the sum of intangibles for the 15 years that the bonds will need to be paid.

ASSIGNMENT 2: USING THE SPREADSHEET FOR DECISION SUPPORT

You will now complete the case by (1) using the spreadsheet model to gather data needed to answer questions from the CFO and university president about the arena's operations, (2) documenting your findings in a memo, and (3) giving an oral presentation if your instructor requires it.

Assignment 2A: Using the Spreadsheet to Gather Data

You have built the spreadsheet to create "what-if" scenarios with the model's input values. The inputs represent the logic of a question, and the outputs provide information needed to answer the question. The scenarios are based on the following questions from the CFO and university president.

Question 1 (base case): What is the shortfall in 2015 and the cumulative shortfall for 15 years in the "base case"? The inputs for this case are shown in Figure 10-8.

Recruiting Effort	2
Student Fee	M
Women's League	N
Interest Rate	3%
Number of Concerts	12
Average Attendance	3,000
Average Ticket Price	$55.00

Source: © Cengage Learning 2015

FIGURE 10-8   Input data for Question 1

Enter the inputs and then observe the outputs in the Summary of Key Results area. Next, you should record the results in a summary area by copying the data from the Summary of Key Results cells and pasting it into the summary area. Highlight the area to be copied, click the Copy button in the Clipboard group of the Home tab, and then click where you want the copy to be placed. Next, click the Paste button's down arrow in the Clipboard group, select Paste Special from the menu, and then select Values in the next window. Click OK. You could also use a second worksheet for this purpose, as shown in Figure 10-9. Note that the results shown in the figure are for illustrative purposes only.

	A	B
1	**BASE CASE:**	
2	**Inputs:**	
3	Recruiting Effort	2
4	Student Fee	M
5	Women's League	N
6	Interest Rate	3%
7	Number of Concerts	12
8	Average Attendance	3,000
9	Average Ticket Price	$ 55.00
10	**Results:**	
11	Shortfall in 2015	$ (2,000,000)
12	Accumulated Shortfall	$ (12,000,000)

Source: Used with permission from Microsoft Corporation

FIGURE 10-9   Inputs for Question 1 and results recorded in summary area

Question 2: What if the arena is built but the university does not substantially upgrade the basketball operation and schedules few concerts? The inputs are shown in Figure 10-10.

Recruiting Effort	1
Student Fee	L
Women's League	N
Interest Rate	3%
Number of Concerts	6
Average Attendance	3,000
Average Ticket Price	$55.00

Source: © Cengage Learning 2015

FIGURE 10-10   Input data for Question 2

Enter the inputs, observe the outputs in the Summary of Key Results area, and then record the results in the summary area.

Question 3: What if the university is aggressive with the basketball operations, does host the women's league, and books more than one concert per month? The inputs are shown in Figure 10-11.

Recruiting Effort	3
Student Fee	O
Women's League	Y
Interest Rate	3%
Number of Concerts	18
Average Attendance	3,000
Average Ticket Price	$55.00

Source: © Cengage Learning 2015

FIGURE 10-11   Input data for Question 3

Enter the inputs, observe the outputs in the Summary of Key Results area, and then record the results in the summary area.

Question 4: What if the project was delayed and the interest rate increased to 4 percent? In other words, what if you adjusted the base case for 4 percent interest? Enter the inputs, observe the outputs in the Summary of Key Results area, and then record the results in the summary area.

Question 5: What is the impact of the women's league? How much would the shortfall change in Year 1 and overall if the university and the women's league reached an agreement? In other words, what if you adjusted the base case to host the women's league in the arena? Enter the inputs, observe the outputs in the Summary of Key Results area, and then record the results in the summary area.

Question 6: What is the impact of the student fee? How much would the shortfall change in Year 1 and overall if the fee was increased significantly? In other words, what if you adjusted the base case to allow the Outrageous fee? Enter the inputs, observe the outputs in the Summary of Key Results area, and then record the results in the summary area.

Question 7: How many concerts per year would be needed to produce a surplus in Year 1? In other words, what if you adjusted the base case to increase the number of concerts? Enter the inputs, observe the outputs in the Summary of Key Results area, and then record the results in the summary area.

Question 8: What is the financial impact of adjusting the base case to employ the maximum basketball recruiting effort? Enter the inputs, observe the outputs in the Summary of Key Results area, and then record the results in the summary area.

Question 9: The university president thinks that the governor would not want the arena to host more than two concerts per month. The president also thinks that the governor would not be upset with a Year 1 shortfall as long as it was less than $1 million and the 15-year total was positive. Assuming an interest rate of 3 percent, what strategy (if any) can you create to achieve this result? Use what-if scenarios with the inputs to achieve the needed result, and then record the results in the summary area.

When you finish gathering data for the preceding nine questions, print the model's worksheet with any set of inputs. Print the summary sheet as well, and then save the spreadsheet for the final time by selecting Save from the File menu.

Assignment 2B: Documenting your Findings and Recommendation in a Memo

Document your findings in a memo that answers the nine questions from the preceding section. The memo should also summarize your general conclusions about how the university president should try to convince the governor to fund the arena. Use the following guidelines to prepare your memo in Microsoft Word:

- Your memo should have proper headings such as Date, To, From, and Subject. You can address the memo to the university president. You should set up the memo as discussed in Tutorial E.
- Briefly outline the situation. You need not provide much background—you can assume that readers are familiar with the situation.
- Answer the nine questions in the body of the memo.

- What general advice can you give to the president that will help him convince the governor to fund the arena? In other words, what combination of basketball operations, student fees, and hosting the women's league and concerts makes the best financial sense?
- Include tables and charts to support your claims, as your instructor specifies. Tutorial E explains how to create a table in Microsoft Word. Tutorial F explains how to create charts in Excel.

ASSIGNMENT 3: GIVING AN ORAL PRESENTATION

Your instructor may request that you explain your analysis and results in an oral presentation. If so, assume that the CFO and university president are impressed by your analysis and want you to give a presentation to summarize the results. Prepare to speak for 10 minutes or less. Use visual aids or handouts that you think are appropriate. Tutorial F explains how to prepare and give an oral presentation.

DELIVERABLES

Assemble the following deliverables for your instructor:

1. Printout of your memo
2. Spreadsheet printouts
3. Flash drive or CD that includes your Word file and Excel file

Staple the printouts together with the memo on top. If you have more than one .xlsx file on your flash drive or CD, write your instructor a note that identifies your spreadsheet model's .xlsx file.

PART 5

INTEGRATION CASES USING ACCESS AND EXCEL

THE COLLEGE GPA ANALYSIS

Decision Support with Access and Excel

PREVIEW

A small college needs to know what factors it can use to predict the future grade point averages (GPAs) of high school students who apply for admission. Administrators will use this knowledge to accept or reject the students' applications. In this case, you will use Microsoft Access and Excel to perform the analysis.

PREPARATION

- Review database and spreadsheet concepts discussed in class and in your textbook.
- Complete any exercises that your instructor assigns.
- Complete any parts of Tutorials B, C, and D that your instructor assigns, or refer to them as necessary.
- Review file-saving procedures for Microsoft Windows programs, as discussed in Tutorial C.
- Refer to Tutorials E or F as necessary.

BACKGROUND

A small New England college draws its 800 students from Rhode Island, Vermont, New Hampshire, and Maine. College administrators want to know how the average GPA for certain groups of their students compare. For example, how do men's GPAs compare with women's GPAs? How do the GPAs of students in sororities and fraternities compare with those of other students? Do GPAs differ depending on the students' home state? Do student GPAs improve each year at the college? Can the current students' data be used to predict the performance of future students and accept or reject the applications of high school seniors?

A database file has been started, and is available for your use. The file will save you time and reduce the amount of data you need to enter. Locate Case 11 in your data files and then select **College.accdb**.

The database has two tables called Student and Accomplishment. These tables are discussed next. Figure 11-1 shows the design of the Student table.

Field Name	Data Type
Student Number	Short Text
State	Short Text
Gender	Short Text
Greek	Short Text
Year	Short Text

Source: Used with permission from Microsoft Corporation

FIGURE 11-1 Design of Student table

Student Number is the primary key field in the table; each student is assigned a unique identification number. The value in the State field is the student's home state. The next field indicates the student's gender. The value in the Greek field indicates whether the student has joined a fraternity or sorority while at school. The value in the Year field shows the student's status as of the end of the current year: freshman, sophomore, junior, or senior. Figure 11-2 shows a few records in the Student table.

Student

Student Number	State	Gender	Greek	Year
1001	Maine	M	Y	SR
1002	Maine	M	Y	SR
1003	Maine	M	Y	SR
1004	Rhode Island	M	N	SR
1005	Vermont	M	N	SR
1006	New Hampshire	M	N	SR
1007	Vermont	F	Y	SR

Source: Used with permission from Microsoft Corporation

FIGURE 11-2 Data records in Student table

For example, student 1001 is a man from Maine. He joined a fraternity during his college career. The student is a senior.

The Accomplishment table shows the hours taken and credits earned by each student as of the end of the current year. The table's design is shown in Figure 11-3. Figure 11-4 shows a few records from the Accomplishment table.

Field Name	Data Type
Student Number	Short Text
Hours	Number
Credits	Number

Source: Used with permission from Microsoft Corporation

FIGURE 11-3 Design of Accomplishment table

Accomplishment

Student Number	Hours	Credits
1001	120	462
1002	120	292
1003	120	381
1004	120	340
1005	120	453
1006	120	425
1007	120	478

Source: Used with permission from Microsoft Corporation

FIGURE 11-4 Data records in Accomplishment table

For example, student 1001 has taken 120 hours of coursework and has earned 462 academic credits.

In Assignment 1, you will gather data to answer a set of questions for administrators, using the Access database and an Excel spreadsheet. In Assignment 2, you will write a memorandum that discusses your findings. In Assignment 3, you may be asked to prepare an oral presentation of your analysis.

ASSIGNMENT 1: USING ACCESS AND EXCEL TO GATHER DATA FOR DECISION SUPPORT

You will create two queries in Access. The query output will be imported into Excel and then analyzed using pivot tables. You will use the student GPA data to identify high school seniors who should be accepted as incoming freshmen at the college.

Creating Two Queries in Access

Create a query to calculate each college student's GPA (credits divided by hours). Format the output field as a standard number with three decimals. Save the query as GPADATA. Your output should look like that in Figure 11-5.

GPADATA

Student Number	State	Gender	Greek	Year	GPA
1001	Maine	M	Y	SR	3.850
1002	Maine	M	Y	SR	2.433
1003	Maine	M	Y	SR	3.175
1004	Rhode Islanc	M	N	SR	2.833
1005	Vermont	M	N	SR	3.775
1006	New Hampsl	M	N	SR	3.542
1007	Vermont	F	Y	SR	3.983

Source: Used with permission from Microsoft Corporation

FIGURE 11-5 Student GPAs

Notice that student 1001's GPA is 3.850, which is 462 credits divided by 120 hours. You will have 800 output records, although only a few are shown in Figure 11-5.

Next, create a query to calculate each college senior's GPA. Format the output field as a standard number with three decimals. Save the query as SRGPA. Your output should look like that in Figure 11-6.

SRGPA

Student Number	State	Gender	GPA
1001	Maine	M	3.850
1002	Maine	M	2.433
1003	Maine	M	3.175
1004	Rhode Islanc	M	2.833
1005	Vermont	M	3.775
1006	New Hampsl	M	3.542
1007	Vermont	F	3.983

Source: Used with permission from Microsoft Corporation

FIGURE 11-6 Senior student GPAs

You will have 200 output records, although only a few are shown in Figure 11-6.

Close the database and exit Access. You will now import your query data into Excel from Access. A spreadsheet file has been started, and is available for your use. Locate Case 11 in your data files and then select **GPA.xlsx**. The file contains a worksheet named Student GPA and a worksheet named Predict GPA.

Importing Query Output into Excel

Import the GPADATA query output into Excel. To begin, select cell A1 of the Student GPA worksheet. Click the Data tab, and then select From Access in the Get External Data group. Specify the filename of the database, and then specify the GPADATA query and where to place the imported data (cell A1). The data is imported into Excel as a data table, which is the format you want. The first few rows of your table should look like the data in Figure 11-7.

	A	B	C	D	E	F
1	Student Number	State	Gender	Greek	Year	GPA
2	1001	Maine	M	Y	SR	3.85
3	1002	Maine	M	Y	SR	2.433333333
4	1003	Maine	M	Y	SR	3.175
5	1004	Rhode Island	M	N	SR	2.833333333
6	1005	Vermont	M	N	SR	3.775
7	1006	New Hampshire	M	N	SR	3.541666667
8	1007	Vermont	F	Y	SR	3.983333333

Source: Used with permission from Microsoft Corporation

FIGURE 11-7 First few rows of the Student GPA worksheet

Import the SRGPA data into the Predict GPA worksheet in the same way. The first few rows of the worksheet should look like the data in Figure 11-8.

	A	B	C	D
1	Student Number	State	Gender	GPA
2	1001	Maine	M	3.85
3	1002	Maine	M	2.433333333
4	1003	Maine	M	3.175
5	1004	Rhode Island	M	2.833333333
6	1005	Vermont	M	3.775
7	1006	New Hampshire	M	3.541666667
8	1007	Vermont	F	3.983333333

Source: Used with permission from Microsoft Corporation

FIGURE 11-8   First few rows of the Predict GPA worksheet

After importing the student data, you need to assign codes to the states as follows: Maine, 1; Rhode Island, 2; Vermont, 3; and New Hampshire, 4. Likewise, you need to assign gender codes of 1 for Male and 2 for Female.

You will want to work with the data as a regular data range, so select cell A1 and then select Convert to Range in the Tools group on the Design tab.

Using Pivot Tables to Gather Data

Create a set of pivot tables to provide data that helps answer the following questions:

1.   Do the average GPAs of men and women differ significantly?
2.   Do the average GPAs of "Greek" students and non-Greeks differ significantly?
3.   Do average GPAs differ significantly by state?
4.   Do average GPAs differ significantly by year? In other words, do student GPAs improve noticeably from freshman year to sophomore year, and so on? Do their GPAs decline noticeably?

For details on creating and using pivot tables, see Tutorial E. A simplified procedure is summarized here using the Gender pivot table as an example. To begin, click cell A1 in the data table. Select the Insert tab, and then click the Pivot Table button from the Tables group. Choose the existing worksheet as the target location. Next, specify a likely pivot table location; H1 would be a good choice. Drag the Gender field label from the pivot table field list to the Rows area in the bottom-right corner. Finally, drag the GPA field label to the Sigma Values area, and change the Value Field Setting from Sum to Average.

Your pivot table should look like the one shown in Figure 11-9.

Source: Used with permission from Microsoft Corporation

FIGURE 11-9   The Gender pivot table

In the figure, notice that the Grand Total row shows the average for all students.

Assuming that a difference of .10 is significant for your analysis, answer the following questions:

1.   Do the average GPAs of men and women differ by more than .10?
2.   Do the average GPAs of Greek students and non-Greeks differ by more than .10?
3.   Does any state's average GPA differ by more than .10 from the overall average GPA?

4. Does the average sophomore's GPA differ by more than .10 from the average freshman's GPA? Does the average junior's GPA differ by more than .10 from the average sophomore's GPA? Does the average senior's GPA differ by more than .10 from the average junior's GPA?

By formula, show the differences next to the four pivot tables. By manual entry or by formula, indicate which differences are significant and which are insignificant.

Using Pivot Tables to Gather More Refined Data

You want to know how male and female students progress from year to year. In other words, for each gender, how do average GPAs change from freshman year to sophomore year, from sophomore year to junior year, and so on? Set up a pivot table for this data. The lower-right corner of the pivot table is shown in Figure 11-10.

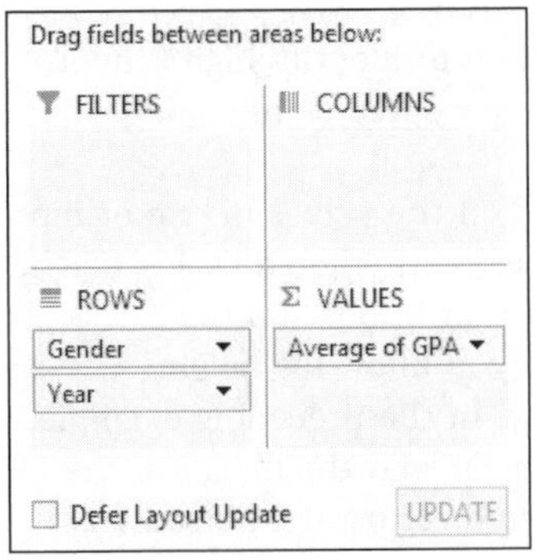

Source: Used with permission from Microsoft Corporation

FIGURE 11-10 Pivot table setup

Enter a note next to the pivot table that states how average GPAs change from year to year when analyzed by students' gender. You can also enter this note using a cell comment. To enter a cell comment, right-click a cell and select Insert Comment from the menu.

You also want to know how male and female student GPAs differ when analyzed by their home states. For example, how do the GPAs of female students from Vermont differ from those of their male counterparts from Vermont? Set up a pivot table to display the data. Use a cell comment or a note next to the pivot table to indicate which states (if any) have significant GPA differences between the genders. For example, if the GPAs of male students from New Hampshire are significantly better than those of their female counterparts from New Hampshire, indicate that finding in your note.

Using College Student GPAs to Identify High School Seniors Who Should Be Accepted as Incoming Freshmen

The Predict GPA worksheet contains records for 50 high school seniors who are applying to the college. The first few records are shown in Figure 11-11.

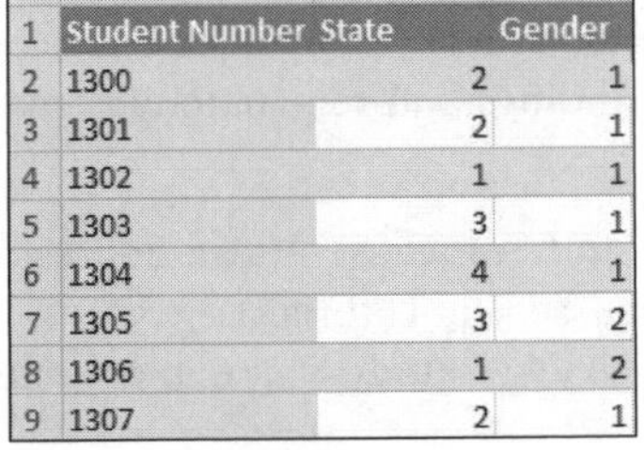

	G	H	I
1	Student Number	State	Gender
2	1300	2	1
3	1301	2	1
4	1302	1	1
5	1303	3	1
6	1304	4	1
7	1305	3	2
8	1306	1	2
9	1307	2	1

Source: Used with permission from Microsoft Corporation

FIGURE 11-11 Applicant data

Earlier in this case, you imported seniors' GPA data into the Predict GPA worksheet. Now you need to convert the data using the coding system for states and genders that was explained previously. You should make these conversions using IF statements in columns to the right of the seniors' data. Then copy the coded data back to columns B and C. Highlight the area to be copied, click the Copy button in the Clipboard group of the Home tab, and then click where you want the copy to be placed. Next, click the Paste button's down arrow in the Clipboard group, select Paste Special from the menu, and then select Values in the next window. Click OK. The first few rows of the Predict GPA worksheet should then look like the data shown in Figure 11-12.

	A	B	C	D	E	F	G	H	I
1	Student Number	State	Gender	GPA			Student Number	State	Gender
2	1001	1	1	3.85			1300	2	1
3	1002	1	1	2.433333333			1301	2	1
4	1003	1	1	3.175			1302	1	1
5	1004	2	1	2.833333333			1303	3	1
6	1005	3	1	3.775			1304	4	1
7	1006	4	1	3.541666667			1305	3	2

Source: Used with permission from Microsoft Corporation

FIGURE 11-12   Data in Predict GPA worksheet

The next step is to predict college GPAs for the high school students, and then to identify high school students whose GPAs are predicted to be 3.00 or greater. These students will be accepted as incoming freshmen.

You can use the TREND function to forecast the value of one variable based on the values of one or more other variables. For example, you might know that a man's weight can sometimes be predicted by his height. If you know the heights and weights of 100 men, you could use the TREND function to develop the relationship between weight and height, and then predict the weight of a man if you know his height.

The syntax of the TREND function is =TREND(known Ys, known Xs, new X). In the preceding example, the 100 known weights are the known Ys, the 100 known heights are the known Xs, and the height of the person whose weight you want to predict is the new X. The TREND function would return the person's predicted weight. Tutorial E discusses the TREND function in more detail.

In the data, the college senior GPAs in column D are the known Ys. The known Xs are the state and gender values in columns B and C. The TREND function "learns" the relationship between the GPAs and the state and gender values. Next, the function uses what it has learned to predict each high school senior's college GPA based on his or her state and gender.

Based on the predicted GPA, you should use an IF statement to accept or reject the applicant. Figure 11-13 shows the values you should get for the first high school applicant.

	A	B	C	D	E	F	G	H	I	J	K
1	Student Number	State	Gender	GPA			Student Number	State	Gender	GPA Prediction	Accept?
2	1001	1	1	3.85			1300	2	1	2.88	Reject

Source: Used with permission from Microsoft Corporation

FIGURE 11-13   Predicted GPA and acceptance decision for a student

Use the COUNTIF function to determine the numbers of students who are accepted and rejected. (The COUNTIF function is discussed in Tutorial E.) Based on the GPA threshold of 3.00 and the overall average college GPA shown in your pivot tables, are the numbers of accepted students and rejections reasonable? Make a note about this question in your Predict GPA spreadsheet.

ASSIGNMENT 2: DOCUMENTING FINDINGS IN A MEMO

In this assignment, you write a memo in Microsoft Word that documents your findings and recommendations. In your memo, observe the following requirements:

- Your memo should have proper headings such as Date, To, From, and Subject. You can address the memo to the college's administrators. Set up the memo as discussed in Tutorial E.
- Briefly outline the situation. However, you need not provide much background—you can assume that readers are familiar with the situation.
- State your findings in the body of the memo. You should answer the questions based on your pivot table data. You should summarize the results of accepting high school seniors based on the GPAs of current students.
- Support your work graphically, following your instructor's guidance. You may need to show results in a table. Tutorial E describes how to create a table in Microsoft Word.

ASSIGNMENT 3: GIVING AN ORAL PRESENTATION

Your instructor may request that you summarize your analysis and results in an oral presentation. Assume that you would be talking to college administrators for 10 minutes or less. Use visual aids or handouts as appropriate. See Tutorial F for guidance on preparing and giving an oral presentation.

DELIVERABLES

Assemble the following deliverables for your instructor:

1. Printout of your memo
2. Spreadsheet and/or database printouts, as required by your instructor
3. Electronic media such as flash drive or CD, which should include your Word file, Access file, and Excel file

Staple the printouts together with the memo on top. If you have .xlsx or .accdb files from other assignments stored on your electronic media, write your instructor a note that identifies the files for this assignment.

THE NONPROFIT DONOR ANALYSIS

Decision Support with Access and Excel

PREVIEW

A nonprofit organization has many donors who contribute small amounts. In this case, you will use Microsoft Access and Excel to analyze donations in previous years and use the data to help predict the level of donations in the current year.

PREPARATION

- Review database and spreadsheet concepts discussed in class and in your textbook.
- Complete any exercises that your instructor assigns.
- Complete any parts of Tutorials B, C, and D that your instructor assigns, or refer to them as necessary.
- Review file-saving procedures for Windows programs, as discussed in Tutorial C.
- Refer to Tutorials E and F as necessary.

BACKGROUND

A nonprofit organization in your hometown is dedicated to feeding and clothing homeless people. Data for donations in prior years has been recorded in an Access database. The organization has a new chief financial officer (CFO) who wants to update the database records in several ways and start a formal budgeting process. You have been called in to help the CFO with these tasks.

The CFO knows which donors have died in the current year (2015) and wants to delete their records from the database. The CFO believes that people who stop giving to a charity cannot be persuaded to resume giving, so she also wants to delete the records of people who have not made donations in the past two years.

The CFO wants to predict the level of donations for the current year based on an analysis of donations in prior years. Donations typically come from people who live nearby, as indicated by donors' zip codes. People of all ages donate. The CFO wants to know if trends exist for donations based on donors' ages or zip codes.

Data for donors and their history of giving is included in the Development.accdb database. The database is available for your use; it will save you time and reduce the amount of data you need to enter. Locate Case 12 in your data files and then select **Development.accdb**.

The tables in the database file are discussed next. Figure 12-1 shows the design of the Biographical table.

Biographical	
Field Name	Data Type
Donor Num	Short Text
Last Name	Short Text
First Name	Short Text
Gender	Short Text
Zip	Number
YOB	Number

Source: Used with permission from Microsoft Corporation

FIGURE 12-1   Design of Biographical table

Each donor has a unique number in the Donor Num field, which is the primary key. The YOB field records each donor's year of birth.

Figure 12-2 shows a few records in the Biographical table.

Biographical						
Donor Num	Last Name	First Name	Gender	Zip	YOB	Click to Add
10	Agostino	Joseph	m	19603	1969	
100	Bowies	Michael	m	19600	1988	
101	Boyce	Charlotte	f	19601	1972	

Source: Used with permission from Microsoft Corporation

FIGURE 12-2   Data records in Biographical table

For example, Donor 10 is a man named Joseph Agostino. He lives in zip code 19603, and he was born in 1969.

Unfortunately, the charity must keep updated information about which donors are alive and which have died during the current year. This data is recorded in the Deceased In Year table. Figure 12-3 shows the design.

Deceased In Year	
Field Name	Data Type
Donor Num	Short Text
Deceased	Yes/No

Source: Used with permission from Microsoft Corporation

FIGURE 12-3   Design of Deceased In Year table

The table contains a record for each donor. The Deceased field has a Yes/No data type to indicate whether a donor died in the current year. A "Yes" value is indicated by a check in the box. Figure 12-4 shows a few records in the Deceased In Year table.

Deceased In Year		
Donor Num	Deceased	Click to Add
1	☐	
10	☐	
100	☐	
101	☐	
102	☐	

Source: Used with permission from Microsoft Corporation

FIGURE 12-4   Data records in Deceased In Year table

For example, Donor 10 has not died during the current year.

The Donation History table shows how much money each donor gave to the organization in each of the prior five years from 2010 through 2014. Figure 12-5 shows the design of the Donation History table, and Figure 12-6 shows a few records in the table.

Donation History	
Field Name	Data Type
Donor Num	Short Text
2010	Currency
2011	Currency
2012	Currency
2013	Currency
2014	Currency

Source: Used with permission from Microsoft Corporation

FIGURE 12-5   Design of Donation History table

Donation History					
Donor Num	2010	2011	2012	2013	2014
1	$375.00	$375.00	$410.00	$445.00	$480.00
10	$750.00	$990.00	$1,150.00	$1,350.00	$1,510.00
100	$40.00	$80.00	$110.00	$100.00	$160.00
101	$750.00	$870.00	$1,070.00	$1,270.00	$1,310.00
102	$750.00	$1,030.00	$1,110.00	$1,150.00	$1,190.00

Source: Used with permission from Microsoft Corporation

FIGURE 12-6   Data records in Donation History table

For example, Donor 10 gave $750 in 2010, $990 in 2011, $1,150 in 2012, $1,350 in 2013, and $1,510 in 2014.

In Assignment 1, you will gather data for the CFO using the Access database and an Excel spreadsheet. In Assignment 2, you will write a memorandum that summarizes your findings. In Assignment 3, you may be asked to prepare an oral presentation of your findings.

ASSIGNMENT 1: USING ACCESS AND EXCEL FOR DECISION SUPPORT

The CFO wants to streamline the donor database, and has four queries in mind to help achieve that purpose. After you create the queries, you will import the Access data into Excel for further analysis and a projection of 2015 giving.

Streamlining the Database

In the first query, you develop data about donors who died during the current year (2015). In the second query, you summarize the data for deceased donors. In the third query, you develop data for donors who have not given recently. In the fourth query, you develop a list of people who are most likely to give in the current year.

Query 1

The CFO wants to know the donor number, name, and year of birth of each prior donor who died during the current year (2015). Create a query that shows this information. The first part of the query output should look like the data in Figure 12-7.

Deceased in this Year				
Donor Num	Last Name	First Name	YOB	Deceased
239	Finnegan	Noah	1932	☑
390	James	Madie	1932	☑
549	Millersville	Angel	1936	☑

Source: Used with permission from Microsoft Corporation

FIGURE 12-7   Data for donors who died in 2015

Save the query as Deceased in this Year.

Query 2

The CFO also wants to see summary data for donors who died during the current year. Create a query that shows how many people died, and sort the data by the donors' year of birth. The first part of the query output should look like the data in Figure 12-8.

Deceased in This Year Summary	
Number In Year	YOB
2	1932
2	1936
1	1937
1	1940

Source: Used with permission from Microsoft Corporation

FIGURE 12-8   Summary data for donors who died in 2015

Save the query as Deceased in This Year Summary.

Query 3

The CFO wants to confirm her belief that people who stop giving to a charity cannot be persuaded to resume giving. She wants you to create a query that shows how much money people gave in 2014 if they gave nothing in the prior year. The first part of the query output should look like the data in Figure 12-9.

Zero Donations

Donor Num	Last Name	First Name	2013	2014	YOB
432	King	Jack	$0.00	$0.00	1930
155	Casper	Jarrod	$0.00	$0.00	1930
239	Finnegan	Noah	$0.00	$0.00	1932

Source: Used with permission from Microsoft Corporation

FIGURE 12-9   2014 donations from people who did not give in 2013

You can scroll the query output to see how frequently people resume giving after a year of not giving. Save the query as Zero Donations.

Query 4

In 2015, the CFO wants to solicit donors who gave money during each year of the period from 2010 to 2014, and who are still alive in 2015. Create a query whose output lists those donors. The first part of the query output should look like the data in Figure 12-10.

Possibles

Donor Num	Last Name	First Name	Gender	Zip	YOB	Deceased	2010	2011	2012	2013	2014
869	Young	Zetta	f	19601	1930	☐	$500.00	$550.00	$675.00	$765.00	$855.00
46	Bailey	Elizabeth	f	19603	1930	☐	$100.00	$150.00	$275.00	$355.00	$460.00
342	Hooper	Bryson	m	19603	1930	☐	$400.00	$475.00	$525.00	$545.00	$635.00
826	Ward	Annabelle	f	19603	1930	☐	$500.00	$525.00	$650.00	$690.00	$765.00

Source: Used with permission from Microsoft Corporation

FIGURE 12-10   Donors to solicit in 2015

The output should be sorted by the donors' year of birth. Save the query as Possibles. Close the database and exit Access.

Importing Data into Excel

Open a new file in Excel and save it as **Development.xlsx**.

Import the Possibles query output into Excel. To begin, click the Data tab, and then select From Access in the Get External Data group. Specify the filename of the Access database, and then specify the Possibles query and where to place the imported data in the worksheet (cell A1 is recommended).

The data is imported into Excel as a data table, which is the format that you want. If cell A1 is not already selected, click in A1. In the Table Style Options group, select Total Row to add a totals row to the bottom of the table. Rename the worksheet Possibles. The first few rows of your worksheet should look like Figure 12-11.

	A	B	C	D	E	F	G	H	I	J	K	L
1	Donor Num	Last Name	First Name	Gender	Zip	YOB	Deceased	2010	2011	2012	2013	2014
2	869	Young	Zetta	f	19601	1930	FALSE	500	550	675	765	855
3	46	Bailey	Elizabeth	f	19603	1930	FALSE	100	150	275	355	460
4	342	Hooper	Bryson	m	19603	1930	FALSE	400	475	525	545	635
5	826	Ward	Annabelle	f	19603	1930	FALSE	500	525	650	690	765

Source: Used with permission from Microsoft Corporation

FIGURE 12-11   Rows in the Possibles worksheet

Using a Data Table, Chart, and Pivot Tables to Gather Data

Use the data table to gather some of the data you need. You will develop pivot tables from the data table and then use the pivot tables to develop data about yearly changes in the rates of donations.

The Possibles worksheet shows donations from everyone who has given to the organization in each of the past five years from 2010 to 2014. You need to determine the total amount of money given in the past five years and calculate yearly changes in the rate of overall giving.

Compute each year's total in the Total row of the data table using the row's Sum function. Then use an Excel expression to compute the rate of increase for each year. The formula for calculating the rate of change is:

(current year's amount less prior year's amount) / (prior year's amount)

Your results should look like the data in Figure 12-12.

	A	B	C	D	E	F	G	H	I	J	K	L
761	649	Powell	Tristan	m	19601	1993	FALSE	80	110	140	125	125
762	129	Bullman	Tyler	m	19600	1993	FALSE	120	170	200	195	180
763	740	Sandquist	Joseph	m	19603	1993	FALSE	40	40	90	85	100
764	139	Butler	Kasandra	f	19600	1994	FALSE	160	170	180	235	265
765	193	Cox	London	f	19601	1994	FALSE	200	210	220	245	320
766	Total							$ 302,480	$370,685	$ 432,470	$ 511,175	$ 589,890
767						Rate of Increase			22%	19%	17%	14%

Source: Used with permission from Microsoft Corporation

FIGURE 12-12   Calculation of total giving and rates of change

In the figure, values in the Total row have been formatted as currency, and the values for the rate of increase are for illustrative purposes only. You should add a comment or type text in the spreadsheet to describe the trends in total giving and how the donation rate has changed over time. In other words, are total donations increasing or decreasing? Is the rate of increase decelerating or accelerating?

Next, show the total amount of donations in a scatter chart. A convenient way to create this chart is to manually enter the totals into the spreadsheet beneath the data table, including appropriate headings. Highlight the data, including the headings. Click the Insert tab, click the Scatter button in the Charts group, and then select the type of scatter chart you want. You can improve the chart's title by clicking it and then entering the text you want. Your chart should look like the one in Figure 12-13.

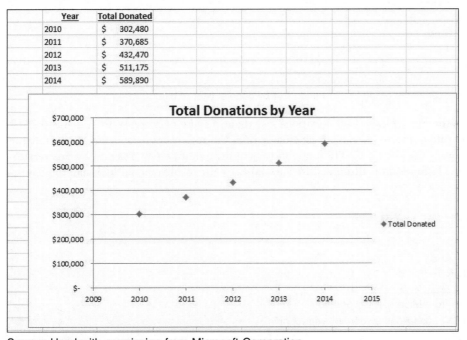

Year	Total Donated
2010	$ 302,480
2011	$ 370,685
2012	$ 432,470
2013	$ 511,175
2014	$ 589,890

Source: Used with permission from Microsoft Corporation

FIGURE 12-13   Scatter chart of total donations for each year

It should be clear from the figure that donors have been generous in recent years. Does giving differ by gender? To answer that question, create a pivot table that shows total donations in 2013 and 2014 by gender. Click cell A2 to tell Excel to base the pivot table on all rows in the data table. Place the pivot table in the existing worksheet; cell N2 would be a good location. Your results should look like the data in Figure 12-14.

N	O	P	Q
Row Labels ▾	**Sum of 2013**	**Sum of 2014**	**Increase%**
f	$ 224,135	$ 257,270	
m	$ 287,040	$ 332,620	
Grand Total	**$ 511,175**	**$ 589,890**	

Source: Used with permission from Microsoft Corporation

FIGURE 12-14 Pivot table that compares male and female giving in 2013 and 2014

Next to the pivot table, use a formula to compute the yearly rate increase in donations for female and male donors. Enter a cell comment or text that indicates whether men's and women's donations are significantly different.

Does giving differ by zip code? To answer that question, create a pivot table that shows total donations in 2013 and 2014 by zip code. Base the pivot table on the entire data table. Place the pivot table in the existing worksheet beneath the first pivot table. Use a formula to compute the yearly rate increase in donations for each of the three zip codes. Enter a cell comment or text that indicates whether donations from the three zip codes are significantly different.

Does giving differ based on a donor's year of birth? To answer that question, create a pivot table that shows 2013 and 2014 total donations by year of birth. Base the pivot table on the entire data table. Place the pivot table in the existing worksheet beneath the zip code pivot table. Use a formula to compute the rate increases in donations for each year.

Use an IF statement to show whether the percentage change in donations from 2013 to 2014 is less than the percentage change from 2012 to 2013. You can hard-code the percentage change from 2012 to 2013 in your IF statement. Copy the formula down the column to apply it to all years. The top part of your pivot table should look like the data in Figure 12-15.

N	O	P	Q	R
Row Labels ▾	**Sum of 2013**	**Sum of 2014**	**Increase%**	
1930	$ 2,355	$ 2,715		
1931	$ 1,540	$ 1,720		
1932	$ 1,615	$ 1,930		

Source: Used with permission from Microsoft Corporation

FIGURE 12-15 Pivot table that compares giving in 2013 and 2014 by donors' year of birth

In how many years was the rate of increased donations less than the rate of increase from 2012 to 2013? In how many years was the rate of increased donations greater than the rate of increase from 2012 to 2013? At the bottom of the pivot table, use the COUNTIF function to count the numbers of Yes and No answers to these questions. (The COUNTIF function is discussed in Tutorial E.) Your results should look like the data in Figure 12-16.

N	O	P	Q	R
1993	$ 720	$ 705	-2.1%	Y
1994	$ 480	$ 585	17.9%	Y
Grand Total	**$ 511,175**	**$ 589,890**		
Increased less than 2012-2013 rate of increase?			Y	30
			N	35

Source: Used with permission from Microsoft Corporation

FIGURE 12-16 Summary data of donations by donors' year of birth

In the figure, the Yes and No values shown (30 and 35) are for illustrative purposes only. Enter a cell comment or text to indicate whether giving is affected significantly by a donor's age. In other words, is the change consistent for most ages?

Does giving differ by some combination of donors' gender and zip code? To answer that question, create a pivot table that shows total donations in 2013 and 2014, organized both by zip code and gender. Base the pivot table on the entire data table, and place the pivot table in the existing worksheet. Use formulas to compute the rate of increase in donations for each of the years. Your results should look like the data in Figure 12-17.

S	T	U	V
Row Labels ▼	Sum of 2013	Sum of 2014	Increase%
⊟ f	$ 224,135	$ 257,270	
19600	$ 82,245	$ 94,650	
19601	$ 74,905	$ 84,830	
19603	$ 66,985	$ 77,790	
⊟ m	$ 287,040	$ 332,620	
19600	$ 98,360	$ 114,070	
19601	$ 89,745	$ 103,400	
19603	$ 98,935	$ 115,150	
Grand Total	$ 511,175	$ 589,890	

Source: Used with permission from Microsoft Corporation

FIGURE 12-17   Pivot table comparing donations by gender and zip code

The table includes six groups of donors to consider (two genders multiplied by three zip codes). Enter a cell comment or text to indicate which groups differ significantly from the average.

Predicting 2015 Donations

You can use the rates of increase computed with the gender–zip code pivot table to generate data for predicted donations. The first step is to copy the data from the donor table in the Possibles worksheet to another worksheet. When you paste this data into the new sheet, the form of the data may be changed to a normal data range by Excel. This change is acceptable for your purposes. Rename the worksheet to Prediction.

Create a formula that shows the rate of increased giving for the first donor. The rate is a function of gender and zip code. The formula should use the AND function, which is discussed in Tutorial C. Figure 12-18 shows the structure of the first part of this formula in cell M2 for the first donor.

M2				*fx*	=IF(AND(D2="m",E2="19600"),0.131,IF(AND(D2="m",E2="19601"),0.123,IF(AND(D2				
D	E	F	G	H	I	J	K	L	M	N	O	P
1	Gender	Zip	YOB	Deceased	2010	2011	2012	2013	2014	Rate	Predicted	Rate2	Pessimistic
2	f	19601	1930	FALSE	500	550	675	765	$ 855	0.133	$ 969	0.103	$ 943
3	f	19603	1930	FALSE	100	150	275	355	$ 460	0.161	$ 534	0.131	$ 520

Source: Used with permission from Microsoft Corporation

FIGURE 12-18   Formula structure for increased donation rate

In the figure, the rate increases shown in the formula (.131 and .123) are for illustrative purposes only. In your formula, you should use the rate increases computed next to the gender-zip code pivot table in the Possibles worksheet. Copy the formula down the column for all donors.

By formula, apply the rate of increase to each donor to generate data that predicts donations in 2015. Copy the formula down the column for all donors.

What if donations are not made at the level expected? You can generate a less optimistic prediction by assuming that the actual rate of increase will be .03 less than the predicted rate you just calculated. For example, if the expected rate of increase is .131, the less optimistic rate would be .131 − .030 = .101. For each donor, revise the formulas to calculate giving at the reduced rate. (In Figure 12-18, this rate is labeled as Rate2.) Copy the formula down the column for all donors. Use the formula to apply the decreased rate to each donor and generate a pessimistic estimate of 2015 giving.

Based on these results, create a pivot table that shows data for predicted giving levels and pessimistic giving levels, organized by gender and zip code. Insert the pivot table in the existing worksheet.

ASSIGNMENT 2: DOCUMENTING FINDINGS IN A MEMO

In this assignment, you write a memo in Microsoft Word that documents your findings. The charity's CFO wants answers to the following questions:

- How did you streamline the database to generate a donor list for 2015? On what basis have changes been made?
- How would you describe donations by the people who will be solicited in 2015? Have donations increased? If so, what is the rate of increase?
- Does giving differ significantly by gender, by zip code, or by year of birth?
- What are your predictions for 2015 giving? How did you develop those predictions?

You should summarize actual and predicted amounts in a table that has the format shown in Figure 12-19.

Gender and Zip	2014 Actual	2015 Predicted	2015 Pessimistic
F-19600			
F-19601			
F-19603			
Total Female			
M-19600			
M-19601			
M-19603			
Total Male			
Grand Total			

Source: © Cengage Learning 2015

FIGURE 12-19   Format of table to insert in memo

You can find the table's contents in your Excel pivot tables of gender and zip code data.

In your memo, observe the following requirements:

- Your memo should have proper headings such as Date, To, From, and Subject. You can address the memo to the CFO of the nonprofit organization. Set up the memo as discussed in Tutorial E.
- Briefly outline the situation. However, you need not provide much background—you can assume that readers are generally familiar with your task.
- In the body of the memo, answer the CFO's questions.
- Insert the table that summarizes actual and predicted amounts. Your instructor will tell you if other graphical support is required in the memo.

ASSIGNMENT 3: GIVING AN ORAL PRESENTATION

Your instructor may request that you summarize your analysis and results in an oral presentation. Assume that you would be talking to managers at the charity for 10 minutes or less. Use visual aids or handouts as appropriate. See Tutorial F for guidance on preparing and giving an oral presentation.

DELIVERABLES

Assemble the following deliverables for your instructor:

1. Printout of your memo
2. Spreadsheet printouts, if required by your instructor
3. Query printouts, if required by your instructor
4. Electronic media such as flash drive or CD, which should include your Word file, Access file, and Excel file

Staple the printouts together with the memo on top. If you have more than one .xlsx file or .accdb file stored on your electronic media, write your instructor a note that identifies the files for this assignment.

PART 6

ADVANCED SKILLS
USING EXCEL

TUTORIAL **E**

GUIDANCE FOR EXCEL CASES

The Microsoft Excel cases in this book require the student to write a memorandum that includes a table. Guidelines for preparing a memo in Microsoft Word and instructions for entering a table in a Word document are provided to begin this tutorial. Also, some of the cases in this book require the use of advanced Excel techniques. Those techniques are explained in this tutorial rather than in the cases themselves:

- Using data tables
- Using pivot tables
- Using built-in functions

You can refer to Sheet 1 of Tutorial E_data.xlsx when reading about data tables. Refer to Sheet 2 when reading about pivot tables.

PREPARING A MEMORANDUM IN WORD

A business memo should include proper headings, such as TO, FROM, DATE, and SUBJECT. If you want to use a Word memo template, follow these steps:

1. In Word, click File.
2. Click New.
3. Click the Memos button in the Office.com or Microsoft Office Online Templates section.
4. Double-click the Contemporary design memo or another memo design of your choice.

The first time you do this, you may need to click Download to install the template. You might also have to search for the memo templates.

ENTERING A TABLE INTO A WORD DOCUMENT

Enter a table into a Word document using the following procedure:

1. Click the cursor where you want the table to appear in the Word document.
2. In the Tables group on the Insert tab, click the Table drop-down menu.
3. Click Insert Table.
4. Choose the number of rows and columns.
5. Click OK.

DATA TABLES

An Excel data table is a contiguous range of data that has been designated as a table. Once you make this designation, the table gains certain properties that are useful for data analysis. (Note that in some previous versions of Excel, data tables were called *data lists*.) Suppose you have a list of runners who have completed a race, as shown in Figure E-1.

	A	B	C	D	E	F
1	**RUNNER#**	**LAST**	**FIRST**	**AGE**	**GENDER**	**TIME (MIN)**
2	100	HARRIS	JANE	O	F	70
3	101	HILL	GLENN	Y	M	70
4	102	GARCIA	PEDRO	M	M	85
5	103	HILBERT	DORIS	M	F	90
6	104	DOAKS	SALLY	Y	F	94
7	105	JONES	SUE	Y	F	95
8	106	SMITH	PETE	M	M	100
9	107	DOE	JANE	O	F	100
10	108	BRADY	PETE	O	M	100
11	109	BRADY	JOE	O	M	120
12	110	HEEBER	SALLY	M	F	125
13	111	DOLTZ	HAL	O	M	130
14	112	PEEBLES	AL	Y	M	63

Source: Used with permission from Microsoft Corporation

FIGURE E-1   Data table example

To turn the information into a data table, highlight the data range, including headings, and click the Insert tab. Then click Table in the Tables group. The Create Table window appears, as shown in Figure E-2.

	A1		*f_x*	RUNNER#		

	A	B	C	D	E	F
1	**RUNNER#**	**LAST**	**FIRST**	**AGE**	**GENDER**	**TIME (MIN)**
2	100	HARRIS	JANE	O	F	70
3	101	HILL	GLENN	Y	M	70
4	102	GARCI				85
5	103	HILBEF				90
6	104	DOAKS				94
7	105	JONES				95
8	106	SMITH				100
9	107	DOE				100
10	108	BRADY				100
11	109	BRADY	JOE	O	M	120
12	110	HEEBER	SALLY	M	F	125
13	111	DOLTZ	HAL	O	M	130
14	112	PEEBLES	AL	Y	M	63

Create Table dialog box: Where is the data for your table? =$A$1:$F$14   ☑ My table has headers   [OK] [Cancel]

Source: Used with permission from Microsoft Corporation

FIGURE E-2   Create Table window

When you click OK, the data range appears as a table. In the Table Style Options group on the Design tab, click the Total Row check box to add a totals row to the data table. You also can select a light style in the Table Styles list to get rid of the contrasting color in the table's rows. Figure E-3 shows the results.

	A	B	C	D	E	F
1	**RUNNER#** ▾	**LAST** ▾	**FIRST** ▾	**AGE** ▾	**GENDER** ▾	**TIME (MIN)** ▾
2	100	HARRIS	JANE	O	F	70
3	101	HILL	GLENN	Y	M	70
4	102	GARCIA	PEDRO	M	M	85
5	103	HILBERT	DORIS	M	F	90
6	104	DOAKS	SALLY	Y	F	94
7	105	JONES	SUE	Y	F	95
8	106	SMITH	PETE	M	M	100
9	107	DOE	JANE	O	F	100
10	108	BRADY	PETE	O	M	100
11	109	BRADY	JOE	O	M	120
12	110	HEEBER	SALLY	M	F	125
13	111	DOLTZ	HAL	O	M	130
14	112	PEEBLES	AL	Y	M	63
15	Total					1242

Source: Used with permission from Microsoft Corporation

FIGURE E-3   Data table example

The headings have acquired drop-down menu tabs, as you can see in Figure E-3.

You can sort the data table records by any field. Perhaps you want to sort by times. If so, click the drop-down menu in the TIME (MIN) heading, and then click Sort Smallest to Largest. You get the results shown in Figure E-4.

	A	B	C	D	E	F
1	RUNNER ▾	LAST ▾	FIRST ▾	AGE ▾	GENDE ▾	TIME (MIN ▾
2	112	PEEBLES	AL	Y	M	63
3	100	HARRIS	JANE	O	F	70
4	101	HILL	GLENN	Y	M	70
5	102	GARCIA	PEDRO	M	M	85
6	103	HILBERT	DORIS	M	F	90
7	104	DOAKS	SALLY	Y	F	94
8	105	JONES	SUE	Y	F	95
9	106	SMITH	PETE	M	M	100
10	107	DOE	JANE	O	F	100
11	108	BRADY	PETE	O	M	100
12	109	BRADY	JOE	O	M	120
13	110	HEEBER	SALLY	M	F	125
14	111	DOLTZ	HAL	O	M	130
15	Total					1242

Source: Used with permission from Microsoft Corporation

FIGURE E-4 Sorting list by drop-down menu

You can see that Peebles had the best time and Doltz had the worst time. You also can sort from Largest to Smallest.

In addition, you can sort by more than one criterion. Assume that you want to sort first by gender and then by time (within gender). You first sort from A to Z. Then you again click the Gender drop-down tab, point to Sort by Color, and then click Custom Sort. In the Sort window that appears, click Add Level and choose Time as the next criterion. See Figure E-5.

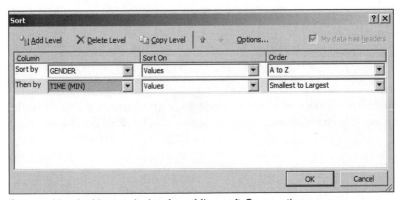

Source: Used with permission from Microsoft Corporation

FIGURE E-5 Sorting on multiple criteria

Click OK to get the results shown in Figure E-6.

	A	B	C	D	E	F
1	RUNNER	LAST	FIRST	AGE	GENDE	TIME (MIN
2	100	HARRIS	JANE	O	F	70
3	103	HILBERT	DORIS	M	F	90
4	104	DOAKS	SALLY	Y	F	94
5	105	JONES	SUE	Y	F	95
6	107	DOE	JANE	O	F	100
7	110	HEEBER	SALLY	M	F	125
8	112	PEEBLES	AL	Y	M	63
9	101	HILL	GLENN	Y	M	70
10	102	GARCIA	PEDRO	M	M	85
11	106	SMITH	PETE	M	M	100
12	108	BRADY	PETE	O	M	100
13	109	BRADY	JOE	O	M	120
14	111	DOLTZ	HAL	O	M	130
15	Total					1242

Source: Used with permission from Microsoft Corporation

FIGURE E-6   Sorting by gender and time (within gender)

You can see that Harris had the best female time and that Peebles had the best male time.

Perhaps you want to see the top *n* listings for some attribute; for example, you may want to see the top five runners' times. Select the Time column's drop-down menu, and select Number Filters. From the menu that appears, click Top 10. The Top 10 AutoFilter window appears, as shown in Figure E-7.

Source: Used with permission from Microsoft Corporation

FIGURE E-7   Top 10 AutoFilter window

This window lets you specify the number of values you want. You might see 10 values as a default setting when the window appears. Figure E-7 shows that the user specified five values. Click OK to get the results shown in Figure E-8.

	A	B	C	D	E	F
1	RUNNER	LAST	FIRST	AGE	GENDE	TIME (MIN
6	107	DOE	JANE	O	F	100
7	110	HEEBER	SALLY	M	F	125
11	106	SMITH	PETE	M	M	100
12	108	BRADY	PETE	O	M	100
13	109	BRADY	JOE	O	M	120
14	111	DOLTZ	HAL	O	M	130
15	Total					675

Source: Used with permission from Microsoft Corporation

FIGURE E-8   Top 5 times

The output contains more than five data records because there are ties at 100 minutes. If you want to see all of the records again, click the Time drop-down menu and click Clear Filter From "TIME (MIN)." The full table of data reappears, as shown in Figure E-9.

	A	B	C	D	E	F
1	RUNNER	LAST	FIRST	AGE	GENDE	TIME (MIN
2	100	HARRIS	JANE	O	F	70
3	103	HILBERT	DORIS	M	F	90
4	104	DOAKS	SALLY	Y	F	94
5	105	JONES	SUE	Y	F	95
6	107	DOE	JANE	O	F	100
7	110	HEEBER	SALLY	M	F	125
8	112	PEEBLES	AL	Y	M	63
9	101	HILL	GLENN	Y	M	70
10	102	GARCIA	PEDRO	M	M	85
11	106	SMITH	PETE	M	M	100
12	108	BRADY	PETE	O	M	100
13	109	BRADY	JOE	O	M	120
14	111	DOLTZ	HAL	O	M	130
15	Total					1242

Source: Used with permission from Microsoft Corporation

FIGURE E-9   Restoring all data to screen

Each of the cells in the Total row has a drop-down menu. The menu choices are statistical operations that you can perform on the totals—for example, you can take a sum, take an average, take a minimum or maximum, count the number of records, and so on. Assume that the Time drop-down menu was selected, as shown in Figure E-10. Note that the Sum operator is highlighted by default.

	A	B	C	D	E	F
1	RUNNER	LAST	FIRST	AGE	GENDE	TIME (MIN
2	100	HARRIS	JANE	O	F	70
3	103	HILBERT	DORIS	M	F	90
4	104	DOAKS	SALLY	Y	F	94
5	105	JONES	SUE	Y	F	95
6	107	DOE	JANE	O	F	100
7	110	HEEBER	SALLY	M	F	125
8	112	PEEBLES	AL	Y	M	63
9	101	HILL	GLENN	Y	M	70
10	102	GARCIA	PEDRO	M	M	85
11	106	SMITH	PETE	M	M	100
12	108	BRADY	PETE	O	M	100
13	109	BRADY	JOE	O	M	120
14	111	DOLTZ	HAL	O	M	130
15	Total					1242
16						None
17						Average
18						Count
19						Count Numbers
20						Max
21						Min
22						Sum
						StdDev
						Var
						More Functions...

Source: Used with permission from Microsoft Corporation

FIGURE E-10   Selecting Time drop-down menu in Total row

By changing from Sum to Average, you find that the average time for all runners was 95.5 minutes, as shown in Figure E-11.

	A	B	C	D	E	F
1	RUNNER	LAST	FIRST	AGE	GENDE	TIME (MIN
2	100 HARRIS	JANE	O	F		70
3	103 HILBERT	DORIS	M	F		90
4	104 DOAKS	SALLY	Y	F		94
5	105 JONES	SUE	Y	F		95
6	107 DOE	JANE	O	F		100
7	110 HEEBER	SALLY	M	F		125
8	112 PEEBLES	AL	Y	M		63
9	101 HILL	GLENN	Y	M		70
10	102 GARCIA	PEDRO	M	M		85
11	106 SMITH	PETE	M	M		100
12	108 BRADY	PETE	O	M		100
13	109 BRADY	JOE	O	M		120
14	111 DOLTZ	HAL	O	M		130
15	Total					95.53846154

Source: Used with permission from Microsoft Corporation

FIGURE E-11   Average running time shown in Total row

PIVOT TABLES

Suppose you have data for a company's sales transactions by month, by salesperson, and by amount for each product type. You would like to display each salesperson's total sales by type of product sold and by month. You can use a pivot table in Excel to tabulate that summary data. A pivot table is built around one or more dimensions and thus can summarize large amounts of data. Figure E-12 shows total sales cross-tabulated by salesperson and by month.

	A	B	C	D	E
1	Name	Product	January	February	March
2	Jones	Product1	30,000	35,000	40,000
3	Jones	Product2	33,000	34,000	45,000
4	Jones	Product3	24,000	30,000	42,000
5	Smith	Product1	40,000	38,000	36,000
6	Smith	Product2	41,000	37,000	38,000
7	Smith	Product3	39,000	50,000	33,000
8	Bonds	Product1	25,000	26,000	25,000
9	Bonds	Product2	22,000	25,000	24,000
10	Bonds	Product3	19,000	20,000	19,000
11	Ruth	Product1	44,000	42,000	33,000
12	Ruth	Product2	45,000	40,000	30,000
13	Ruth	Product3	50,000	52,000	35,000

Source: Used with permission from Microsoft Corporation

FIGURE E-12   Excel spreadsheet data

You can create pivot tables and many other kinds of tables with the Excel PivotTable tool. To create a pivot table from the data in Figure E-12, follow these steps:

1. Starting in the spreadsheet in Figure E-12, click a cell in the data range, and then click the Insert tab. In the Tables group, choose PivotTable. You see the screen shown in Figure E-13.

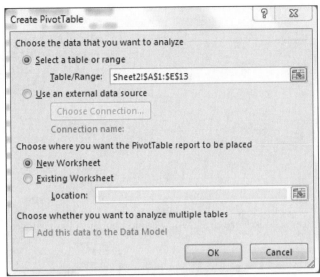

Source: Used with permission from Microsoft Corporation

FIGURE E-13   Creating a pivot table

2. Make sure New Worksheet is checked under "Choose where you want the PivotTable report to be placed." Click OK. The screen shown in Figure E-14 appears. If it does not, right-click in a cell in the pivot table area, and click PivotTable Options from the menu that appears. Click the Display tab and then check the Classic PivotTable layout.

Source: Used with permission from Microsoft Corporation

FIGURE E-14   PivotTable design screen

The data range's column headings are shown in the PivotTable Fields section on the right side of the screen. From there, you can click and drag column headings into the Row, Column, and Value areas that appear in the spreadsheet.

3. If you want to see the total sales by product for each salesperson, drag the Name field to the Drop Column Fields Here area in the spreadsheet. You should see the result shown in Figure E-15.

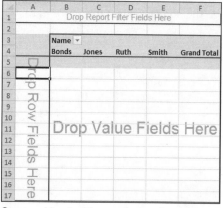

Source: Used with permission from Microsoft Corporation

FIGURE E-15   Column fields

4. Next, drag the Product field to the Drop Row Fields Here area. You should see the result shown in Figure E-16.

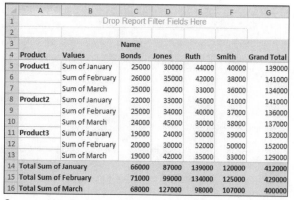

Source: Used with permission from Microsoft Corporation

FIGURE E-16   Row fields

5. Finally, drag the month fields (January, February, and March) individually to the Drop Value Fields Here area to produce the finalized pivot table. You should see the result shown in Figure E-17.

			Name				
Product	Values		Bonds	Jones	Ruth	Smith	Grand Total
Product1	Sum of January		25000	30000	44000	40000	139000
	Sum of February		26000	35000	42000	38000	141000
	Sum of March		25000	40000	33000	36000	134000
Product2	Sum of January		22000	33000	45000	41000	141000
	Sum of February		25000	34000	40000	37000	136000
	Sum of March		24000	45000	30000	38000	137000
Product3	Sum of January		19000	24000	50000	39000	132000
	Sum of February		20000	30000	52000	50000	152000
	Sum of March		19000	42000	35000	33000	129000
Total Sum of January			66000	87000	139000	120000	412000
Total Sum of February			71000	99000	134000	125000	429000
Total Sum of March			68000	127000	98000	107000	400000

Source: Used with permission from Microsoft Corporation

FIGURE E-17   Data items

If your table's values are organized differently, drag the Values heading to the right of the Product heading, as shown in Figure E-17.

By default, Excel adds all of the sales for each salesperson by month for each product. At the bottom of the pivot table, Excel also shows the total sales for each month for all products.

Refer to the pivot table window and note the four small panes in the lower-right corner. The Values pane lets you easily change from the default Sum operator to another one (Min, Max, Average, and so on). Click the drop-down arrow, select Value Field Settings, and then select the desired operator.

BUILT-IN FUNCTIONS

You might need to use some of the following functions when solving the Excel cases elsewhere in this text:

- MIN, MAX, AVERAGE, COUNTIF, ROUND, ROUNDUP, RANDBETWEEN, TREND, and PMT

The syntax of these functions is discussed in this section. The following examples are based on the runner data shown in Figure E-18.

	A	B	C	D	E	F	G
1	RUNNER#	LAST	FIRST	AGE	GENDER	HEIGHT (IN)	TIME (MIN)
2	100	HARRIS	JANE	O	F	60	70
3	101	HILL	GLENN	Y	M	65	70
4	102	GARCIA	PEDRO	M	M	76	85
5	103	HILBERT	DORIS	M	F	64	90
6	104	DOAKS	SALLY	Y	F	62	94
7	105	JONES	SUE	Y	F	64	95
8	106	SMITH	PETE	M	M	73	100
9	107	DOE	JANE	O	F	66	100
10	108	BRADY	PETE	O	M	73	100
11	109	BRADY	JOE	O	M	71	120
12	110	HEEBER	SALLY	M	F	59	125
13	111	DOLTZ	HAL	O	M	76	130
14	112	PEEBLES	AL	Y	M	76	63

Source: Used with permission from Microsoft Corporation

FIGURE E-18   Runner data used to illustrate built-in functions

The data is the same as that shown in Figure E-1, except that Figure E-18 includes a column for the runners' height in inches.

MIN and MAX Functions

The MIN function determines the smallest value in a range of data. The MAX function returns the largest. Say that we want to know the fastest time for all runners, which would be the minimum time in column G. The MIN function computes the smallest value in a set of values. The set of values could be a data range, or it could be a series of cell addresses separated by commas. The syntax of the MIN function is as follows:

- MIN(set of data)

To show the minimum time in cell C16, you would enter the formula shown in the formula bar in Figure E-19:

C16			fx	=MIN(G2:G14)			
	A	B	C	D	E	F	G
1	RUNNER#	LAST	FIRST	AGE	GENDER	HEIGHT	TIME (MIN)
2	100	HARRIS	JANE	O	F	60	70
3	101	HILL	GLENN	Y	M	65	70
4	102	GARCIA	PEDRO	M	M	76	85
5	103	HILBERT	DORIS	M	F	64	90
6	104	DOAKS	SALLY	Y	F	62	94
7	105	JONES	SUE	Y	F	64	95
8	106	SMITH	PETE	M	M	73	100
9	107	DOE	JANE	O	F	66	100
10	108	BRADY	PETE	O	M	73	100
11	109	BRADY	JOE	O	M	71	120
12	110	HEEBER	SALLY	M	F	59	125
13	111	DOLTZ	HAL	O	M	76	130
14	112	PEEBLES	AL	Y	M	76	63
15							
16	MINIMUM TIME:		63				

Source: Used with permission from Microsoft Corporation

FIGURE E-19   MIN function in cell C16

(Assume that you typed the label "MINIMUM TIME:" into cell A16.) You can see that the fastest time is 63 minutes.

To see the slowest time in cell G16, use the MAX function, whose syntax parallels that of the MIN function, except that the largest value in the set is determined. See Figure E-20.

	A	B	C	D	E	F	G
	RUNNER#	LAST	FIRST	AGE	GENDER	HEIGHT	TIME (MIN)
1	RUNNER#	LAST	FIRST	AGE	GENDER	HEIGHT	TIME (MIN)
2	100	HARRIS	JANE	O	F	60	70
3	101	HILL	GLENN	Y	M	65	70
4	102	GARCIA	PEDRO	M	M	76	85
5	103	HILBERT	DORIS	M	F	64	90
6	104	DOAKS	SALLY	Y	F	62	94
7	105	JONES	SUE	Y	F	64	95
8	106	SMITH	PETE	M	M	73	100
9	107	DOE	JANE	O	F	66	100
10	108	BRADY	PETE	O	M	73	100
11	109	BRADY	JOE	O	M	71	120
12	110	HEEBER	SALLY	M	F	59	125
13	111	DOLTZ	HAL	O	M	76	130
14	112	PEEBLES	AL	Y	M	76	63
15							
16	MINIMUM TIME:		63		MAXIMUM TIME:		130

G16 = =MAX(G2:G14)

Source: Used with permission from Microsoft Corporation

FIGURE E-20   MAX function in cell G16

AVERAGE, ROUND, and ROUNDUP Functions

The AVERAGE function computes the average of a set of values. Figure E-21 shows the use of the AVERAGE function in cell C17:

C17 = =AVERAGE(G2:G14)

	A	B	C	D	E	F	G
1	RUNNER#	LAST	FIRST	AGE	GENDER	HEIGHT	TIME (MIN)
2	100	HARRIS	JANE	O	F	60	70
3	101	HILL	GLENN	Y	M	65	70
4	102	GARCIA	PEDRO	M	M	76	85
5	103	HILBERT	DORIS	M	F	64	90
6	104	DOAKS	SALLY	Y	F	62	94
7	105	JONES	SUE	Y	F	64	95
8	106	SMITH	PETE	M	M	73	100
9	107	DOE	JANE	O	F	66	100
10	108	BRADY	PETE	O	M	73	100
11	109	BRADY	JOE	O	M	71	120
12	110	HEEBER	SALLY	M	F	59	125
13	111	DOLTZ	HAL	O	M	76	130
14	112	PEEBLES	AL	Y	M	76	63
15							
16	MINIMUM TIME:		63		MAXIMUM TIME:		130
17	AVERAGE TIME:		95.53846				

Source: Used with permission from Microsoft Corporation

FIGURE E-21   AVERAGE function in cell C17

Notice that the value shown is a real number with many digits. What if you wanted to have the value rounded to a certain number of digits? Of course, you could format the output cell, but doing that changes only what is shown on the screen. You want the cell's contents actually to *be* the rounded number. Therefore, you need to use the ROUND function. Its syntax is:

- ROUND(number, number of digits)

Figure E-22 shows the rounded average time (with two decimal places) in cell G17.

Source: Used with permission from Microsoft Corporation

FIGURE E-22 ROUND function used in cell G17

To achieve this output, cell C17 was used as the value to be rounded. Recall from Figure E-21 that cell C17 had the formula =AVERAGE(G2:G14). The following ROUND formula would produce the same output in cell G17: =ROUND(AVERAGE(G2:G14),2). In this case, Excel evaluates the formula "inside out." First, the AVERAGE function is evaluated, yielding the average with many digits. That value is then input to the ROUND function and rounded to two decimal places.

The ROUNDUP function works much like the ROUND function. ROUNDUP's output is always rounded up to the next value. For example, the value 4 would appear in a cell that contained the following formula: =ROUNDUP(3.12,0). In Figure E-22, if the formula in cell G17 had been =ROUNDUP(AVERAGE(G2:G14),0), the value 96 would have been the result. In other words, 95.54 rounded up with no decimal places becomes 96.

COUNTIF Function

The COUNTIF function counts the number of values in a range that meet a specified condition. The syntax is:

- COUNTIF(range of data, condition)

The condition is a logical expression such as "=1", ">6", or "=F". The condition is shown with quotation marks, even if a number is involved.

Assume that you want to see the number of female runners in cell C18. Figure E-23 shows the formula used.

Source: Used with permission from Microsoft Corporation

FIGURE E-23 COUNTIF function used in cell C18

The logic of the formula is: Count the number of times that "F" appears in the data range E2:E14.

As another example of using COUNTIF, assume that column H shows the rounded ratio of each runner's height in inches to the runner's time in minutes (see Figure E-24).

	H2		f_x	=ROUND(G2/F2,2)			
	A	B	C	D	E	F	G	H
1	RUNNER#	LAST	FIRST	AGE	GENDER	HEIGHT	TIME (MIN)	RATIO
2	100	HARRIS	JANE	O	F	60	70	1.17
3	101	HILL	GLENN	Y	M	65	70	1.08
4	102	GARCIA	PEDRO	M	M	76	85	1.12
5	103	HILBERT	DORIS	M	F	64	90	1.41
6	104	DOAKS	SALLY	Y	F	62	94	1.52
7	105	JONES	SUE	Y	F	64	95	1.48
8	106	SMITH	PETE	M	M	73	100	1.37
9	107	DOE	JANE	O	F	66	100	1.52
10	108	BRADY	PETE	O	M	73	100	1.37
11	109	BRADY	JOE	O	M	71	120	1.69
12	110	HEEBER	SALLY	M	F	59	125	2.12
13	111	DOLTZ	HAL	O	M	76	130	1.71
14	112	PEEBLES	AL	Y	M	76	63	0.83
15								
16	MINIMUM TIME:		63		MAXIMUM TIME:		130	
17	AVERAGE TIME:		95.53846		ROUNDED AVERAGE:		95.54	
18	NUMBER OF FEMALES:		6					

Source: Used with permission from Microsoft Corporation

FIGURE E-24 Ratio of height to time in column H

Assume that all runners whose height in inches is less than their time in minutes will get an award. How many awards are needed? If the ratio is less than 1, an award is warranted. The COUNTIF function in cell G18 computes a count of ratios less than 1, as shown in Figure E-25.

	G18		f_x	=COUNTIF(H2:H14,"<1")			
	A	B	C	D	E	F	G	H
1	RUNNER#	LAST	FIRST	AGE	GENDER	HEIGHT	TIME (MIN)	RATIO
2	100	HARRIS	JANE	O	F	60	70	1.17
3	101	HILL	GLENN	Y	M	65	70	1.08
4	102	GARCIA	PEDRO	M	M	76	85	1.12
5	103	HILBERT	DORIS	M	F	64	90	1.41
6	104	DOAKS	SALLY	Y	F	62	94	1.52
7	105	JONES	SUE	Y	F	64	95	1.48
8	106	SMITH	PETE	M	M	73	100	1.37
9	107	DOE	JANE	O	F	66	100	1.52
10	108	BRADY	PETE	O	M	73	100	1.37
11	109	BRADY	JOE	O	M	71	120	1.69
12	110	HEEBER	SALLY	M	F	59	125	2.12
13	111	DOLTZ	HAL	O	M	76	130	1.71
14	112	PEEBLES	AL	Y	M	76	63	0.83
15								
16	MINIMUM TIME:		63		MAXIMUM TIME:		130	
17	AVERAGE TIME:		95.53846		ROUNDED AVERAGE:		95.54	
18	NUMBER OF FEMALES:		6		RATIOS<1:		1	

Source: Used with permission from Microsoft Corporation

FIGURE E-25 COUNTIF function used in cell G18

RANDBETWEEN Function

If you wanted a cell to contain a randomly generated integer in the range from 1 to 9, you would use the formula =RANDBETWEEN(1,9). Any value between 1 and 9 inclusive would be output by the formula. An example is shown in Figure E-26.

A2			fx	=RANDBETWEEN(1,9)
	A	B	C	D	E
1	Position				
2	9				

Source: Used with permission from Microsoft Corporation

FIGURE E-26   RANDBETWEEN function used in cell A2

Assume that you copied and pasted the formula to generate a column of 100 numbers between 1 and 9. Every time a value was changed in the spreadsheet, Excel would recalculate the 100 RANDBETWEEN formulas to change the 100 random values. Therefore, you might want to settle on the random values once they are generated. To do this, copy the 100 values, click Paste Special, and then click Values to put the values in the same range. The contents of the cells will change from formulas to literal values.

TREND Function

The TREND function can be used to estimate a variable's value based on the values of other variables. For example, you might know the heights, genders, and weights for 20 people. Correlations exist among these three characteristics. You also have height and gender data for 10 other people, and you want to estimate their weights based on the data you have. The data is shown in Figure E-27.

	A	B	C	D	E	F	G	H	I
1	Person	Height	Gender	Weight		Person	Height	Gender	Pred Weight
2	101	70	1	190		130	71	1	
3	102	60	2	110		131	61	2	
4	103	72	1	200		132	70	1	
5	104	62	2	120		133	63	2	
6	105	66	1	175		134	65	1	
7	106	66	2	140		135	65	2	
8	107	64	1	170		136	67	1	
9	108	70	2	155		137	70	2	
10	109	62	1	150		138	61	1	
11	110	66	2	150		139	68	2	
12	111	68	1	186					
13	112	68	2	200					
14	113	70	1	200					
15	114	62	2	100					
16	115	72	1	210					
17	116	63	2	110					
18	117	71	1	200					
19	118	64	2	130					
20	119	70	1	170					
21	120	61	2	120					

Source: Used with permission from Microsoft Corporation

FIGURE E-27   Data for people's heights, genders, and weights

The TREND function requires numerical values. In the data, the code for a male is 1 and the code for a female is 2. Height values are measured in inches and weight values are in pounds. For example, person 101 is a male who is 5 feet, 10 inches tall and weighs 190 pounds.

You can use the TREND function to examine a set of data and "learn" the relationship between two or more variables. In this example, the TREND function learns how the heights and genders of 20 people correlate to their weights. Then, given 10 other people's heights and genders, the TREND function applies what it knows to estimate their weights.

The syntax for the TREND function is:

- =TREND(known Ys, known Xs, new Xs)

In the example, the known Ys are the known weights for 20 people, the known Xs are the related heights and genders, and the new Xs are heights and genders of 10 people for whom you want estimated weights. The formula is shown in Figure E-28.

- =TREND($D$2:$D$21, $B$2:$C$21, G2:H2)

Cells D2 to D21 hold the known weights for 20 people. Cells B2 to C21 hold the values of the two predictor variables (height and gender) for those 20 people. Cells G2 and H2 are the predictor variables for person 130, for whom you want a predicted weight. The predicted weight formula is in cell I2.

I2	▼ : × ✓ ƒx	=TREND($D$2:$D$21,$B$2:$C$21,G2:H2)							
	A	B	C	D	E	F	G	H	I
1	Person	Height	Gender	Weight		Person	Height	Gender	Pred Weight
2	101	70	1	190		130	71	1	200
3	102	60	2	110		131	61	2	114
4	103	72	1	200		132	70	1	194
5	104	62	2	120		133	63	2	126

Source: Used with permission from Microsoft Corporation

FIGURE E-28   Calculation of predicted weight for person 130

When you copy the formula down the cells in column I for the 10 people, you calculate weight predictions for all of them. By using absolute addressing, the only address changes are the predictor height and gender values for the 10 people.

PMT Function

The PMT function calculates a loan payment. The syntax is:

- =PMT(interest rate, number of periods, initial loan principal)

As an example, assume that you have a 6 percent, 30-year loan for $100,000. The calculation of the monthly payment is shown in Figure E-29.

B5	▼ : × ✓ ƒx	=PMT(B1/12,B2*12,B3) * -1		
	A	B	C	D	E
1	Annual rate:	6%			
2	Years:	30			
3	Principal:	$ 100,000			
4					
5	Monthly Payment:	$599.55			

Source: Used with permission from Microsoft Corporation

FIGURE E-29   Calculation of monthly loan payment

The formula is in cell B5. The monthly interest rate is the annual rate in cell B1 divided by 12. The number of months covered by the loan is the number of years (see cell B2) multiplied by 12. The loan principal is in cell B3. The PMT function returns a negative number, so the expression is multiplied by –1.

Loan payments for the year are computed by multiplying the monthly payment by 12.

PART 7

PRESENTATION SKILLS

TUTORIAL **F**

GIVING AN ORAL PRESENTATION

Giving an oral presentation in class lets you practice the presentation skills you will need in the workplace. The presentations you create for the cases in this textbook will be similar to professional business presentations. You will be expected to present objective, technical results to your organization's stakeholders, and you will have to support your presentation with visual aids commonly used in the business world. During your presentation, your instructor might assign your classmates to role-play an audience of business managers, bankers, or employees. They might also provide feedback on your presentation.

Follow these four steps to create an effective presentation:

1. Plan your presentation.
2. Draft your presentation.
3. Create graphics and other visual aids.
4. Practice delivering your presentation.

PLANNING YOUR PRESENTATION

When planning an oral presentation, you need to know your time limits, establish your purpose, analyze your audience, and gather information. This section explores each of these elements.

Knowing Your Time Limits

You need to consider your time limits on two levels. First, consider how much time you will have to deliver your presentation. For example, what are the key points in your material that can be covered in 10 minutes? The element of time is the primary constraint of any presentation. It limits the breadth and depth of your talk, and the number of visual aids that you can use. Second, consider how much time you will need for the process of preparing your presentation—drafting your presentation, creating graphics, and practicing your delivery.

Establishing Your Purpose

After considering your time limits, you must define your purpose: what you need to say and to whom you will say it. For the Access cases in this book, your purpose will be to inform and explain. For instance, a business's owners, managers, and employees may need to know how the company's database is organized and how they can use it to fill in forms and create reports. In contrast, for the Excel cases, your purpose will be to recommend a course of action based on the results of your business model. You will make the recommendations to business owners, managers, and bankers based on the results of inputting and running various scenarios.

Analyzing Your Audience

Once you have established the purpose of your presentation, you should analyze your audience. Ask yourself: What does my audience already know about the subject? What do the audience members want to know? What do they need to know? Do they have any biases or personal agendas that I should consider? What level of technical detail is best suited to their level of knowledge and interest?

In some Access cases, you will make a presentation to an audience that might not be familiar with Access or with databases in general. In other cases, you might be giving your presentation to a business owner who started to work on a database but was not able to finish it. Tailor the presentation to suit your audience.

For the Excel cases, you are most often interpreting results for an audience of bankers or business managers. In those instances, the audience will not need to know the detailed technical aspects of how you generated your results. But what if your audience consists of engineers or scientists? They will certainly be more interested in the structure and rationale of your decision models. Regardless of the audience, your listeners need to know what assumptions you made prior to developing your spreadsheets because those assumptions might affect their opinion of your results.

Gathering Information

Because you will have just completed a case as you begin preparing your oral presentation, you will already have the basic information you need. For the Access cases, you should review the main points of the case and your goals. Make sure you include all of the points you think are important for the audience to understand. In addition, you might want to go beyond the requirements and explain additional ways in which the database could be used to benefit the organization, now or in the future.

For the Excel cases, you can refer to the tutorials for assistance in interpreting the results from your spreadsheet analysis. For some cases, you might want to use the Internet or the library to research business trends or background information that can support your presentation.

DRAFTING YOUR REPORT AND PRESENTATION

When you have completed the planning stage, you are ready to begin drafting the presentation. At this point, you might be tempted to write your presentation and then memorize it word for word. Even if you could memorize your presentation verbatim, however, your delivery would sound unnatural because people use a simpler vocabulary and shorter sentences when they speak than when they write. For example, read the previous paragraph out loud as if you were presenting it to an audience.

In many business situations, you will be required both to submit a written report of your work and give a PowerPoint presentation. First, write your report, and then design your PowerPoint slides as a "brief" of that report to discuss its main points. When drafting your report and the accompanying PowerPoint slides, follow this sequence:

1. Write the main body of your report.
2. Write the introduction to your report.
3. Write the conclusion to your report.
4. Prepare your presentation (the PowerPoint slides) using your report's main points.

Writing the Main Body

When you draft your report, write the body first. If you try to write the opening paragraph first, you might spend an inordinate amount of time attempting to craft your words perfectly, only to revise the introduction after you write the body of the report.

Keeping Your Audience in Mind

To write the main body, review your purpose and your audience profile. What are the main points you need to make? What are your audience's needs, interests, and technical expertise? It is important to include some technical details in your report and presentation, but keep in mind the technical expertise of your audience.

Remember that the people reading your report or listening to your presentation have their own agendas—put yourself in their places and ask, "What do I need to get out of this presentation?" For example, in the Access cases, an employee might want to know how to enter information on a form, but the business owner might be more interested in generating queries and reports. You need to address their different needs in your presentation. For example, you might say, "And now, let's look at how data entry associates can input data into this form."

Similarly, in the Excel cases, your audience will consist of business owners, managers, bankers, and perhaps some technical professionals. The owners and managers will be concerned with profitability, growth, and customer service. In contrast, the bankers' main concern will be repayment of a loan. Technical professionals will be more concerned with how well your decision model is designed, along with the credibility of the results. You need to address the interests of each group.

Using Transitions and Repetition in your Presentation

During your presentation, remember that the audience is not reading the text of your report, so you need to include transitions to compensate. Words such as *next, first, second,* and *finally* will help the audience follow the sequence of your ideas. Words such as *however, in contrast, on the other hand,* and *similarly* will help the audience follow shifts in thought. You can use your voice to convey emphasis.

Also consider using hand gestures to emphasize what you say. For instance, if you list three items, you can use your fingers to tick off each item as you discuss it. Similarly, if you state that profits will be flat, you can make a level motion with your hand for emphasis.

You may be speaking behind a podium or standing beside a projection screen, or both. If you feel uncomfortable standing in one place and you can walk without blocking the audience's view of the screen, feel free to move around. You can emphasize a transition by changing your position. If you tend to fidget, shift, or rock from one foot to the other, try to anchor yourself. A favorite technique of some speakers is to come from behind the podium and place one hand on it while speaking. They get the anchoring effect of the podium while removing the barrier it places between them and the audience. Use the stance or technique that makes you feel most comfortable, as long as your posture or actions do not distract the audience.

As you draft your presentation, repeat key points to emphasize them. For example, suppose your main point is that outsourcing labor will provide the greatest gains in net income. Begin by previewing that concept, and state that you will demonstrate how outsourcing labor will yield the biggest profits. Then provide statistics that support your claim, and show visual aids that graphically illustrate your point. Summarize by repeating your point: "As you can see, outsourcing labor does yield the biggest profits."

Relying on Graphics to Support Your Talk

As you write the main body, think of how to integrate graphics into your presentation. Do not waste words with a long description if a graphic can bring instant comprehension. For instance, instead of describing how information from a query can be turned into a report, show the query and a completed report. Figures F-1 and F-2 illustrate an Access query and the resulting report.

| Order Query 1 | | | | | |
Customer Name	City	Product Name	Qty	Price per Unit	Total
Applewood Restaurant	Martinsburg	Frozen Alligator on a Stick	20	$27.99	$559.80
Applewood Restaurant	Martinsburg	Nogales Chipotle Sauce	15	$11.49	$172.35
Applewood Restaurant	Martinsburg	Mom's Deep Dish Apple Pie	12	$12.49	$149.88
Fresh Catch Fishery	Salem	Brumley's Seafood Cocktail Sauce	24	$4.79	$114.96
Fresh Catch Fishery	Salem	NY Smoked Salmon	21	$21.99	$461.79
Fresh Catch Fishery	Salem	Mama Mia's Tiramisu	15	$17.99	$269.85
Jimmy's Crab House	Elkton	Frozen Alligator on a Stick	12	$27.99	$335.88
Jimmy's Crab House	Elkton	Brumley's Seafood Cocktail Sauce	24	$4.79	$114.96
Jimmy's Crab House	Elkton	Mama Mia's Tiramisu	18	$17.99	$323.82
Jimmy's Crab House	Elkton	Mom's Deep Dish Apple Pie	36	$12.49	$449.64

Source: Used with permission from Microsoft Corporation

FIGURE F-1 Access query

Customer Name	City	Product Name	Qty	Price per Unit	Total
Applewood Restaurant	Martinsburg	Frozen Alligator on a Stick	20	$27.99	$559.80
Applewood Restaurant	Martinsburg	Nogales Chipotle Sauce	15	$11.49	$172.35
Applewood Restaurant	Martinsburg	Mom's Deep Dish Apple Pie	12	$12.49	$149.88
Fresh Catch Fishery	Salem	Brumley's Seafood Cocktail Sauce	24	$4.79	$114.96
Fresh Catch Fishery	Salem	NY Smoked Salmon	21	$21.99	$461.79
Fresh Catch Fishery	Salem	Mama Mia's Tiramisu	15	$17.99	$269.85
Jimmy's Crab House	Elkton	Frozen Alligator on a Stick	12	$27.99	$335.88
Jimmy's Crab House	Elkton	Brumley's Seafood Cocktail Sauce	24	$4.79	$114.96
Jimmy's Crab House	Elkton	Mama Mia's Tiramisu	18	$17.99	$323.82
Jimmy's Crab House	Elkton	Mom's Deep Dish Apple Pie	36	$12.49	$449.64

Title: May 2013 Orders--Fine Foods, Inc. — Thursday, July 04, 2013, 7:51:38 AM

Total Orders $2,952.93

Page 1 of 1

Source: Used with permission from Microsoft Corporation

FIGURE F-2 Access report

Also consider what kinds of graphic media are available and how well you can use them. Your employer will expect you to be able to use Microsoft PowerPoint to prepare your presentation as a slide show. Luckily, many college freshmen are required to take an introductory course that covers Microsoft Office and PowerPoint. If you are not familiar with PowerPoint, several excellent tutorials on the Web can help you learn the basics.

Anticipating the Unexpected

Even though you are only drafting your report and presentation at this stage, eventually you will answer questions from the audience. Being able to handle questions smoothly is the mark of a business professional. The first steps to addressing audience questions are being able to anticipate them and preparing your answers.

You will not use all the facts you gather for your report or presentation. However, as you draft your report, you might want to jot down those facts and keep them handy, in case you need them to answer questions from the audience. PowerPoint has a Notes section where you can include notes for each slide and print them to help you answer questions that arise during your presentation. You will learn how to print notes for your slides later in the tutorial.

The questions you receive depend on the nature of your presentation. For example, during a presentation of an Excel decision model, you might be asked why you are not recommending a certain course of action, or why you left it out of your report. If you have already prepared notes that anticipate such questions, you will probably remember your answers without even having to refer to the notes.

Another potential problem is determining how much technical detail you should display in your slides. In one sense, writing your report will be easier because you can include any graphics, tables, or data you want. Because you have a time limit for your presentation, the question of what to include or leave out becomes more challenging. One approach to this problem is to create more slides than you think you need, and then use the Hide Slide option in PowerPoint to "hide" the extra slides. For example, you might create slides that contain technical details you do not think you will have time to present. However, if you are asked for more details on a particular technical point, you can "unhide" a slide and display the detailed information needed to answer the question. You will learn more about the Hide Slide and Unhide Slide options later in the tutorial.

Writing the Introduction

After you have written the main body of your report and presentation, you can develop the introduction. The introduction should be only a paragraph or two, and it should preview the main points you will cover.

For some of the Access cases, you might want to include general information about databases: what they can do, why they are used, and how they can help a company become more efficient and profitable. You will not need to say much about the business operation because the audience already works for the company.

For the Excel cases, you might want to include an introduction of the general business scenario and describe any assumptions you used to create and run your decision support models. Excel is used for decision support, so you should describe the decision criteria you selected for the model.

Writing the Conclusion

Every good report or presentation needs a good ending. Do not leave the audience hanging. Your conclusion should be brief—only a paragraph or two—and it should give your presentation a sense of closure. Use the conclusion to repeat your main points or, for the Excel cases, to recap your findings and recommendations.

On many occasions, information learned during a business project reveals new opportunities for other projects. Your conclusion should provide closure for the immediate project, but if the project reveals possibilities for future improvements, include them in a "path forward" statement.

CREATING GRAPHICS

Visual aids are a powerful means of getting your point across and making it understandable to your audience. Visual aids come in a variety of forms, some of which are more effective than others. The integrated graphics tools in Microsoft Office can help you prepare a presentation with powerful impact.

Choosing Presentation Media

The media you use will depend on the situation and the media you have available, but remember: *You must maintain control of the media or you will lose the attention of your audience.*

The following list highlights the most common media used in a classroom or business conference room, along with their strengths and weaknesses:

- **PowerPoint slides and a projection system**—These are the predominant presentation media for academic and business use. You can use a portable screen and a simple projector hooked up to a PC, or you can use a full multimedia center. Also, although they are not yet universal in business, touch-sensitive projection screens (for example, Smart Board™ technology) are gaining popularity in college classrooms. The ability to project and display slides, video and sound clips, and live Web pages makes the projection system a powerful presentation tool. *Negatives:* Depending on the complexity of the equipment, you might have difficulties setting it up and getting it to work properly. Also, you often must darken the room to use the projector, and it may be difficult to refer to written notes during your presentation. When using presentation media, you must be able to access and load your PowerPoint file easily. Make sure your file is available from at least two sources that the equipment can access, such as a thumb drive, CD, DVD, or online folder. If your presentation has active links to Web pages, make sure that the presentation computer has Internet access.
- **Handouts**—You can create handouts of your presentation for the audience, which once was the norm for many business meetings. Handouts allow the audience to take notes on applicable slides. If the numbers on a screen are hard to read from the back of the room, your audience can refer to their handouts. With the growing emergence of "green" business practices, however, unnecessary paper use is being discouraged. Many businesses now require reports and presentation slides to be posted at a common site where the audience can access them later. Often, this site is a "public" drive on a business network. *Negatives:* Giving your audience reading material may distract their attention from your presentation. They could read your slides and possibly draw wrong conclusions from them before you have a chance to explain them.

- **Overhead transparencies**—Transparencies are rarely used anymore in business, but some academics prefer them, particularly if they have to write numbers, equations, or formulas on a display large enough for students to see from the back row in a lecture hall. *Negatives:* Transparencies require an overhead projector, and frequently their edges are visually distorted due to the design of the projector lens. You have to use special transparency sheets in a photocopier to create your slides. For both reasons, it is best to avoid using overheads.
- **Whiteboards**—Whiteboards are common in both the business conference room and the classroom. They are useful for posting questions or brainstorming, but you should not use one in your presentation. *Negatives:* You have to face away from your audience to use a whiteboard, and if you are not used to writing on one, it can be difficult to write text that is large enough and legible. Use whiteboards only to jot down questions or ideas that you will check on after the presentation is finished.
- **Flip charts**—Flip charts (also known as easel boards) are large pads of paper on a portable stand. They are used like whiteboards, except that you do not erase your work when the page is full— you flip over to a fresh sheet. Like whiteboards, flip charts are useful for capturing questions or ideas that you want to research after the presentation is finished. Flip charts have the same negatives as whiteboards. Their one advantage is that you can tear off the paper and take it with you when you leave.

Creating Graphs and Charts

Strictly speaking, charts and graphs are not the same thing, although many graphs are referred to as charts. Usually charts show relationships and graphs show change. However, Excel makes no distinction and calls both entities *charts*.

Charts are easy to create in Excel. Unfortunately, the process is so easy that people frequently create graphics that are meaningless, misleading, or inaccurate. This section explains how to select the most appropriate graphics.

You should use pie charts to display data that is related to a whole. For example, you might use a pie chart when breaking down manufacturing costs into Direct Materials, Direct Labor, and Manufacturing Overhead, as shown in Figure F-3. (Note that when you create a pie chart, Excel 2013 will convert the numbers you want to graph into percentages of 100.)

LCD TV Manufacturing Cost

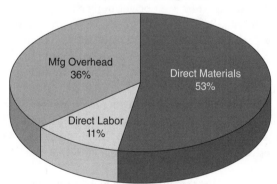

Source: © Cengage Learning 2015
FIGURE F-3   3D Pie chart: appropriate use

You would *not*, however, use a pie chart to display a company's sales over a three-year period. For example, the pie chart in Figure F-4 is meaningless because it is not useful to think of the period "as a whole" or the years as its "parts."

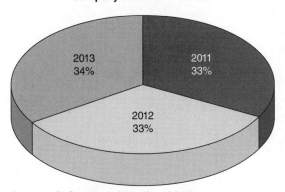

Source: © Cengage Learning 2015
FIGURE F-4 3D Pie chart: inappropriate use

You should use vertical bar charts (also called column charts) to compare several amounts at the same time, or to compare the same data collected for successive periods of time. The same type of company sales data shown incorrectly in Figure F-4 can be compared correctly using a vertical bar chart (see Figure F-5).

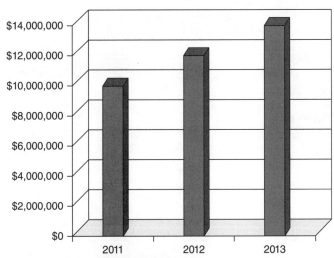

Source: © Cengage Learning 2015
FIGURE F-5 3D Column chart: appropriate use

As another example, you might want to compare the sales revenues from several different products. You can use a clustered bar chart to show changes in each product's sales over time, as in Figure F-6. This type of bar chart is called a "clustered column" chart in Excel.

When building a chart, include labels that explain the graphics. For instance, when using a graph with an x- and y-axis, you should show what each axis represents so your audience does not puzzle over the graphic while you are speaking. Figures F-6 and F-7 illustrate the necessity of good labels.

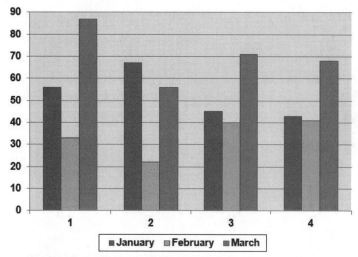

Source: © Cengage Learning 2015

FIGURE F-6   Clustered column graph without title or axis labels

Source: © Cengage Learning 2015

FIGURE F-7   3D Clustered column graph with title and axis labels

In Figure F-6, the graph has no title and neither axis is labeled. Are the amounts in units or dollars? What elements are represented by each cluster of bars? In contrast, Figure F-7 provides a comprehensive snapshot of product sales, which would support a talk rather than create confusion. Note also how the 3D chart style adds visual depth to the chart. Using the 3D chart, the audience can more easily discern that February sales were lower across all product categories.

Another common pitfall of visual aids is charts that have a misleading premise. For example, suppose you want to show how sales are distributed among your inventory, and their contribution to net income. If you simply take the number of items sold in a given month, as displayed in Figure F-8, the visual fails to give your audience a sense of the actual dollar value of those sales. It is far more appropriate and informative to graph the net income for the items sold instead of the number of items sold. The graph in Figure F-9 provides a more accurate picture of which items contribute the most to net income.

Number of Items Sold in June

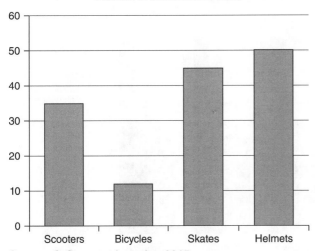

Source: © Cengage Learning 2015

FIGURE F-8   Graph of number of items sold that does not reflect generated income

Net Income by Product in June

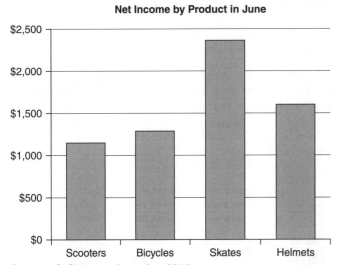

Source: © Cengage Learning 2015

FIGURE F-9   Graph of net income by item sold

You should also avoid putting too much data in a single comparative chart. For example, assume that you want to compare monthly mortgage payments for two loan amounts with different interest rates and time frames. You have a spreadsheet that computes the payment data, as shown in Figure F-10.

	A	B	C	D	E	F	G
1	**Calculation of Monthly Payment**						
2	Rate	6.00%	6.10%	6.20%	6.30%	6.40%	6.50%
3	Amount	$ 100,000	$ 100,000	$ 100,000	$ 100,000	$ 100,000	$ 100,000
4	Payment (360 Payments)	$ 599	$ 605	$ 612	$ 618	$ 625	$ 632
5	Payment (180 Payments)	$ 843	$ 849	$ 854	$ 860	$ 865	$ 871
6	Amount	$ 150,000	$ 150,000	$ 150,000	$ 150,000	$ 150,000	$ 150,000
7	Payment (360 Payments)	$ 899	$ 908	$ 918	$ 928	$ 938	$ 948
8	Payment (180 Payments)	$ 1,265	$ 1,273	$ 1,282	$ 1,290	$ 1,298	$ 1,306

Source: Used with permission from Microsoft Corporation

FIGURE F-10   Calculation of monthly payment

In Excel, it is possible (but not advisable) to capture all of the information in a single clustered column chart, as shown in Figure F-11.

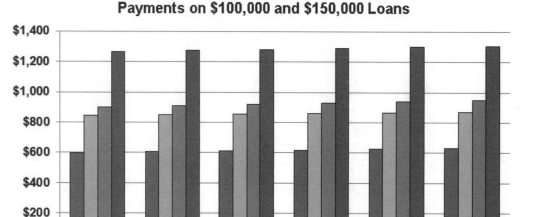

Source: © Cengage Learning 2015

FIGURE F-11   Too much information in one chart

The chart contains a great deal of information. Putting the $100,000 and $150,000 loan payments in the same "cluster" may confuse the readers. They would probably find it easier to understand one chart that summarizes the $100,000 loan (see Figure F-12) and a second chart that covers the $150,000 loan.

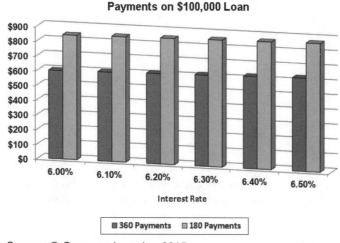

Source: © Cengage Learning 2015

FIGURE F-12   Good balance of information and visual depth

You could then augment the charts with text that summarizes the main differences between the payments for each loan amount. In that fashion, the reader is led step by step through the analysis.

Excel 2007 and later versions no longer have a Chart Wizard; instead, the Insert tab includes a Charts group. Once you create a chart and click it, two chart-specific tabs appear under the Chart Tools heading on the Ribbon to assist you with chart design and formatting. Excel 2013 also adds three menu buttons to the right of the chart: Chart Elements, Chart Styles, and Chart Filters (see Figure F-13). Click each button to see a menu that helps you edit your chart. If you are unfamiliar with the charting tools in Excel, ask your instructor for guidance or refer to the many Excel tutorials on the Web.

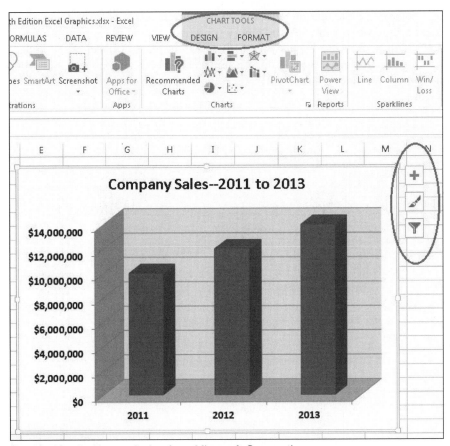

Source: Used with permission from Microsoft Corporation

FIGURE F-13   Chart tools and new menu buttons in Excel 2013

Creating PowerPoint Presentations

PowerPoint presentations are easy to create. When you open PowerPoint, it starts a new presentation for you. You can select from many different themes, styles, and slide layouts by clicking the Design tab. If none of PowerPoint's default themes suit you, you can download theme "templates" from Microsoft Office Online. When choosing a theme and style for your slides, such as background colors or graphics, fonts, and fills, keep the following guidelines in mind:

- In older versions of PowerPoint, users were advised to avoid pastel backgrounds or theme colors and to keep their slide backgrounds dark. Because of the increasing quality of graphics in both computer hardware and projection systems, most of the default themes in PowerPoint will project well and be easy to read.
- If your projection screen is small or your presentation room is large, consider using boldface type for all of your text to make it readable from the back of the room. If you have time to visit the presentation site beforehand, bring your PowerPoint file, project a slide on the screen, and look at it from the back row of the audience area. If you can read the text, the font is large enough.
- Use transitions and animations to keep your presentation lively, but do not go overboard with them. Swirling letters and pinwheeling words can distract the audience from your presentation.

- It is an excellent idea to animate the text on your slides with entrance effects so that only one bullet point appears at a time when you click the mouse (or when you tap the screen using a touch-sensitive board). This approach prevents your audience from reading ahead of the bullet point being discussed and keeps their attention on you. Entrance effects can be incorporated and managed using the Add Animation button in PowerPoint 2013, as shown in Figures F-14 and F-15.

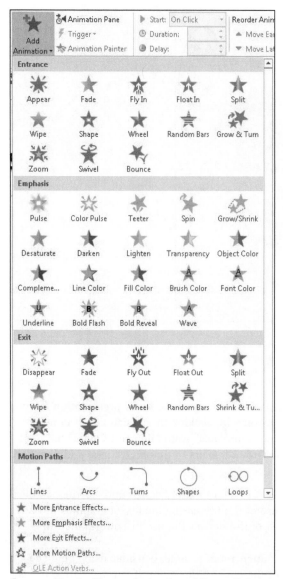

Source: Used with permission from Microsoft Corporation

FIGURE F-14   The Add Animation button on the Ribbon in PowerPoint 2013

Source: Used with permission from Microsoft Corporation

FIGURE F-15   Add Entrance Effect window

N O T E—DIFFERENCES IN POWERPOINT ANIMATION TOOLS IN LAST THREE VERSIONS

The structure of the animation tools changed considerably from PowerPoint 2007 to the 2010 version. The Custom Animation button and pane were removed, and most of the custom animation tools were incorporated using the Add Animation button in PowerPoint 2010. You can still use an animation pane to organize and edit your animations within a slide. PowerPoint 2013 has the same Animations tab and Advanced Animations group as PowerPoint 2010.

- Consider creating PowerPoint slides that have a section for your notes. You can print the notes from the Print dialog box by choosing Notes Pages from the Print menu, as shown in Figure F-16. Each slide will be printed at half its normal size, and your notes will appear beneath each slide, as shown by the print preview on the right side of Figure F-16.

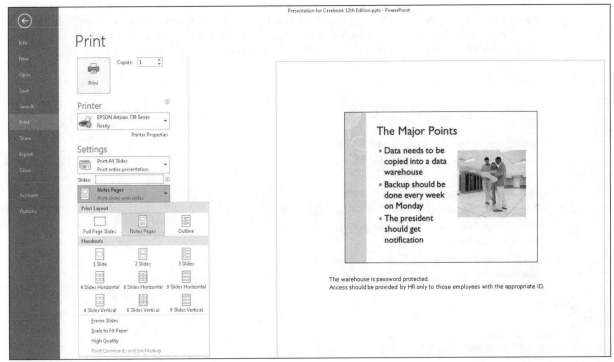

Source: Used with permission from Microsoft Corporation

FIGURE F-16   Printing notes page and slide print preview in PowerPoint 2013

- Finally, you should check your PowerPoint slides on a projection screen before your presentation. Information that looks good on a computer display may not be readable on the projection screen.

Using Visual Aids Effectively

Make sure you choose the visual aids that will work most effectively, and that you have enough without using too many. How many is too many? The amount of time you have to speak will determine the number of visual aids you should use, as will your target audience. A good rule of thumb is to allow at least one minute to present each PowerPoint slide. Leave a minimum of two minutes for audience questions after a 10-minute presentation, and allow up to 25 percent of your total presentation time to address questions after longer presentations. (For example, for a 20-minute presentation, figure on taking five minutes for questions.) For a 10-minute talk, try to keep the body of your presentation to eight slides or less. Your target audience will also influence your selection of visual aids. For instance, your slides will need more graphics and animation if you are addressing a group of teenagers than if you are presenting to a board of directors. Remember to use visual aids to emphasize your main points, not to detract from them.

Review each of your slides and visual aids to make sure it meets the following criteria:

- The font size of the text is large enough to read from the back of the presentation area.
- The slide or visual aid does not contain misleading graphics, typographical errors, or misspelled words—the quality of your work is a direct reflection on you.
- The content of your visual aid is relevant to the key points of your presentation.
- The slide or visual aid does not detract from your message. Your animations, pictures, and sound effects should support the text. Your visuals should look professional.
- A visual aid should look good in the presentation environment. If possible, rehearse your PowerPoint slides beforehand in the room where you will give the presentation. Make sure you can read your slides easily from the back row of seats in the room. If you have a friend who can sit in, ask her or him to listen to your voice from the back row of seats. If you have trouble projecting your voice clearly, consider using a microphone for your presentation.
- All numbers should be rounded unless decimals or pennies are crucial. For example, your company might only pay fractions of a cent per Web hit, but this cost may become significant after millions of Web hits.
- Slides should not look too busy or crowded. Many PowerPoint experts have a "6 by 6" rule for bullet points on a slide, which means you should include no more than six bullet points per slide and no more than six words per bullet point. Also avoid putting too many labels or pictures on a slide. Clip art can be "cutesy" and therefore has no place in a professional business presentation. A well-selected picture or two can add emphasis to the theme of a slide. For examples of a slide that is too busy versus one that conveys its points succinctly, see Figures F-17 and F-18.

Major Points

- Data needs to be copied into a data warehouse
- Backup should be done every week on Monday
- The president should get notification
- The vice president should get notification
- The data should be available on the Web
- Web access should be on a secure server
- HR sets passwords
- Only certain personnel in HR can set passwords
- Users need to show ID to obtain a password
- ID cards need to be the latest version

Source: Used with permission from Microsoft Corporation
FIGURE F-17 Busy slide

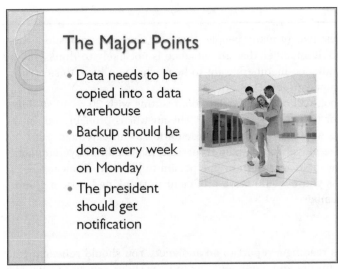

Source: Used with permission from Microsoft Corporation

FIGURE F-18 Slide with appropriate number of bullet points and a supporting photo

You may find that you have created more slides than you have time to present, and you are unsure of which slides you should delete. Some may have data that an audience member might ask about. Fortunately, PowerPoint lets you "hide" slides; these hidden slides will not be displayed in Slide Show view unless you "unhide" them in Normal view. Hiding slides is an excellent way to keep detailed data handy in case your audience asks to see it. Figure F-19 shows how to hide a slide in a PowerPoint presentation. Right-click the slide you want to hide, and then click Hide Slide from the menu to mark the slide as hidden in the presentation. To unhide the slide, right-click it and then click Unhide Slide from the menu. Click the slide to display it in Slide Show view.

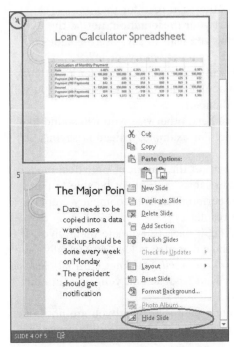

Source: Used with permission from Microsoft Corporation

FIGURE F-19 Hiding a slide in PowerPoint

PRACTICING YOUR DELIVERY

Surveys indicate that public speaking is the greatest fear of many people. However, fear or nervousness can be channeled into positive energy to do a good job. Remember that an audience is not likely to think you are nervous unless you fidget or your voice cracks. Audience members want to hear what you have to say, so think about them and their interests—not about how you feel.

Your presentations for the cases in this textbook will occur in a classroom setting with 20 to 40 students. Ask yourself: Am I afraid when I talk to just one or two of my classmates? The answer is probably no. In addition, they will all have to give presentations as well. Think of your presentation as an extended conversation with several classmates. Let your gaze move from person to person, making brief eye contact with each of them randomly as you speak. As your focus moves from one person to another, think to yourself: I am speaking to one person at a time. As you become more proficient in speaking before a group, your gaze will move naturally among audience members.

Tips for Practicing Your Delivery

Giving an effective presentation is not the same as reading a report to an audience. You should rehearse your message well enough so that you can present it naturally and confidently, with your slides or other visual aids smoothly intermingled with your speaking. The following tips will help you hone the effectiveness of your delivery:

- Practice your presentation several times, and use your visual aids when you practice.
- Show your slides at the right time. Luckily, PowerPoint makes this easy; you can click the slide when you are ready to talk about it. Use cues as necessary in your speaker's notes.
- Maintain eye and voice contact with the audience when using the visual aid. Do not turn your back on your audience. It is acceptable to turn sideways to glance at your slide. A popular trick of experienced speakers is to walk around and steal a glance at the slide while moving to a new position.
- Refer to your visual aids in your talk and use hand gestures where appropriate. Do not ignore your own visual aid, but do not read it to your audience—they can read for themselves.
- Keep in mind that your slides or visual aids should support your presentation, not *be* the presentation. Do not try to crowd the slide with everything you plan to say. Use the slides to illustrate key points and statistics, and fill in the rest of the content with your talk.
- Check your time, especially when practicing. If you stay within the time limit when practicing, you will probably finish a minute or two early when you actually give the presentation. You will be a little nervous and will talk a little faster to a live audience.
- Use numbers effectively. When speaking, use rounded numbers; otherwise, you will sound like a computer. Also make numbers as meaningful as possible. For example, instead of saying "in 83 percent of cases," say "in five out of six cases."
- Do not extrapolate, speculate, or otherwise "reach" to interpret the output of statistical models. For example, suppose your Excel model has many input variables. You might be able to point out a trend, but often you cannot say with mathematical certainty that if a company employs the inputs in the same combination, it will get the same results.
- Some people prefer recording their presentation and playing it back to evaluate themselves. It is amazing how many people are shocked when they hear their recorded voice—and usually they are not pleased with it. In addition, you will hear every *um, uh, well, you know*, throat-clearing noise, and other verbal distraction in your speech. If you want feedback on your presentation, have a friend listen to it.
- If you use a pointer, be careful where you wave it. It is not a light saber, and you are not Luke Skywalker. Unless you absolutely have to use one to point at crucial data on a slide, leave the pointer home.

Handling Questions

Fielding questions from an audience can be tricky because you cannot anticipate all of the questions you might be asked. When answering questions from an audience, *treat everyone with courtesy and respect*. Use the following strategies to handle questions:

- Try to anticipate as many questions as possible, and prepare answers in advance. Remember that you can gather much of the information to prepare those answers while drafting your presentation. The Notes section under each slide in PowerPoint is a good place to write anticipated questions and your answers. Hidden slides can also contain the data you need to answer questions about important details.
- Mention at the beginning of your talk that you will take questions at the end of the presentation, which helps prevent questions from interrupting the flow and timing of your talk. In fact, many PowerPoint presentations end with a Questions slide. If someone tries to interrupt, say that you will be happy to answer the question when you are finished, or that the next graphic answers the question. Of course, this point does not apply to the company CEO—you *always* stop to answer the CEO's questions.
- When answering a question, a good practice is to repeat the question if you have any doubt that the entire audience heard it. Then deliver your answer to the whole audience, but make sure you close by looking directly at the person who asked the question.
- Strive to be informative, not persuasive. In other words, use facts to answer questions. For instance, if someone asks your opinion about a given outcome, you might show an Excel slide that displays the Solver's output; then you can use the data as the basis for answering the question. In that light, it is probably a good idea to have computer access to your Excel model or Access database if your presentation venue permits it, but avoid using either unless you absolutely need it.
- If you do not know the answer to a question, it is acceptable to say so, and it is certainly better than trying to fake the answer. For instance, if someone asks you the difference between the Simplex LP and GRG solving methods in Excel Solver, you might say, "That is an excellent question, but I really don't know the answer—let me research it and get back to you." Then follow up after the presentation by researching the answer and contacting the person who asked the question.
- Signal when you are finished. You might say that you have time for one more question. Wrap up the talk yourself and thank your audience for their attention.

Handling a "Problem" Audience

A "problem" audience or a heckler is every speaker's nightmare. Fortunately, this experience is rare in the classroom: Your audience will consist of classmates who also have to give presentations, and your instructor will be present to intervene in case of problems.

Heckling can be a common occurrence in the political arena, but it does not happen often in the business world. Most senior managers will not tolerate unprofessional conduct in a business meeting. However, fellow business associates might challenge you in what you perceive as a hostile manner. If so, remain calm, be professional, and rely on facts. The rest of the audience will watch to see how you react—if you behave professionally, you make the heckler appear unprofessional by comparison and you'll gain the empathy of the audience.

A more common problem is a question from an audience member who lacks technical expertise. For instance, suppose you explained how to enter data into an Access form, but someone did not understand your explanation. Ask the questioner what part of the explanation was confusing. If you can answer the question briefly and clearly, do so. If your answer turns into a time-consuming dialogue, offer to give the person a one-on-one explanation after the presentation.

Another common problem is receiving a question that you have already answered. The best solution is to give the answer again, as briefly as possible, using different words in case your original answer confused the person. If someone persists in asking questions that have obvious answers, you might ask the audience, "Who would like to answer that question?" The questioner should get the hint.

PRESENTATION TOOLKIT

You can use the form in Figure F-20 for preparation, the form in Figure F-21 for evaluation of Access presentations, and the form in Figure F-22 for evaluation of Excel presentations.

Preparation Checklist

Facilities and Equipment

☐ The room contains the equipment that I need.
☐ The equipment works and I've tested it with my visual aids.
☐ Outlets and electrical cords are available and sufficient.
☐ All the chairs are aligned so that everyone can see me and hear me.
☐ Everyone will be able to see my visual aids.
☐ The lights can be dimmed when/if needed.
☐ Sufficient light will be available so I can read my notes when the lights are dimmed.

Presentation Materials

☐ My notes are available, and I can read them while standing up.
☐ My visual aids are assembled in the order that I'll use them.
☐ A laser pointer or a wand will be available if needed.

Self

☐ I've practiced my delivery.
☐ I am comfortable with my presentation and visual aids.
☐ I am prepared to answer questions.
☐ I can dress appropriately for the situation.

Source: © Cengage Learning 2015

FIGURE F-20 Preparation checklist

Evaluating Access Presentations

Course: _____   Speaker: _____   Date: _____

Rate the presentation by these criteria:
4=Outstanding   3=Good   2=Adequate   1=Needs Improvement
N/A=Not Applicable

Content

_____ The presentation contained a brief and effective introduction.

_____ Main ideas were easy to follow and understand.

_____ Explanation of database design was clear and logical.

_____ Explanation of using the form was easy to understand.

_____ Explanation of running the queries and their output was clear.

_____ Explanation of the report was clear, logical, and useful.

_____ Additional recommendations for database use were helpful.

_____ Visuals were appropriate for the audience and the task.

_____ Visuals were understandable, visible, and correct.

_____ The conclusion was satisfying and gave a sense of closure.

Delivery

_____ Was poised, confident, and in control of the audience

_____ Made eye contact

_____ Spoke clearly, distinctly, and naturally

_____ Avoided using slang and poor grammar

_____ Avoided distracting mannerisms

_____ Employed natural gestures

_____ Used visual aids with ease

_____ Was courteous and professional when answering questions

_____ Did not exceed time limit

Submitted by: _____

Source: © Cengage Learning 2015

FIGURE F-21   Form for evaluation of Access presentations

Evaluating Excel Presentations

Course: _____   Speaker: _____   Date: _____

Rate the presentation by these criteria:
4=Outstanding   3=Good   2=Adequate   1=Needs Improvement
N/A=Not Applicable

Content

_____ The presentation contained a brief and effective introduction.

_____ The explanation of assumptions and goals was clear and logical.

_____ The explanation of software output was logically organized.

_____ The explanation of software output was thorough.

_____ Effective transitions linked main ideas.

_____ Solid facts supported final recommendations.

_____ Visuals were appropriate for the audience and the task.

_____ Visuals were understandable, visible, and correct.

_____ The conclusion was satisfying and gave a sense of closure.

Delivery

_____ Was poised, confident, and in control of the audience

_____ Made eye contact

_____ Spoke clearly, distinctly, and naturally

_____ Avoided using slang and poor grammar

_____ Avoided distracting mannerisms

_____ Employed natural gestures

_____ Used visual aids with ease

_____ Was courteous and professional when answering questions

_____ Did not exceed time limit

Submitted by: _____

Source: © Cengage Learning 2015

FIGURE F-22   Form for evaluation of Excel presentations

INDEX

Note: Page numbers in **boldface** indicate key terms.

G

H

I